Vesna Tomašić, Bruno Zelić (Eds.)
Environmental Engineering

Also of Interest

Industrial Biotechnology.
Benvenuto, 2019
ISBN 978-3-11-053639-3, e-ISBN 978-3-11-053662-1

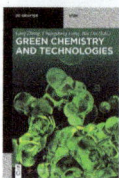

Green Chemistry and Technologies.
Zhang, Gong, Bin (Eds.), 2018
ISBN 978-3-11-047861-7, e-ISBN 978-3-11-047931-7

Environmental Toxicology.
Botana (Ed.), 2018
ISBN 978-3-11-044203-8, e-ISBN 978-3-11-044204-5

Physical Sciences Reviews.
e-ISSN 2365-659X

Environmental Engineering

Basic Principles

Edited by
Vesna Tomašić and Bruno Zelić

DE GRUYTER

Editors

Prof. Vesna Tomašić
University of Zagreb
Faculty of Chemical Engineering & Technology
Marulicev trg 19
10000 Zagreb
Croatia
vtomas@fkit.hr

Prof. Bruno Zelić
University of Zagreb
Faculty of Chemical Engineering & Technology
Marulicev trg 19
10000 Zagreb
Croatia
bzelic@fkit.hr

ISBN 978-3-11-046801-4
e-ISBN (PDF) 978-3-11-046803-8
e-ISBN (EPUB) 978-3-11-046807-6

Library of Congress Control Number: 2018949255

Bibliographic information published by the Deutsche Nationalbibliothek
The Deutsche Nationalbibliothek lists this publication in the Deutsche Nationalbibliografie;
detailed bibliographic data are available on the Internet at http://dnb.dnb.de.

© 2018 Walter de Gruyter GmbH, Berlin/Boston
Cover image: Fotosearch/Getty Images Plus
Typesetting: Integra Software Services Pvt. Ltd.
Printing and binding: CPI books GmbH, Leck

www.degruyter.com

Preface

Progress that was made in the last centuries, concerning human living conditions is more than evident. We live longer and better in a state of slight obsession where we always want more and more of new available products, food, accessorise, etc. that can make our lives more easy and comfortable. Somewhere in that rush of today's lives we forgot or we rarely stop to think what the cost of such strong urbanisation and industrialization is. What is the cost of taking a new plastic bag in every store that we enter? Or throwing food and plastic in to the garbage without recycling? Or driving a car in the store that is 5 min away from our home? Or taking a 20 minutes shower every day with drinkable water? Besides the fact we stopped appreciating things due to their accessibility, what is the real cost of this and other everyday actions that we all do? Limited natural resources, polluted air and water, extinction of species, global warming, etc. are just some consequences of something that we love call modern life. But, how modern this destruction really is? How modern will be if tomorrows designers will have to focus on creating coolest gas masks because air will not be good for breathing without previous purification? Or creating a new canister that will be easy to transport from point A to point B because luxury of taking a shower in a bath will be just that – luxury?

Since there is always but and since there is always hope, some things are changing. Slowly, but they still do and perhaps we can call them awareness awakening? Awareness that if we continue going in this direction one day we will run out of path and there will be nothing on horizon. Awareness that it is the final moment to put environment in to the focus. To change our selfish human perspective from us to nature.

There is a saying *"You can't teach old dog new tricks"* so perhaps the best way to start setting things right are new generations. To transfer the knowledge and information and sow the seeds of the awareness among engineering and science students hoping that one day that little seed will grow in to something strong and powerful. This textbook will, hopefully, be a good base or a good soil for that seed growth. It is organized in to eight different areas important to all present and future environmental engineers together with introduction, modelling of environmental processes and risk assessment. All that is necessary for new human direction when it comes to nature and putting her back in to the centre.

The first chapter gives short introduction to basic environmental engineering principles and puts whole idea behind this new science field in to the concept. It presents the main areas of interest that are more deeply explored trough Chapters 2-9.

The second chapter sets the stage with one of the basic aspects of environment – Ecology. Main focus of this field is to define diversity, distribution and number of particular organisms, as well as cooperation and competition between organisms, both within and among ecosystems. Basically, makes connections between organisms and their environment since it is not possible for one to work without the other.

https://doi.org/10.1515/9783110468038-201

In order to understand ecology, we have to understand the very base: environment and living organisms.

Trough Chapters 3-5, entitled – Environmental Chemistry, Environmental Microbiology and Environmental Geology mentioned bases are covered. Chapter 3 presents the chemical nature of environmental problems and processes. Main focus is set on physical-chemical phenomena that occur in water, soil and atmosphere trying to summarize them on the basic environmental chemistry which is strictly related with chemical equilibrium phenomena. Because nothing in nature functions as a single but as a part of bigger picture, overcoming this principles a foundation for future construction of more complex environmental engineering problems is set, i.e. human impact on air, water and soil. Chapter 4 focuses on the dominant form of life on planet Earth – microorganisms. And although for many people they are out of sight and therefore out of mind almost every surface is colonized by them. From the environment point of view, environmental microbiology is concerned with the study of microorganisms in the soil, water and air and their application in bioremediation processes for biotreatment of water, wastewater, solid waste, soil and gas. And finally in Chapter 5 the importance of geological environment is presented since many natural hazards, which have great impact on humans and their environment, are caused by geological settings. On the other hand, human activities have great impact on the physical environment, especially in the last decades due to dramatic human population growth. Natural disasters often hit densely populated areas causing tremendous death toll and material damage. Demand for resources enhanced remarkably, as well as waste production. Exploitation of mineral resources deteriorates huge areas of land, produce enormous mine waste and pollute soil, water and air.

Water, soil and air, their quality and protection are stressed down in Chapter 1 as most important issues in environmental engineering. Before exploring them more extensively in Chapters 7-9, Chapter 6 focuses on the engineering base of mentioned processes – transport phenomena. The main idea is to understand and then describe the specific process using phenomenological equations that rely on three elementary physical processes: momentum, energy and mass transport. In the end, if that is achieved those processes, using mathematical models (described in Chapter 10) can be predicted, designed, optimized, developed; some catastrophes can be avoided, multiple solutions for a certain problem can be presented, etc...

Following the three main environmental issues, Chapter 7 focuses on air pollution. Air pollution is not just an environmental but a social problem which leads to a multitude of adverse effects on human health and standard of human life, state of the ecosystems and global change of climate. The chapter deals with the complexities of the air pollution and presents an overview of different technical processes and equipment for air pollution control, as well as basic principles of their work. As for air, Chapter 8 introduces water and wastewater characterization and gives an overview of the most important instrumental techniques used in water analysis. The presence of new pollutants (mostly of synthetic organic origin), the possibility of their characterization and modern removal techniques are particularly emphasized. The chapter describes the basic principles of biological water treatment, as well as modern processes involving membrane separation processes and advanced oxidation processes. The last issue, soil treatment is described in Chapter 9. As a conditionally renewable natural resource, soil has a decisive influence on sustainable development of global economy, especially on sustainable agriculture and environmental protection. In recent decades, a growing interest prevails for non-production soil functions, primarily those relating to environmental protection. It especially refers to protection of natural resources whose quality depends directly on soil and soil management. Soil contamination is correlated with the degree of industrialization and intensity of agrochemical usage and is one of the most dangerous forms of soil degradation with the consequences that are reflected in virtually the entire biosphere, primarily at heterotrophic organisms and also at mankind as a food consumer.

As mentioned previously, by understanding chemical nature and transport phenomena together with microorganism behaviour many mentioned issues can be described, predicted, emerged problems, etc. can be solved by using mathematical models. In the Chapter 10 a short introduction into models (types of the models and their application) with accent on mathematical modelling in environmental engineering is given. Some basic advices on how to approach mathematical modelling, how to use different types of models and how to apply different methods for solving of mathematical models will be discussed on the examples describing the environmental processes.

The final chapter, Chapter 11, introduces risk assessment and examines the basic concepts of risk and hazard, their causes, consequences and probabilities. The methodology of environmental risk assessment including effect assessment, exposure assessment and risk estimation/characterisation is overviewed. Some of the commonly used techniques in risk assessments are introduced and illustrated by examples Environmental risk assessment should provide input to the risk management so well-informed decisions can be made, to protect human health and the environment.

Vesna Tomašić
Bruno Zelić

Contents

Aleksandra Sander, Jasna Prlić Kardum, Gordana Matijašić and Krunoslav Žižek

Karolina Maduna and Vesna Tomašić

Danijela Ašperger, Davor Dolar, Krešimir Košutić, Hrvoje Kušić, Ana Lončarić
Božić and Marija Vuković Domanovac

Ivica Kisić, Željka Zgorelec and Aleksandra Percin

Anita Šalić, Ana Jurinjak Tušek and Bruno Zelić

Hrvoje Kušić and Ana Lončarić Božić

List of contributing authors

Danijela Ašperger
University of Zagreb
Faculty of Chemical Engineering and
Technology
Marulićev trg 19
HR-10000 Zagreb
Croatia

Tomislav Bolanča
University of Zagreb
Faculty of Chemical Engineering and
Technology
Marulićev trg 19
HR-10000 Zagreb
Croatia
tbolanca@fkit.hr

Felicita Briški
University of Zagreb
Faculty of Chemical Engineering and
Technology
Marulićev trg 19
HR-10000 Zagreb
Croatia

Davor Dolar
University of Zagreb
Faculty of Chemical Engineering and
Technology
Marulićev trg 19
HR-10000 Zagreb
Croatia

Ana Jurinjak Tušek
University of Zagreb
Faculty of Food Technology and Biotechnology
Pierottijeva 6
HR-10000 Zagreb
Croatia

Jasna Prlić Kardum
University of Zagreb
Faculty of Chemical Engineering and
Technology
Marulićev trg 19
HR-10000 Zagreb
Croatia

Ivica Kisić
University of Zagreb
Faculty of Agriculture
Department of General Agronomy
Svetošimunska cesta 25
HR-10000 Zagreb
Croatia
ikisic@agr.hr

Krešimir Košutić
University of Zagreb
Faculty of Chemical Engineering and
Technology
Marulićev trg 19
HR-10000 Zagreb
Croatia
kkosutic@fkit.hr

Zoran Kovač
University of Zagreb
Faculty of Mining, Geology and Petroleum
Engineering
Pierottijeva 6
HR-10000 Zagreb
Croatia

Hrvoje Kušić
University of Zagreb
Faculty of Chemical Engineering and
Technology
Marulićev trg 19
HR-10000 Zagreb
Croatia
hkusic@fkit.hr

https://doi.org/10.1515/9783110468038-202

Ana Lončarić Božić
University of Zagreb
Faculty of Chemical Engineering and
Technology
Marulićev trg 19
HR-10000 Zagreb
Croatia
abozic@fkit.hr

Karolina Maduna
University of Zagreb
Faculty of Chemical Engineering and
Technology
Marulićev trg 19
HR-10000 Zagreb
Croatia
kmaduna@flkit.hr

Gordana Matijašić
University of Zagreb
Faculty of Chemical Engineering and
Technology
Marulićev trg 19
HR-10000 Zagreb
Croatia

Zlatko Mihaljević
University of Zagreb
Department of Biochemistry and Molecular
Biology
Ante Kovačića 1
HR-10000 Zagreb
Croatia
zlatko.mihaljevic@biol.pmf.hr

Marta Mileusnić
University of Zagreb
Faculty of Mining, Geology and Petroleum
Engineering
Pierottijeva 6
HR-10000 Zagreb
Croatia

Zoran Nakić
University of Zagreb
Faculty of Mining, Geology and Petroleum
Engineering

Pierottijeva 6
HR-10000 Zagreb
Croatia
zoran.nakic@rgn.hr

Mirjana Novak Stankov
University of Zagreb
Faculty of Chemical Engineering and
Technology
Marulićev trg 19
HR-10000 Zagreb
Croatia
mirnovak@fkit.hr

Krešimir Pavlić
University of Zagreb
Faculty of Mining, Geology and Petroleum
Engineering
Pierottijeva 6
HR-10000 Zagreb
Croatia

Aleksandra Percin
University of Zagreb
Faculty of Agriculture
Department of General Agronomy
Svetošimunska cesta 25
HR-10000 Zagreb
Croatia

Anita Šalić
University of Zagreb
Faculty of Chemical Engineering and
Technology
Marulićev trg 19
HR-10000 Zagreb
Croatia

Aleksandra Sander
University of Zagreb
Faculty of Chemical Engineering and
Technology
Marulićev trg 19
HR-10000 Zagreb
Croatia

Ivančica Ternjej
University of Zagreb
Department of Biochemistry and Molecular
Biology
Ante Kovačića 1
HR-10000 Zagreb
Croatia
ivancica.ternjej@biol.pmf.hr

Šime Ukić
University of Zagreb
Faculty of Chemical Engineering and
Technology
Marulićev trg 19
HR-10000 Zagreb
Croatia
sukic@fkit.hr

Marija Vuković Domanovac
University of Zagreb
Faculty of Chemical Engineering and
Technology

Marulićev trg 19
HR-10000 Zagreb
Croatia
mvukovic@flkit.hr

Željka Zgorelec
University of Zagreb
Faculty of Agriculture
Department of General Agronomy
Svetošimunska cesta 25
HR-10000 Zagreb
Croatia

Krunoslav Žižek
University of Zagreb
Faculty of Chemical Engineering and
Technology
Marulićev trg 19
HR-10000 Zagreb
Croatia
kzizek@fkit.hr

Anita Šalić and Bruno Zelić

1 Introduction to environmental engineering

Abstract: Nowadays we can easily say that environmental engineering is truly an interdisciplinary science. Combining biology, ecology, geology, geography, mathematics, chemistry, agronomy, medicine, economy, etc. environmental engineering strives to use environmental understanding and advancements in technology to serve mankind by decreasing production of environmental hazards and the effects of those hazards already present in the soil, water, and air. Major activities of environmental engineer involve water supply, waste water and solid management, air and noise pollution control, environmental sustainability, environmental impact assessment, climate changes, etc. And all this with only one main goal – to prevent or reduce undesirable impacts of human activities on the environment. To ensure we all have tomorrow.

Keywords: environment, engineering, pollution, environmental issues, humans

> *"We won't have a society if we destroy the environment"*
>
> *Margaret Mead*

1.1 A new engineering field?

The last two to three decades have confirmed a growing understanding that no part of our world is immune from environmental consequences of human activities. There is a pressing need for international cooperation and solidarity to preserve our forests, safekeeping our water and oceans, and stabilize the Earth's atmosphere. The strong interest in protecting the environment has placed new responsibilities, especially for the engineers. Therefore, environmental engineers got under a huge spotlight. But the question is, *Is this really a new concept*? Did the humans of twentieth and twenty-first centuries become more aware of their environment, protection of themselves, animals, plants, and plant populations from the effects of adverse environmental factors, including toxic chemicals and wastes, pathogenic bacteria and global effects such as warming, ozone layer depletion, and weather-pattern change? Or this goes way back in to the history?

   Although the first description of title "environmental engineer" appears somewhere around 1960s, the history of practice itself goes long in to the history. Since the early days of formation of the first communities (small group of people living on permanent settlements) a concern about clean water, waste separation, and sewage system had to be developed. With the growth of population, magnitude of those simple actions increased with addition of concern about air and soil quality.

https://doi.org/10.1515/9783110468038-001

Scoping trough the written history of those early ages some examples of good environmental engineering practice can be found. Like Roman aqueducts that provided constant and clean water to the large cities of Empire or development of sewers system in some Indian cities 5,000 years ago. But true jump was made parallel with rapid industrialization in eighteenth century. With the increase of population in the cities, great migrations, increased exploitation of nature goods, increase demand for the large quantities of food and fresh water in one place, disposal of waste, appearance of massive epidemic of water transmitted diseases like cholera etc. all lead to significant concerned about public health.

The first environmental engineer is said to have been Joseph Bazalgette who designed the first large-scale sewerage system in London in the mid-nineteenth century. This engineering venture is considered the beginning era of modern environmental engineering. Since then, the main idea – to improve natural environment for human habitation – remains the same but the amplitude of work severely intensified.

1.2 New era – todays environmental engineering

Since the modern environmental engineering is relatively a new science maybe the best way to start is by defining it. Simply by scoping the Internet several definitions of environmental engineering can be found:

> *The application of science and engineering knowledge and concepts to care for and/or restore our natural environment and/or solve environmental problems.* (unknown)

> *The application of science and engineering principles to improve the natural environment (air, water, and/or land resources) to provide healthy water, air, and land for human habitation (house or home) and for other organisms, and to remediate polluted sites.* (unknown)

> *The branch of engineering concerned with the application of scientific and engineering principles for protection of human populations from the effects of adverse environmental factors; protection of environments, both local and global, from potentially deleterious effects of natural and human activities; and improvement of environmental quality.* (The American Academy of Environmental Engineers)

> *Is that branch of engineering that is concerned with protecting the environment from the potentially deleterious effects of human activity, protecting human populations from the effects of adverse environmental actors and improving environmental quality for human health and wellbeing.* (Peavy, 1985 [1])

> *Environmental engineering deals with design and construction of the processes and equipment intended to lessen the impact of man's activities on the environment.* (unknown)

Summarizes, the main goal of environmental engineering or better environmental engineers is to restore, to protect, to improve, and to provide. In order to do that, like any engineer they have to follow some basic principles like planning, design, construction and operation of equipment, systems, and structures for the benefit of society (Figure 1.1).

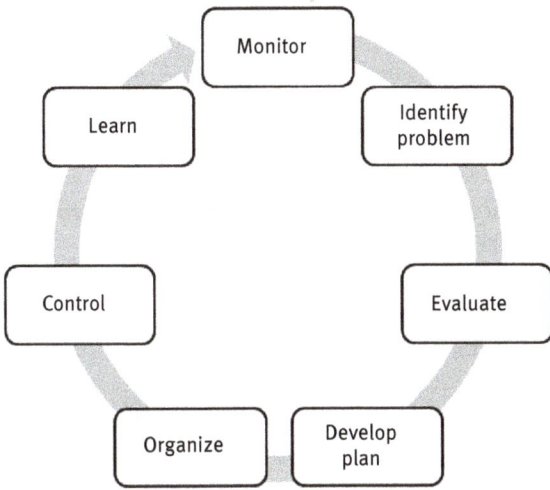

Figure 1.1: Basic principles for general problem management (inspired by [2]).

The basic step or better, one of the mayor responsibilities of environmental engineering is to **monitor** and potentially predict or even prevent harmful events in the environment. Using the previous knowledge environmental monitoring should ensures the management of natural resources contributing to sustainable development through processes and activities which characterize and monitor the quality of the environment. One of good examples is a strong pressure on upcoming industries to not only satisfy economic and technical sustainability principles but also social and ecological. Also there is a request for existing processes to adapt this new paradigm (Figure 1.2).

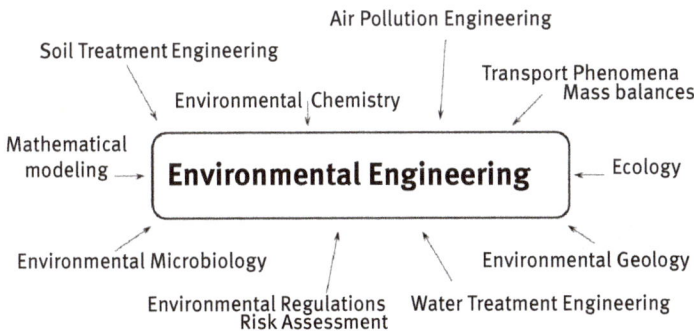

Figure 1.2: Suitability diagram.

According to Tonelli et al. [3] the need to reduce or contain the ecological footprint of the industry will affect the whole industrial system. In stepping to a low-carbon, resource-efficient approach, industry has to be considered not only as part of the problem but as part of the solution.

Leading companies are preparing for this transformation on three fronts:
–   rapidly reducing the resource- and energy-intensity in producing existing goods
–   investigating the options for a thorough redesign of the industrial system
–   radically rethinking business models.

Finally, a redesigned industrial system should
–   add the same value with a reduction of 25% on input materials and energy
–   make use of the 90% of discarded extracted materials
–   use benign materials that can be reused according to "cradle-to-cradle" concept
–   refurbish and reuse sophisticated long-lasting components
–   mimic and nurture the environmental niches.

Let's go back to the basic principles. Usually predictions and prevention are not possible and if a problem appears, then the cycle of the events is launched (Figure 1.1) in order to detect the presence of pollutants and tracking them back to their source. Usually, finding the source can present a significant challenge. For instance, the source of contamination in a lake could be anywhere within several thousands of acres of land surrounding the lake and its tributaries. Contamination of oceans can present even greater challenges in identifying the source. In order to, *identify* the problem, *evaluate* the situation, *develop* a plan and then potently *resolve* the situation requires extensive knowledge (Figure 1.3) of the chemistry and biology of the potential contaminants as well as the industrial or agricultural processes that might lead to their release (Chapters 2–6). And then, depending on area of pollution many other sciences get involved in finding a solution.

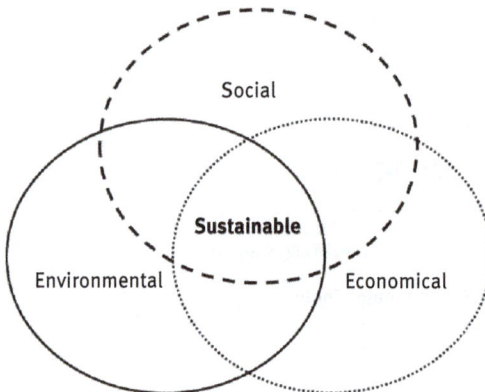

Figure 1.3: Diagram of environment engineering fundamentals.

This all means that a role of environmental engineer is also to coordinate other engineers to find best solutions for existing problems and to make rapid advancement in cleaning up the environment with eco-friendly techniques.

Usually many environmental problems cannot be stopped completely but rather reduced and during the process there are many ethical question involved. One of the most unpopular decisions that have to be made is to or not to shut down business that is a source of environmental pollutions because of the potential for severe economic consequences. That's why; nowadays environmental engineers often work with businesses to determine ways to avoid or reduce the production of pollutants or to separate them so they can be disposed of in a safe manner (Chapter 11).

At the end, one of the most valuable outcomes of the cycle (besides the protection and problem resolving) is obtained **knowledge** that can be utilized for future goals (Chapter 10).

As examples for proposed system, air resource, solid waste and waste water management diagrams are presented in Figures 1.4–1.6.

1.3 Environmental issues

The next important question is *What are exactly environmental issues that today's environmental engineer has to face*? Usually they are divided in to three major groups:
– air quality
– soil quality
– water quality

and each of them present important part of the everyday life, not only for humans but for all other organisms on Planet – plants, microorganisms, animals, etc.

Although it my seam that nowadays nature is rebelling more and more simply by watching number of catastrophes (tsunami in Indonesia, earthquakes in Italy, soil movement in South Amerika, etc.) it is still little percentage in total pollution amount in comparison to human impact. We can easily say that today there is serious environmental crisis due to global industrialization. Large quantities of water are polluted as well as air, forests are destroyed due to increased need for wood, land for food production or building new factories or simply urbanization, amount of solid waste is increasing rapidly with special accent on chemical and toxic material and chemicals, global warming, destruction of ozone layer are just some of the problems.

Generally, factors impacting environment can be divided in two groups [4–6]:
1. *Natural changes*
   a. Natural processes
   b. Natural disasters

2. *Anthropogenic changes*
   c. *Industrialization* – although the industrialization is the base of human society development, in parallel it presents global environmental problem. Depletion of resources and creation of enormous waste is just some of the problems.

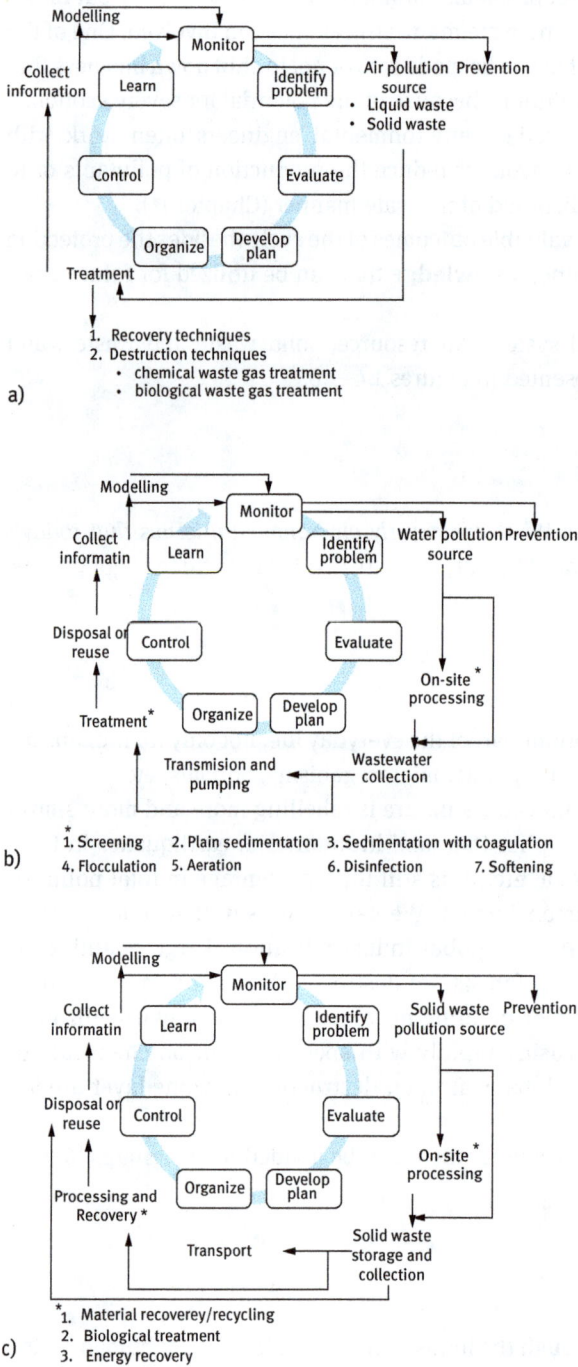

Figure 1.4: Management systems for (a) air resource (b) water pollution and (c) solid waste pollution.

d. *Change in ecological balance* – topic that opens two important areas. First, wrong management of the biosphere leads to visible disturbance in ecosystem (mostly referring to extinction of wildlife and habitual destruction (i.e. destruction of rain forests or coral reefs)). Many species have extinct and many are almost there. Second problem is rapid genetic engineering development where genetically modified organisms can be harmful and even toxic to wildlife.

e. *Global warming* – leads to rising temperatures of the ocean and earth surface. Year 2012 was characterized as the warmest year ever, after that year 2013 got the same title, as well as 2014, 2015, etc. As a consequence, polar ice caps are melting, sea level is rising flooding the area around sea where a great number of humane population lives which will lead to massive migration in the future etc.

f. *Overpopulation* – is considered one of the most critical environmental problems. By year 2050, the global human population is expected to grow by 2 billion people, thereby reaching a level of 9.6 billion people. This will lead to major problem in water, food and energy supply.

g. *Natural resources depletion* – is considered second crucial environmental problem. Since the starting of industrial era, the natural resources are constantly utilized for the production of one or more products for human consumption. One of the major natural resource that is being exploited rapidly are fossil fuels so in near future humans will have to think more about using renewable sources of energy.

h. *Deforestation* – refers to clearing of green cover and making that land available for residential, industrial or commercial purpose. The process results in decreasing rainfall, increasing global temperature, loss of top soil, modification of climatic conditions, etc.

i. *Air pollution* – release of different air pollutants in the atmosphere and environment causes many of other environmental issues like greenhouse effect, ozone layer destruction, smog formation, etc.

j. *Water pollution* – it is believed that in near future water will be a rare commodity and that many economic and political problems will emerge from that since wars are expected for this natural resource.

k. *Waste disposal* – as mentioned, rapid industrialization created large amounts of waste. Plastic, packaging, electronic waste, nuclear waste, etc. are becoming serious environmental problem.

l. Some other issues: *noise pollution, radiation pollution, soil erosion, climate change, ocean acidification, ozone layer depletion, acid rains, public health issues, etc.*

Unfortunately, all of these problems are global since they affect every individual, organization, community and country, and by becoming environmental stewards, it keeps the economy moving, which is necessary for growth and long-term viability and to solve them cooperation of all sectors is necessary.

Since air, soil and water are defined as main areas of environmental engineering interest, few additional questions are addressed.

1.3.1 Air quality – Why it is such a problem?

Air pollution may be defined as the presence in the atmosphere of any substance (or combination of substances) that is harmful to human health and welfare; offensive or objectionable to man, either externally or internally; or which by its presence will directly or indirectly adversely affect the welfare of man. It can be visible or invisible and perhaps the most familiar and obvious form of air pollution is smog hanging over cities.

Air pollutants fall into two main categories:
1. those that are present in many areas because they are the products of daily-life activities such as transportation, power generation, space and water heating, and waste incineration, and
2. those generated by activities such as chemical, manufacturing, and agricultural processing whose pollutant byproducts tend to be localized in nearby areas or are spread long distances by tall stacks and prevailing winds.

They are also categorized by their emission characteristics:
1. point sources, such as power plants, incinerators, and large processing plants;
2. area sources, such as space and water heating in buildings; and
3. mobile sources, mainly cars and trucks, but also lawn mowers and blowers and airplanes [7].

Besides these anthropogenic causes of air pollution, there are also natural sources like dust, methane, smoke and carbon monoxide from wildfires, volcano activity, etc.

Air pollution presents a problem after exposer and depends on how pollutant is hazard. The most common sources of air pollution include particulates, ozone, nitrogen dioxide and sulphur dioxide. It can affect an individual, a certain groups or entire populations. It is a significant risk factor for a number of pollution-related diseases. Some of them are new cases of asthma, exacerbate (worsen) a previously existing respiratory illness, and provoke development or progression of chronic illnesses including lung cancer, chronic obstructive pulmonary disease, and emphysema. Air pollutants also negatively and significantly harm lung development, creating an additional risk factor for developing lung diseases later in life [8, 9]. Individual reactions to air pollutants depend on the type of pollutant a person is exposed to, the degree of exposure, and the individual's health status and genetics [10]. Also the World Health Organization (WHO) estimated in 2014 that air pollution every year causes the premature death of some 7 million people worldwide [11]. India has the highest death rate due to air pollution. India also has more deaths from asthma than any other nation according to the WHO. In December 2013, air pollution was estimated to kill 500,000 people in China each year. There is a positive correlation between pneumonia-related deaths and air pollution from motor vehicle emissions.

Besides humans, air pollution effects agriculture. Yield reductions on a national scale were estimated to be about 5% in USA, with the economic benefit of reducing ozone concentrations by 40% estimated to be about 3 billion dollars annually. The ozone exposure corresponding to a 10% yield loss is exceeded over most of Europe, indicating the large potential for substantial effects on crop production. Detailed continent wide estimates of actual production losses have not yet been made, although individual filtration studies at Mediterranean sites, where ozone exposures are among the highest in Europe, have shown yield losses above 20% on sensitive crops. Situation is even more critical in developing countries like valley of Mexico – 40% yield lost in sensitive cultivar, Nile delta – 30%, Pakistan – 40%, etc. [12]

Also, by each year economy costs are increasing. According to the calculations, global air pollution-related healthcare costs are projected to increase from USD 21 billion (using constant 2010 USD and PPP exchange rates) in 2015 to USD 176 billion 2005 in 2060. By 2060, the annual number of lost working days, which affect labour productivity, are projected to reach 3.7 billion (currently around 1.2 billion) at the global level. And the annual global welfare costs associated with the premature deaths from outdoor air pollution are projected to rise from USD 3 trillion in 2015 to USD 18–25 trillion in 2060 [13].

Besides those impacts air pollution is directly connected with smog, acid rains, eutrophication, depletion of ozone layer and global warming.

So, if air pollution is present are there solutions for it? Some of them are
1. *use public mode of transportation, walk, drive bike or carpool*
2. *conserve energy*
3. *understand the concept of Reduce, Reuse and Recycle*
4. *emphasis on clean energy resources*
5. *use energy efficient devices*
6. *use hybrid cars*
7. *utilize alternative fuels.*

For more details about air pollution environmental engineering see Chapter 7.

1.3.2 Soil quality – What problems emerge from soil pollution?

Soil contamination is mainly located close to waste landfills, industrial/commercial activities diffusing heavy metals, oil industry, military camps, and nuclear power plants. As European society has grown wealthier, it has created more and more rubbish. Each year in the EU, 3 billion tonnes of solid wastes are thrown away (some 90 million tonnes of them are hazardous). This amounts to about 6 tonnes of solid waste for every man, woman and child [14, 15]. In the US, the army alone has estimated that over 1.2 million tonnes of soils have been contaminated with explosives, and the impact of explosives contamination in other countries in the world is of similar magnitude [16].

Based just on mentioned numbers it is obvious that soil pollution probably represents the most faced problem in environmental pollution. Generally, it is because

1. Soil is a point of concentration and recovery of toxic compounds, chemicals, salts, radioactive materials or disease causing agents, which have adverse effects on plant growth and animal health.
2. Soil pollutants can contaminate water: water infiltration is the movement of water from the soil surface into the soil profile and soil is a valuable resource that support cultures and plant life.
3. Soil pollutants have an adverse effect on the physical, chemical and biological properties of the soil and reduce its productivity.

Main causes of soil pollution are as follows
1. industrial activity, especially since the amount of mining and manufacturing has increased;
2. agricultural activities, pesticides and fertilizers which are full of chemicals that are not fully degradable in nature and are widely utilized around the world;
3. waste disposal, where there is also a large amount of industrial and municipal waste that is dumped directly into landfills without any treatment and
4. accidental oil spills, where oil leaks can happen during storage and transport of chemicals [17].

As a historical example, serious problem was caused by widespread application of the pesticide DDT to control agricultural pests in the years following World War II. It was used in urban aerial sprays to control urban mosquito, gypsy moth, Japanese beetle and other insects in the 1940s. While the agricultural benefits were outstanding and crop yields increased dramatically thus reducing world hunger substantially, and malaria was controlled better than it ever had been, numerous species were brought to the verge of extinction due to the impact of the DDT on their reproductive cycles. By 1972, DDT was banned from the United States due to widespread development of resistance to DDT and evidence that DDT use was increasing preterm births and also harming the environment [18]. DDT was found to cause behavioural anomalies and eggshell thinning in populations of bald eagles and peregrine falcons.

So back to the question from title, *What problems emerge from soil pollution?* Main effects of soil pollution are
1. effect on health of humans
2. effect on growth of plants
3. decreased soil fertility
4. toxic dust.

Contaminated or polluted soil directly affects human health through direct contact with soil or via inhalation of soil contaminants which have vaporized. The implication of soils to human health is direct such as ingestion, inhalation, skin contact and dermal absorption. Some epidemiological examples include geohelminth infection and potentially harmful elements via soil ingestion, cancers caused by the inhalation of fibrous minerals, hookworm disease and podoconiosis caused by skin contact with soils [19]. Elliott et al. [20] have found small excess risks of congenital anomalies and low and very low birth weights in populations living near landfill sites. Different contaminants have different effects on human health and the environment depending on their properties. The contaminant effect depends on its potential for dispersion, solubility in water or fat, bioavailability, carcinogenicity and so forth [15].

Not unexpectedly, soil contaminants can also have significant deleterious consequences for ecosystems [21]. An example is already mentioned DDT pesticide. Soil contaminates can alter metabolism of some microorganisms resulting in eradication of some of the primary food chain, which in turn could have major consequences for predator or consumer species. Negative effect is also present in plant metabolism. Result is often reduced crop yields. This has a secondary effect upon soil conservation, since the languishing crops cannot shield the Earth's soil from erosion.

As for air pollution, there are some clean-up options to reduce already made mess. Most of them are based on remediation strategies:
1. *excavate soil and take it to a disposal site*
2. *aeration of soils at the contaminated site* (enhances the risk of creating air pollution)
3. *thermal remediation*
4. *bioremediation*
5. *extraction of groundwater* or soil vapour with an active electromechanical system
6. *containment of the soil contaminants* (such as by capping or paving over in place)
7. *phytoremediation*, or using plants (such as willow) to extract heavy metals
8. *mycoremediation*, or using fungus to metabolize contaminants and accumulate heavy metals and
9. *remediation of oil contaminated sediments with self-collapsing air microbubbles* [22].

For more details about soil treatment environmental engineering see Chapter 9.

1.3.3 Water quality – Are water wars our future?

Water occupies two-third of Earth's surface and that makes it not only a valuable resource for humans but entire ecosystem. By the growth of human population it will be utilized in more and more activates associated with humans. That will undoubtedly lead to reduction of overall water quality and poorer water quality usually means water pollution. Besides that, water pollution can be defined in many ways. Usually, it means one or more substances have built up in water to such an extent that they cause problems for animals or people.

Factors that contribute to water pollution can be categorized into two different groups:

1. *Point sources* that are the easiest to identify and control. Some of sources are factories, sewage system, power plants, underground coalmines, oil wells, etc.
2. *Non-point sources* are ambiguously defined and harder to control. Some of the sources are:
   - when rain or snow moves through the ground and picks up pollutants as it moves towards a major body of water,
   - the runoff of fertilizers from farm animals and crop land,
   - air pollutants getting washed or deposited to earth and
   - storm water drainage from lawns, parking lots, and streets [23, 24].

Regardless the source, in the last decades it became more then obvious that water is becoming a serious environmental problem. Due to environmental impact, drought and flooding are present in many parts of the world. Both of these events strongly impact food supplies and economy. On the other hand, being a universal solvent, water is a major source of infection. According to WHO 80% diseases are water borne. Infectious diseases, like cholera, typhoid fever and other diseases gastroenteritis, diarrhoea, vomiting, skin and kidney problem are just some of them [25].

Drinking water in various countries does not meet WHO standards. Water pollution causes approximately 14,000 deaths per day, mostly due to contamination of drinking water by untreated sewage in developing countries [26].

Not only humans but, as mentioned, entire biosystem is affected by water pollution. Biomass and diversity of communities are to be effected when large amount of toxic materials are released into the streams, lakes and coastal waters in the ocean. This waste can increase secondary productivity while altering the character of the aquatic community. Most fishes especially the species desired as food by man are among the sensitive species that disappear with the least intense pollution.

Direct damage to plants and animals nutrition also affects human health. Plants nutrients including nitrogen, phosphorus and other substances that support the growth of aquatic plant life could be in excess causing algal bloom and excessive weed growth. This makes water to have odour, taste and sometimes colour. Ultimately, the ecological balance of water is altered [26].

Additionally, approximately 97.5% of all water is either salt water or water that has become polluted. Of the remaining 2.5%, nearly 70% is frozen in glaciers and the polar ice caps. Less than 0.01% of all water worldwide is available for human use in lakes, rivers, reservoirs and easily accessible aquifers. Combining all of this with all that is mentioned previously and the fact that conflicts over water, both within countries and between countries, are sharply increasing human future is becoming slightly shaken. Fortunately, up to now few of these conflicts have led to violence [27]. But who knows what will happen tomorrow.

For more details about water and waste water treatment engineering see Chapter 8.

1.4 What is the future of environmental engineering?

It is said that *environmental engineering has a proud history and a bright future.* It is a career that may be challenging, enjoyable, personally satisfying and monetarily rewarding [28]. Predictions are that employment of environmental engineers will grow by 15% from 2012 to 2022, which is faster than the average for all occupations. If we think beyond 2022 and further in to the future, it is expected that technology development will continue growing bringing new challenges for the environment. Population will also grow which will enhance the need for resources, food and water supplies, cities will get overpopulated, etc. Also, "going green" move became more and more popular so new approaches to develop new "green" methods will be needed. Accent on prevention rather than controlling will be the main stream of the future. A new even more dangerous contaminates will probably emerge. A climate change will become more and more obvious.

All of this and more in the end will need proper coordination and management to avoid pollution or eco-damage. And this is where environmental engineering will build bright and prospers future.

1.5 Conclusion

At the end of this first chapter we can just hope that we managed to awaken interest for this broad field. Field, which will rapidly change with development of human society. One can just hope that it will keep the step, walk shoulder to shoulder one next to another, otherwise, if it stutters a little it will have deep consequences on environment and society.

Also, following chapters of this book will give wider picture of problems environmental engineering is facing and hopefully give enough tools to fight all challenges that the future is bringing.

References

[1]   Peavy HS, Rowe DR, Tchobanoglous G. Environmental engineering. UK: McGraw-Hill; 1985.
[2]   Osgood N. 1.040 project management. Cambridge, USA: Massachusetts Institute of Technology: MIT OpenCourseWare, 2004 Spring. https://ocw.mit.edu. License: Creative Commons BY-NC-SA.

[3]   Tonelli F, Evans S, Taticchi P. Industrial sustainability: challenges, perspectives, actions. Int J Business Innovation Res. 2013;7:143–63.

[4]   Pereira JC. Environmental issues and international relations, a new global (dis)order – the role of international relations in promoting a concerted international system. Rev Bras Polít Int. 2015;58:191–209.

[5]   Salati E, Dos Santos AA, Klanin I. Relevant environmental issues. Estudios Advancados. 2007;21:107–27.

[6]   Anand SV. Global environmental issues. Sci Rep – UK. 2013;2:632–40.

[7]   Pfafflin JR, Ziegler EN. Encyclopaedia of environmental science and engineering, Volume 1 A-L. Boca Raton, Florida: CRC Press Taylor & Francis Group 2006.

[8]   Laumbach RJ, Kipen HM. Respiratory health effects of air pollution: update on biomass smoke and traffic pollution. J Allergy Clin Immunol. 2012;129:3–13.

[9]   Del Donno M, Verdur A, Olivieri D. Air pollution and reversible chronic respiratory diseases. Monaldi Arch Chest Dis. 2002;57:164–66.

[10] Vallero DA. Fundamentals of air pollution. USA: Elsevier Academic Press; 2014.

[11] WHO, 7 million premature deaths annually linked to air pollution". New relies, 25 March 2014. (http://www.who.int/mediacentre/news/releases/2014/air-pollution/en/) Accessed 1 January 2018.

[12] Marshal F, Ashmore M, Hinchcliffe F. A hidden threat to food production: air pollution and agriculture in the developing world, GATEKEEPER SERIES No. 73

[13] Policy highlights: The economic consequences of outdoor air pollution. OECD better policies for better lives. Paris, France: OECD, 1–20.

[14] Eurostat, Environmental data centre on waste, (http://epp.eurostat.ec.europa.eu/portal/page/portal/waste/introduction) Accessed 1. January 2018.

[15] Panagos P, Van Liedekerke M, Yigini Y, Montanarella L. Contaminated sites in Europe: review of the current situation based on data collected through a European network. J Environ Public Health. 2013;2013:1–11.

[16] Lewis TA, Newcombe DA, Crawford RL. Bioremediation of soils contaminated with explosives. J Environ Manage. 2004;70:291–307.

[17] Ranieri E, Bombardelli F, Gikas P, Chiaia B. Soil pollution prevention and remediation. Appl Environ Soil Sci. 2016;2016:1–2.

[18] Shabbira A, DiStasioa S, Zhaob J, Cardozob CP, Wolff MS, Caplana AJ. Differential effects of the organochlorine pesticide DDT and its metabolite p,pV-DDE on p-glycoprotein activity and expression. Toxicol Appl Pharmacol. 2005;203:91–98.

[19] Abrahams PW. Soils: their implications to human health. Sci Total Environ. 2002;291:1–32.

[20] Elliott P, Briggs D, Morris S, De Hoogh C, Hurt C, Kold Jensen T, et al. Risk of adverse birth outcomes in populations living near landfill sites. Bmj. 2001;323:363–68.

[21] Hogan M, Patmore L, Latshaw G, Seidman H. Computer modelling of pesticide transport in soil for five instrumented watersheds, prepared for the U.S. Environmental Protection Agency Southeast Water laboratory. Ga A, edited by. Sunnyvale, California: ESL Inc.; 1973.

[22] Agarwal A, Zhou Y, Liu Y. Remediation of oil contaminated sand with self-collapsing air microbubbles. Environ Sci Pollut Res Int. 2016;23:23876–83.

[23] Moss B. Water pollution by agriculture. Phil Trans Royal Society B. 2008;363:659–66.

[24] Hogan CM. Water pollution. Encyclopedia of earth. Washington, DC: National Council on Science and the Environment; 2010.

[25] Juneja T, Chauhdary A. Assessment of water quality and its effect on the health of residents of Jhunjhunu district, Rajasthan: a cross sectional study. J Public Health Epidemiol. 2013;5: 186–91.

[26] Owa FD. Water pollution: sources, effects, control and management. Mediterr J Soc Sci. 2013;4:65–68.
[27] Levy BS, Sidel VW. Water rights and water fights: preventing and resolving conflicts before they boil over. Am J Public Health. 2011;101:778–80.
[28] Weiner R, Matthews RA. Environmental engineering. 4th ed. USA: Elsevier Science; 2003.

Ivančica Ternjej and Zlatko Mihaljević

2 Ecology

Abstract: Ecology is a science that studies the mutual interactions between organisms and their environment. The fundamental subject of interest in ecology is the individual. Topics of interest to ecologists include the diversity, distribution and number of particular organisms, as well as cooperation and competition between organisms, both within and among ecosystems. Today, ecology is a multidisciplinary science. This is particularly true when the subject of interest is the ecosystem or biosphere, which requires the knowledge and input of biologists, chemists, physicists, geologists, geographists, climatologists, hydrologists and many other experts. Ecology is applied in a science of restoration, repairing disturbed sites through human intervention, in natural resource management, and in environmental impact assessments.

Keywords: elements of ecology, environment, habitat, biodiversity

2.1 Introduction

The phrase ecology is used often today and has a wide meaning. Ecology is referred to as a science, ecology as nature, ecology as an idea and ecology as a movement. In this chapter, ecology is used to refer to ecology as a science that studies the mutual interactions between organisms and their environment. For practical reasons, ecology is typically divided into plant ecology, animal ecology and human ecology. One of the first to describe the significance of ecological observations was renowned natural historian Charles Darwin, in his book *On the Origin of the Species* [1]. Darwin explained numerous examples of evolution (adaptation, natural selection, fighting for survival and extinction of the species) which included an entire range of mutual relations between organisms and their interactions with the conditions of their environment. The word ecology (Greek: *oikos* = house, home, habitat; *logos* = study of, knowledge) was first coined by the German biologist Ernst Haeckel in his book *Generelle Morphologie der Organismen* [2].

The fundamental subject of interest in ecology is the individual, the plant or animal organism. Individuals of plants or animals do not live isolated, but are typically grouped, and such groups of similar individuals are called species. Though biologists

This article has previously been published in the journal Physical Sciences Reviews. Please cite as: Ternjej, I., Mihaljevic, Z. Ecology. *Physical Sciences Reviews* [Online] **2017**, *2* (10). DOI: 10.1515/psr-2016-0116

https://doi.org/10.1515/9783110468038-002

have not agreed on a single definition of the species, most accept that species are groups of individuals that share similar morphological and genetic traits, and individuals of a species are able to reproduce with one another and create fertile offspring. Today, ecology is a multidisciplinary science: in order to provide answers to many of the questions this field of science deals with, cooperation is needed from many different fields. This is particularly true when the subject of interest is the ecosystem or biosphere, which requires the knowledge and input of biologist, chemists, physicists, geologists, geographists, climatologists, hydrologists and many other experts.

Humans, in addition to consuming massive amounts of energy, have polluted nature with industrial waste, harmful gases, pesticides, heavy metals and organic compounds, thereby drastically reducing the diversity of plant and animal species, which has started to change the ecological balance. The altered environment negatively impacts human health and is even gradually threatening human survival (climate change, greenhouse effect, destruction of the Earth's ozone layer). Industry – the foundation for global prosperity – has created significant global threats to humankind's opportunities for continued survival. Therefore, future development should be based on protecting the environment, in the form of sustainable development and management. This implies that the use of natural resources must be tolerable, allowing those resources to regenerate and remain in ecological balance. If we master the fundamental principles of ecology, we will be more easily able to create the conditions for the rational use of natural resources of the biosphere, and to improve relations between humans and their natural environment.

2.2 Conditions for life

Many factors influence the distribution and survival of species. Generally, they can be divided into factors of the non-living environment – abiotic – and factors of the living environment – biotic factors. The habitat, or biotope, as the area where a given species lives, is characterised by a specific complex set of ecological factors. None of these factors acts alone, and so organisms are distributed only in those areas where the gradients of specific factors in play overlap in a suitable way. The range of fluctuation of a single ecological factor, within which the life of a certain organism is possible, is called ecological valence (Shelford's law of tolerance) (Figure 2.1). However, the effect of one ecological factor may not be within the limits of the ecological valence, and organism could not survive. For each ecological factor, there is an optimal range, and a minimum and maximum. At both the minimum and maximum, the intensity of ecological factor corresponds to the stress zone, or area pessimum (from lat. pessimus – very bad). Optimum is that quantity or intensity of the ecological factor whose intensity is most suitable for the organism and the consequence is his maximal growth and development. The concept of the limiting effect of an ecological factor was introduced in ecology by V. E. Shelford [3], and it is valid for all ecological factors and for all organisms.

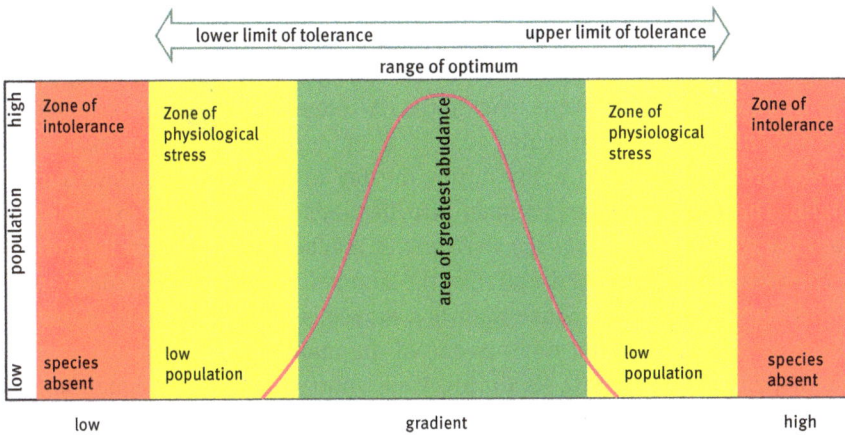

Figure 2.1: A schematic representation of the ecological valence [inspired by [4].

Organisms with a wide range of tolerance for an ecological factor are called eurivalent species, while organisms with a narrow range of tolerance for an ecological factor are called stenovalent species. What is interesting is that one single life factor at a minimum can exclude a certain species from a given habitat, regardless of the fact that all other factors may be at optimum ranges.

2.2.1 Abiotic factors

Temperature and moisture are key ecological factors that determine the distribution and development of plants, and based on these factors, we can differentiate different vegetation zones on Earth. For animals, vegetation is an important factor as shelter and a source of food. In areas where ecological conditions are unfavourable, and the plant cover is poorly developed, the diversity and abundance of animals is also lower, due to the scarcity of suitable shelters and available food. Many animals are distributed only with certain temperature limits that are suitable for their survival, and the very distribution of climate areas on Earth becomes a barrier to their broader dispersal. Monkeys and parrots are primarily distributed in the tropical and subtropical zones; the polar bear, walrus and some diving birds are distributed in the polar regions. Another important ecological factor is light. The rhythmic change of light is a dependent factor for many of the daily and seasonal phenomena and activities in plants and animals. Day length is a factor that determines when leaves begin to fall in many deciduous tree species in moderate climate zones. There are long- and short-day organisms. For example, barley is a long-day plant and blooms under long-day light, while in tobacco, long days prevent blooming. Constant daylight is a reason why plants in the tropics bloom year round, and why polar plants fail to thrive in other parts of the world, as they are adapted for long days. Trout spawns at a temperature of 6 – 7 °C (winter), though spawning can be achieved experimentally in periods of higher temperature (summer) just by reducing the amount of daylight. Shortening daylight also stimulates migrating birds to fly south. How organisms react to the rhythmic changes in the intensity of light – day, night – is of particular ecological importance. Among mammals in temperate forests 30–40 % are active during the day and 60–70 % are active during the night. The distribution of animals also depends on light. For example, in Norway, the wood mouse (*Apodemus sylvaticus*) reaches the maximum latitude of 62°N. The reason for this is the duration of the summer nights, which in that part of the world lasts only 5 hours. Since the wood mouse is a nocturnal animal (active and feeds at night), a shorter night is one of the main reasons why this animal cannot disperse further north, as during such short nights, this animal simply cannot feed enough to ensure its survival.

Specific behaviour patterns are also tied to changes in photoperiod and temperature, such as bird migrations.

In aquatic ecosystems, primary producers (phytoplankton, macrophytes and sea grasses) are limited to the depth of light penetration, water temperature and the quantity of available nutrients, while for animals, the quantity of metabolic gases (oxygen and carbon dioxide), salinity and pressure (in the sea) are important factors.

2.2.2 Biotic relationships

Population

The living area of a species consists of a multitude of intertwining abiotic factors that affect the organism. However, organisms of the same species, and those of other species, also come into contact and affect one another. A population is a group of individuals of the same species in a given area. Each population has their own dynamics and properties: a size of the population, spatial distribution, age structure, sex ratio, birth, death, immigration and emigration rates. The spatial distribution of individuals in the environment could be caused by the distribution of resources or by social relationships among the individuals.

Mutual interactions and relationships between individuals of the same species are called intraspecific relations. If a certain resource essential for the life of the population is limited, this will result in competition among the population members (intraspecific competition), which can limit the population size. When ecologists investigate endangered species, the size of the surviving population is of crucial importance in the design of proper management. Population size also depends on other population characteristics: natality, mortality, survivorship, reproduction potential and dispersion. Populations are increased through the birth of new individuals (natality) and the arrival of new individuals (immigration), while they are reduced through death (mortality) and those leaving the population (emigration).

Population growth can be exponential or logistic. Exponential population growth is characteristic in invertebrates and some vertebrates (rodents) while colonising new habitats, when the two most important resources – space and food – are not yet limiting (Figure 2.2). But no real population can for long continue increasing exponentially and are instead stabilised at a certain level.

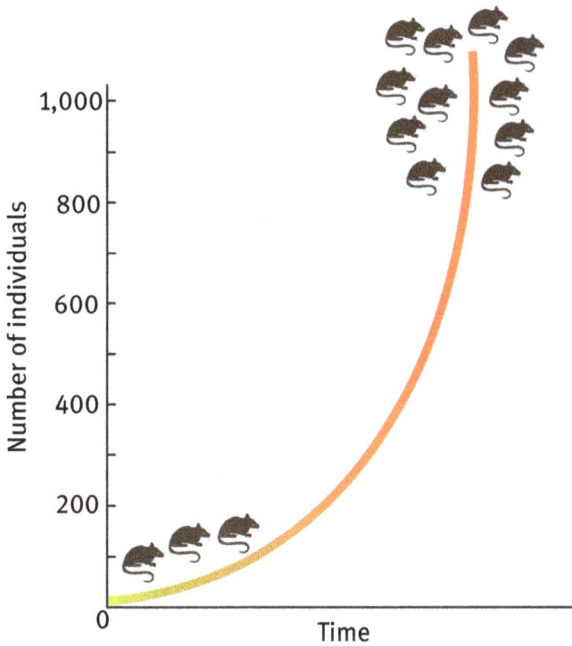

Figure 2.2: The exponential growth curve [inspired by [5].

Logistic growth is caused by the limitation of resources in the habitat. It is characterised by the carrying capacity (K) of the environment, which indicates the maximum population size that the habitat can support with its resources. For example, sheep brought to the Australian island of Tasmania in 1800 experienced a rapid population expansion to almost two million individuals. The result was massive competition for food and by 1885, their numbers had declined and stabilised, to the level of about 1.6 million individuals (Figure 2.3).

The terms r-selection and K-selection have been used by ecologists to describe the growth and reproductive strategies of various organisms. The r and K classification was originally proposed by R. McArthur and O.E. Wilson [6]. R-selected species live in variable, unstable habitats, while K-selected species reside in stable environments. Small body size, early reproductive maturity, high fecundity and low parental

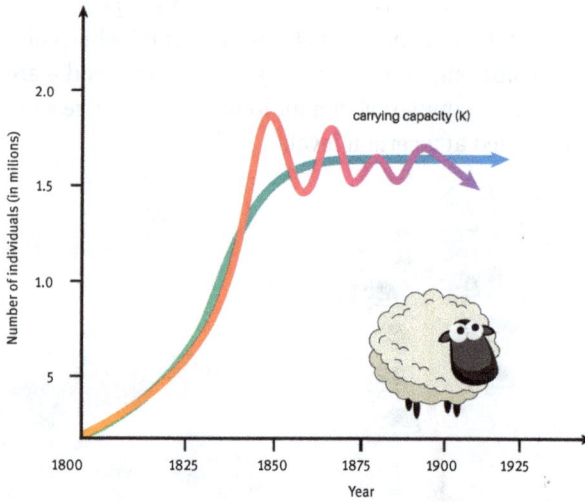

Figure 2.3: The logistic growth curve. When the carrying capacity is reached, there is no further increase in density [inspired by [5].

investment are main characteristics of the r-selection species. Flies and most invertebrates are an example of an r-selected species. The extreme opposite is K-selection species with low fecundity and high parental investment. Most mammals are examples of K-selected species.

2.2.2.1 Interspecific relationships

Relationships between different species (populations) are called interspecific relationships and imply an entire series of mutual influences, such as competition, predation, parasitism, mutualism and commensalism. For example, one species may feed on another, and is, therefore, dependent on it.

Competition

Species can compete for the same food source or space, and thus they are in competition. An interesting example is the barnacles *Chthalamus stellatus* and *Balanus balanoides*. Both species inhabit the tidal zone, though *Chthalamus* always is found in higher areas and *Balanus* in lower areas of that zone. Connell [7] conducted experiments with these barnacles: he moved the rocks housing *Chthalamus* to lower areas and those housing *Balanus* to higher areas, and he physically removed the *Balanus* from the rocks in the lower areas and *Chthalamus* from the rocks in the higher areas. Connell found that *Chthalamus* was able to survive normally in the lower level on the rocks he had placed there. However, *Balanus* did not fare well on the rocks in the higher zone, as this species is less tolerant to drying out. It is interesting that Connell also took photographs of larvae of both species equally colonising the lower zone. However, as *Balanus* individuals grow, their base expands more rapidly and they literally expel the *Chthalamus* larvae from the rocks. On the one hand, the ecological factors act on the vertical distribution of these species; while on the other hand, during colonisation of rocks the acting factor is competition.

Competition can be so great that one species completely expels another from a habitat. This is called competitive exclusion. Many allochthonous (alien) species can also expel autochthonous (indigenous) species through this process of competitive exclusion. In many clean rivers still in a relatively natural state, the introduction of rainbow trout (*Oncorhynchus mykiss*) has resulted in significant reductions of populations of the native brook trout (*Salmo trutta*). The introduction of alien and invasive species has particularly impacted the Dinaric karst rivers of the Adriatic basin, which abound in endemic species [8]. Another example is the introduction of the American grey squirrel (*Sciurus carolinensis*) in Great Britain between 1876 and continued to 1915 or later. By the mid-1920s, the population of the native red squirrel (*Sciurus vulgaris*) had declined drastically due to disease. The spread of introduced species co-occurred with the declining population of the indigenous species. It has been confirmed that the grey squirrel spread in those habitats not inhabited by the domestic red squirrel, and also into those where the domestic squirrel population was in decline due to disease. The introduced species was stronger in competition for food sources and habitat, and also more resistant to disease, and therefore drove out the native species in a great part of Britain. The domestic red squirrel has maintained large populations only in certain isolated localities, which had not been exposed to the epidemic (e. g. on the Isle of Wight off the south-east coast of Great Britain) [9].

There are countless examples of competition. However, most species having the same needs often share the same habitat. This is considered coexistence, indicating that they have somehow succeeded in overcoming or reducing the competition between them. One way different species use the resource is to use it at different times. In forests, predator bird species, such as falcons and buzzards, hunt by day, while owls hunt by night. Both groups feed on small mammals. The macrozooplankton of Visovac Lake in Krka National Park (Croatia) consists of four species of water fleas. Of these, two species are in direct competition for food, and both belong to the same order of water fleas (*Daphnia*) [10]. The species *Daphnia longispina* is small and feeds primarily on algae from 15–20 µm in size. The second water flea species, *Daphnia cucullata*, feeds on green algae of the same size. However, unlike the first species, this species can also feed on slightly larger amounts of small organic particles. The annual cycle of these two crustacean species is shown in Figure 2.4.

In this case, the spatial and temporal distribution of these species is primarily determined by food sources. It can be noted that *D. cucullata* develops the densest populations in June, when the species *D. longispina* is not present yet in the macrozooplankton. Namely, this species begins to appear in July, and when it does, it slowly forces out the first species. During July, their populations are separated spatially: *D. cucullata* is found at depths of 10 and 20 m, while *D. longispina* is

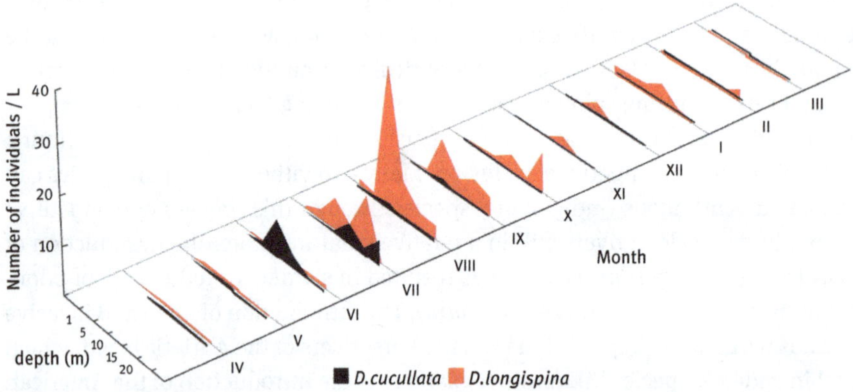

Figure 2.4: Niche separation in space and time between two *Daphnia* species in Visovac Lake.

found at depths of 5 and 15 m. However, for the remainder of the year, only *D. longispina* is present in the macrozooplankton.

As explained in these examples, different species can coexist by separating their ecological niches, either spatially or temporally. Most often, the term ecological niche implies the position of a species in the community in terms of food, which can be considered its microhabitat. It could be said that the habitat of a species is its "address," while its niche is its "profession"! In that way, species living at the same "address" need not have the same "profession." However, this understanding of the ecological niche has been gradually evolving over the past 30 years. Today, many biologists feel that the ecological niche is not determined only by the feeding methods, but also by all the other ecological factors to which a species has adapted.

There is yet another way to avoid competition. In Asia Minor, the distribution ranges of two passerine bird species overlap: the western rock nuthatch (*Sitta neumayer*) and the eastern rock nuthatch (*Sitta tephronata*). Both species feed on insects in the soil. The size of the beak of each species determines the size of prey it can catch. Both nuthatches have a beak 26 mm in length, and therefore, it could be assumed that they will feed on insects of the same size, and that they will be in direct competition with one another. However, this is not the case in nature. In those areas where their distribution ranges overlap, the western rock hatch develops a beak 24 mm in length, while the beak of the eastern rock nuthatch grows to 29 mm! In areas where they do not overlap, and where they are not in competition, both species have a beak 26 mm in length [11]. This phenomenon is called character displacement. It is seen in the fact that species that come into competition in overlapping range areas will change those characteristics that create the competition between them.

Is there competition among plants? Plants resolve possible competitors through allelopathy. This is a phenomenon in which some plants secrete chemical compounds into the soil to inhibit the growth of other plants. For example, walnut or sage secretes compounds into the soil that will disable the group of other plant species to with a radius of several metres from the plant.

Predation

Another important biotic factor that can influence the distribution of a given species is predation. This is the relationship in which one organism, the predator, consumes another, the prey. The prey is alive when the predator first attacks! Predators can overhunt and completely eliminate prey species from a certain area, or they can disable range expansion and colonisation of new areas by potential prey. Though there is theoretical support for such a possibility, there is no actual proof that this occurs in nature. Such extreme events are only known to occur in cases of introductions of species that are not native to an area.

The influence of a predator on the prey population is evidently detrimental. However, predator effects on a prey population are not always easily predictable. There are two reasons for this: prey individuals that are killed or damaged do not always correspond to the random pattern of the entire prey population. Predators tend to select the prey that is easiest to catch: the young (and naïve), the old (and weak) and the sick (and helpless). Second, predation can impact the growth, survival and reproduction of prey. In other words, while predation is harmful to the prey individuals that are captured and consumed, it can be favourable to those individuals not captured. The case of the Kaibab deer is well known [12]. The initial population of 4,000 of these animals on the Kaibab Plateau, located in northern Arizona in the United States along the northern border of the Grand Canyon, grew to 100,000 in just 20 years (1905–1924), as humans hunted and eliminated their natural predators (the cougar and coyotes), Figure 2.5. This massive population growth resulted in great competition, primarily for food resources, and within the next 15 years, the population

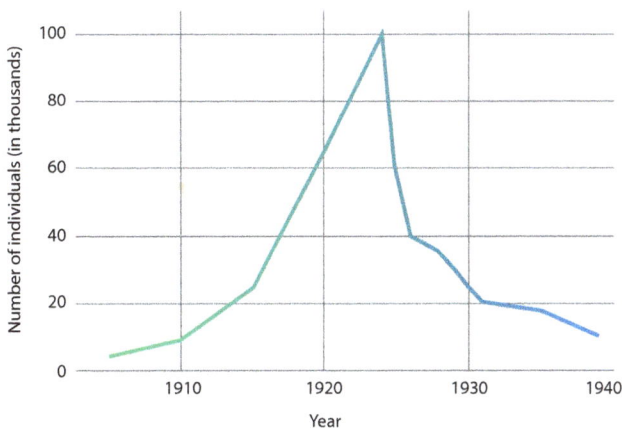

Figure 2.5: Changes in abundance of Kaibab deer after elimination of natural predators [inspired by www.biologycorner.com/worksheets/kaibab_key.html, Keeton Biological science].

declined to 10,000 individuals. Therefore, it could be concluded that the relationship between predator and prey is what maintains the natural balance!

It is also important to note which prey individuals are under the effect of predators and during which life stage. Young individuals cannot yet contribute reproductively to the population, and therefore the effects of predation of this segment of the prey population are less than if the predator catches a reproductively mature individual. The impact of a predator on prey is not constant, and is instead limited to certain time periods. For example, the boa constrictor (*Boa constrictor*) will rest for a while after eating its prey, while it digests. The success rate of hunting prey is not always 100 %, as prey may escape. Also, some predators take the time to learn how to hunt, and their efficacy will change with age: the young tiger or lion are less experienced and effective than adult individuals.

Parasitism

Parasitism is a relationship that is similar to predation. It could be said that parasitism is a special form of predation, in which the predator is most often much smaller than its prey (host). Unlike "classical" predators, the parasite typically does not kill its prey. This is a relationship in which the parasite has a benefit, while the host typically experiences more or less damage. As in predation, parasites are one means of regulating the natural balance. They maintain their host populations to a certain level, to ensure that they will not be left without a food source. However, introduced parasites can cause significant changes in the distribution of potential host populations. Lake trout (*Salvelinus namayscush*) is a widely distributed species throughout northern North America, but it is not adapted to the presence of parasitic lampreys (*Petromyzon* and *Entopshenus*). These lampreys attach to fish and feed off their tissue and bodily fluids. Niagara Falls is a natural barrier preventing the entry of these parasites into the Great Lakes, where large populations of lake trout thrived. However, with the construction of the Welland Canal, connecting Lakes Ontario and Erie, the lampreys spread throughout all the Great Lakes and caused substantial declines in the lake trout populations [13].

A similar effect was seen following the introduction of the parasite that causes malaria in birds (*Plasmodium*) to the Hawaiian Islands. The endemic bird fauna of these islands suffered significant population declines of some species, range contractions and even extinctions [14].

Mutualism

Mutualism is a relationship in which both species benefit. This relationship can imply a close physical interaction between members. Typically, one member of such a pair becomes completely dependent on the other, and for this species, this mutualism is obligatory. In extreme cases, in their interactions, these organisms begin to function as a single individual, as in the case of fungi and algae in lichens. One such example of obligate mutualism is the protozoans and bacteria in the digestive tracts of certain animals, such as ruminants or termites. Protozoans in the digestive tract of termites break down wood into compounds the host can digest further. Neither of these organisms would be able to survive on their own. Mycorrhiza is another form of mutualism. This is a symbiosis between the roots of a vascular plant (tree) and fungi. The fungi supply the tree with nutrients, while the plant supplies the fungi with carbon. However, mutualism may also be facultative (not obligatory), in which mutualists can survive normally without one another. The most common example of this form of mutualism is the hermit crabs and certain sessile anemones. The hermit crab settles in an empty snail shell, upon which one or more anemones can settle. The anemones defend the crab with their stinging cells, and since they are sessile organisms, the crab serves as their "means of transport." However, both of these organisms can survive without the other.

A special form of mutualism can also be seen in the relationship between certain plants and insects (or birds) as pollinators. Coevolution has gone very far here, as can be witnessed in the form of flowers, colouration and special mechanisms developed by plants and their pollinators. The most unusual examples of mutualism can be seen in the relationship between ants and certain plants. The South American species of acacia (*Acacia*) have thorns in which some ant species of the genus *Pseudomyrmex* built their nests [15]. The plant contains the protein-rich Beltian bodies (resembling berries) and the sweet fluids that the ants feed on. In return, they protect the plant from attack by other herbivorous insects and vertebrates. They also eliminate the surrounding vegetation, thus reducing competition with other plants. This specialisation has resulted in identical distribution ranges for these two species.

Commensalism

Commensalism is a relationship in which one species benefits, and the second neither benefits nor suffers. The Adriatic Sea is home to small soft-bodied crabs of the family Pinnotheridae, which are in commensal relations with numerous bivalves, ascidians and the pipes of the polychaete worms (Polychaeta). Among the many genera and species, the most common are the crabs that inhabit the noble pen shell and the crabs that inhabit oyster and mussel shells. The bivalves are not disturbed by these small crabs, which are seeking shelter from predators. Another example of commensalism is epiphytic plants. These are mostly tropical plants characteristic as living on the branches of other plants. They use only space on which organic material and water has collected, and have no other impact on the host plant.

2.2.3 Community

No organism in nature lives in complete isolation. A collection of organisms interacting directly or indirectly in a habitat is a community. The concept of community is based on the efforts of botanists to group certain plant formations. In the early twentieth century, American scientist Frederic Clements [16] proposed that the plant units, which he called the "*association*," are strictly determined and fixed for a long period of time. Today, the term "*community*" involves a group of all living organisms in the same habitat where among them there are relations of mutual dependence and biological connections. Their interaction is important for the functioning of the community.

2.2.3.1 Community structure

The community is characterised not only by a mixture of species, but also by physical features. On land, the vertical structure of the community is mostly determined by plants. Their height, branching and the amount of leaves, determine temperature, moisture and light vertical gradient (Figure 2.6). Plants make the framework of the community in which animal species are adapted to. A well-developed forest ecosystem contains several layers of vegetation. There are the canopy layer, the understory, the shrub layer, ground layer and the forest floor. The most important role in the fixation of sun energy has a canopy. Aquatic systems, like lakes and oceans, are also vertically stratified primarily by penetration of light.

Figure 2.6: Forest vertical structure and light vertical gradient.

The horizontal pattern of community in space is variable. Usually, it includes differences due to latitude and altitude. Increasing latitude and altitude means more extreme conditions, and therefore the composition and appearance of the community changes.

2.2.3.2 Community sustainability

In the early twentieth century, two American scientists, Frederic Clements and Henry Gleason, prompted the question of community sustainability (equilibria).

Clements [16, 17] developed an *"organism concept of community."* In the concept, each species in the community representing an interacting component of the whole. The main processes shaping the community are mutualism and coevolution (predator–prey or competition interactions). On the other hand, Gleason [18, 19] supported *"the individualistic continuum concept."* The community very rarely (almost never) reaches a state of equilibrium. Communities, therefore, are not highly regulated species groups, but rather random populations of plants and animals, in which each individual responds to changes in the environment specifically. Biological factors, such as competition and predation, have no significant impact on community sustainability or equilibria.

Many studies have been conducted in order to prove one of these theories in the last 50 years. Some have shown that predation and competition play an important role in regulating the community. Rapid change in the physical environment can strongly affect the composition and structure of the community. Climate change, weather or natural disasters can also affect communities significantly.

2.2.3.3 Biodiversity

Biological diversity (biodiversity) is the variability of life on Earth. It includes genetic variation, ecosystem variation and species variation [20] within an area, biome or the whole planet.

Diversity is estimated on the basis of species richness (number of species) and species abundance (the number of individuals per species). For estimation of biodiversity, many indexes are used, the most common is Shanon index:

$$H = - \sum_{i=1}^{S} (p_i)(\log_2 p_i)$$

where H is a diversity of species, S is the number of species, and p_i is the proportion of individuals in total sample belonging to the ith species.

The number of species is changing over time (Figure 2.7). The oscillations of diversity and extinction rates are a consequence of changes in a number of factors: biotic and abiotic. It is believed that the biggest impact has a climate change [21]. Human activities disrupt the natural balance. This caused the extinction of many organisms, especially in the last century.

2.2.4 Ecosystem

The concept of the community includes only living organisms in a habitat but ignores the physical environment: nutrients and energy that allow living organisms survival. The ecosystem is a spatial or organisational unit of organisms and their physical environment. The environment is consisting of biotic and abiotic components linked together through nutrient cycles and energy flow. Simply said, all ecosystems (terrestrial and aquatic) have three important parts: producers, consumers and the non-living environment (Figure 2.8).

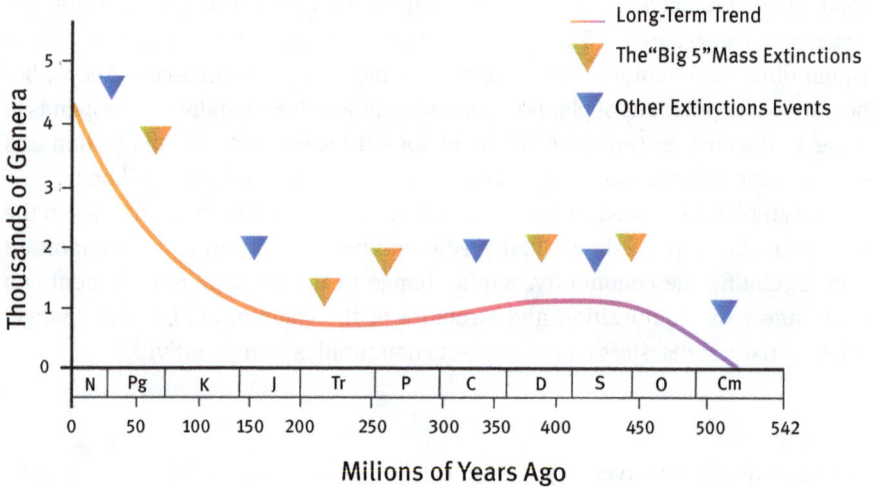

Figure 2.7: Apparent marine fossil diversity during the Phanerozoic; N – Neogene, P – Paleogene, K – Cretaceous, J – Jurassic, Tr – Triassic, P – Permian, C – Carboniferus, D – Devonian S – Silurian, O – Ordovician, Cm – Cambrian, inspired by [22].

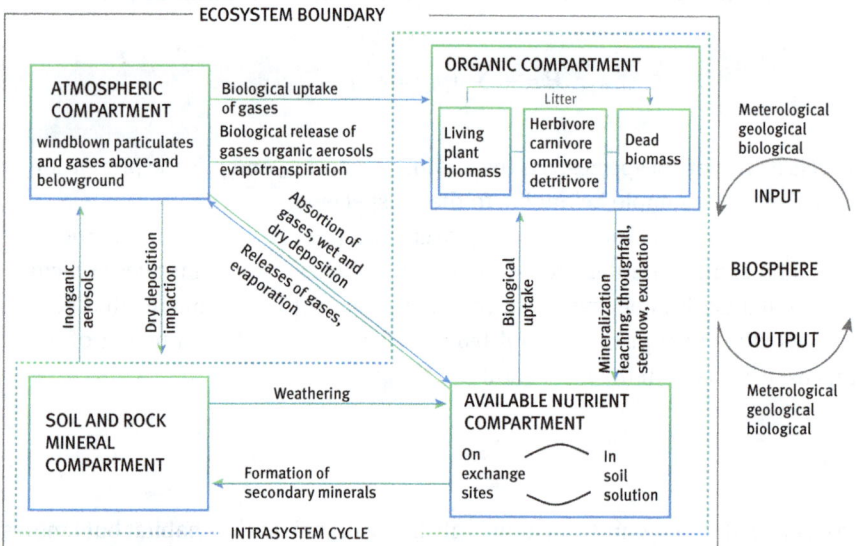

Figure 2.8: Schematic representation of the ecosystem; inspired by [23].

The energy that flows through ecosystems is obtained primarily from the sun. It generally enters the system through the process of photosynthesis. The energy accumulated in the process is called primary production. Total annual primary production on Earth is 162.41×10^9 t. Approximately 1/3 of it occurs in the sea, and the rest on the land. Among the most productive ecosystems in the world are wetlands; however, their share in global production is small because they cover a very small part of the earth's surface.

In terrestrial systems, production is determined by temperature and precipitation, because those directly affect the level of photosynthesis. In the oceans, production is limited by the penetration of light. The phytoplankton (and plants) are located mainly in the surface layers of aquatic ecosystems, where there is enough light. However, the primary production also depends on the amount of nutrients: nutrients from deeper parts must be transported in the surface layers.

The energy stored in plants is going through ecosystem in series of steps called food chains. A food chain shows how the organisms are related to each other by the food they eat. Each level of a food chain represents a different trophic level, but all of them starting from producer organisms (plants) and ending at apex predator species (carnivore). One kind of plant can be a source of food for many herbivores, but also many carnivores can hunt the same prey. So actually, in the ecosystem, many food chains interconnected in food webs.

Organisms in the food chain can be grouped according to the type of food they consume, in trophic levels. In general, all organisms that feed on other organisms are called consumers or secondary producer. Consumers can be divided into subgroups, each of which represents a single trophic level: those which feed on plants (primary producers) are called herbivores or primary consumers. Carnivorous (meat-eating) or predators feed on herbivores (plant eaters) and are called secondary consumers. Almost every habitat, including freshwater ecosystems, includes consumers who do not fit entirely into this simple scheme of trophic levels. These are consumers – decomposers or detritivores, they feed on dead remains of other organisms (detritus). Their role may at first glance do not seem so important, but they are an important link in the flow of energy and nutrient circulation in the ecosystem. Decomposers are involved in the degradation of dead plants and animals, returning nutrients in the cycle and thus allowing growth and development of primary producers.

The amount of energy that is transferred from one trophic level to another is very difficult to determine. When summed, biomass, or energy on each trophic level, constructs an ecological pyramid. It is also called Elton's pyramid, according to Elton [24]. In general, biomass and energy of producers must be greater than the biomass and energy of secondary producer: base of the pyramid determines the size of each of the next trophic level (Figure 2.9).

According to the second law of thermodynamics, as energy transfers between different trophic levels, there is a loss in the form of heat which limits the number of links in the chain. Highly productive systems do not support long food chains; however, they have a lot of species. This means that their food web is very complex. It is also known that ecosystems which are subject to rapid changes in abiotic factors

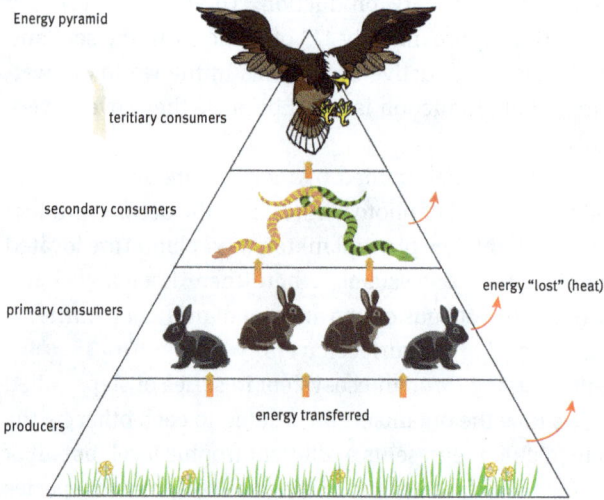

Figure 2.9: Energy pyramid.

have shorter food chains in relation to those stable. Highly stratified systems such as forests or pelagic ocean food chains are longer than those less stratified (meadows, tundra, benthal of a stream).

2.2.4.1 Biogeochemical cycles

All nutrients are transported through the ecosystem from the non-living to the living components in cyclic paths called biogeochemical cycles. The most important biogenic elements such as carbon, nitrogen, phosphorus and sulfur are bound in living organisms in the food chains and, through the process of decomposition of dead organisms, again returned to the lithosphere. Then the whole cycle begins again (Figure 2.10(a)). The main driver of all is the solar energy and is unusually important role for all biogenic elements has a water cycle (Figure 2.10(b)).

Carbon, hydrogen, oxygen, nitrogen, phosphorus and sulfur make 95 % of the biomass on Earth and are called macroelements. Macroelements together with calcium, silicon, magnesium, sodium, potassium, chlorine and iron (micro elements) consist 99 % of the dry weight of living organisms. Other biogenic elements (manganese, molybdenum, copper, vanadium, nickel) appear in much lower concentrations and a group of trace elements.

2.2.4.2 Ecosystem biodiversity and stability

One of the basic questions in ecology is: how many species is necessary for the normal functioning of the ecosystem? Is there a minimum or, if they are redundant species in the system? If there are redundant species: which species are necessary for the maintenance of the ecosystem?

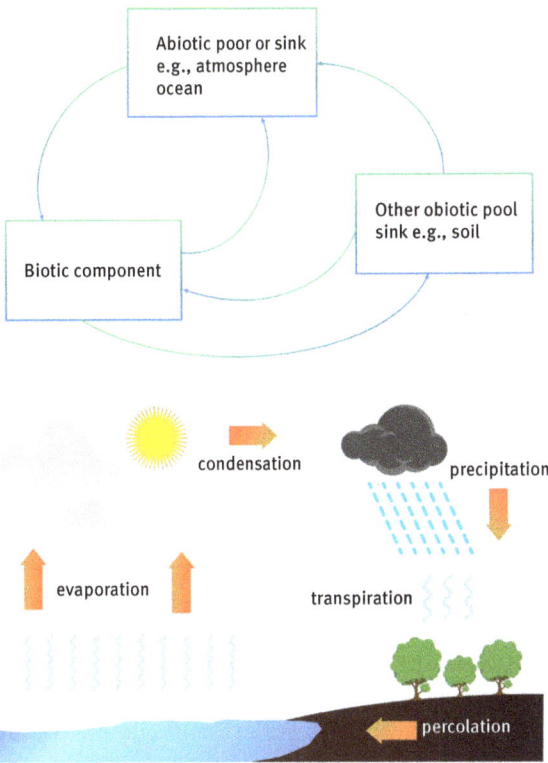

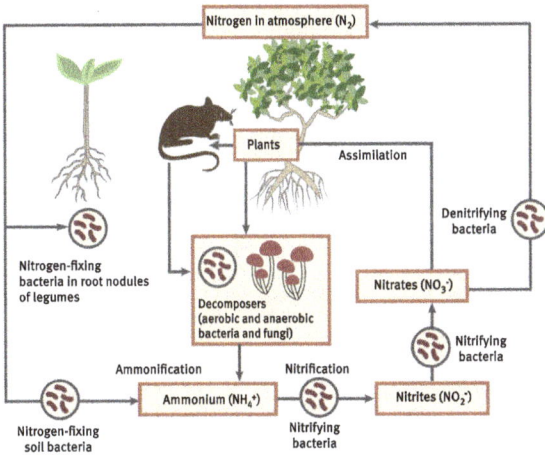

Figure 2.10: (a) nutrients cycle; (b) water cycle; (c) nitrogen cycle.

If we assume that all species are essential for the functioning of the ecosystem, then the loss of any causes imbalance: leads to degradation or collapse of the whole system (Figure 2.11). Maybe a diversity of species in ecosystem has no influence on its function, so the loss of any species causes random and unpredictable consequences.

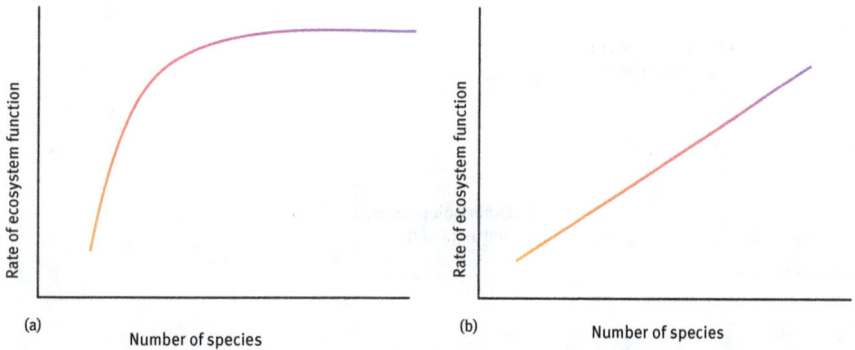

Figure 2.11: The potential relationship between the number of species in the ecosystem and its functioning (primary production, decomposition, circulation of materials); (a) redundant species hypothesis; (b) hypothesis dependent species [25].

These models were tested in the laboratory and *in situ*. The first laboratory research on this subject was conducted by John Lawton with a group of researchers at the Imperial College in London [26]. They found that ecosystems with a large number of species have more efficient primary production and developed a complex plant cover, but also accumulate more nutrients. Their results support the second hypothesis: that the reduction of biodiversity causes poorer functioning ecosystem. However, it is possible that further addition of species in a system causes a certain saturation point so that this experiment does not answer the question posed at the beginning. In any case, this research demonstrates that we can reject the third hypothesis: the functioning of the ecosystem is not random in relation to biodiversity.

The group of authors from the University of Minnesota conducted similar studies [27]. Their results were similar to Lawton's research: greater biodiversity – a better functioning of the ecosystem. However, they found that all species in the system are not equally valuable. The value indicates how much the loss of a certain species impacts the entire ecosystem, but also that some species depend on each other. For example, plants belonging to family (Fabaceae) legumes are very important for the nutrient cycling in the ecosystem, since their roots have symbiotic bacteria that can bind elementary nitrogen. Circulation of nitrogen is one of the basic biogenic cycles that is important for all organisms. By fixing nitrogen, these plants provide more nutrients in the soil, which in turn results in greater production of other plants, and animals, which are connected to them in the food chain. Therefore, the loss of legume species can have a major impact on the production of the entire system. These species are often called and "keystone species," in terms of their importance to the functioning of a system.

Another issue is of extraordinary importance when it comes to biodiversity and ecosystem function: whether biodiversity influences the stability of the ecosystem?

British ecologist Charles Elton during the 1950s observed that more complex and richer types of systems are less susceptible to the influence of sudden changes.

However, the development of mathematical modelling and the use of simulation in the population ecology did not to reach the same results. Models showed that the populations of some species even in diverse ecosystems are subject to fluctuations caused by external factors (disease, drought, etc.). It is possible that regardless of variations in individual species, the functioning of the entire ecosystem remains unchanged. So, the population of a certain species may fall or rise, but ecosystem survives because there are a lot of other populations that still maintain the function of the system.

2.2.5 Succession

Communities in an ecosystem are changing over time. Such changes are called succession. Actually, it would be more accurate to say that the succession is a series of changes in communities within a certain time period. Succession occurs in both terrestrial and aquatic systems. We distinguish between so-called primary and secondary successions. Primary succession takes place on previously completely bare surfaces (without life). Secondary succession happens to already populated habitats after a disturbance.

On August 28, 1883, at 10:02 pm there was an extraordinary disaster. The Indonesian island of Krakatoa exploded in one of the largest volcanic eruptions in recent history. Most of the island was blown off in the explosion. At the site of the former island of Krakatau was born two smaller islets: Anak Krakatau and Rakata. The island of Rakata was completely sterilised: the entire island was covered with a layer of lava 30–60 m thick, and all life on it was destroyed. The first expedition came to Rakata 8 months after the explosion and found only one living creature on the island: one spider! Scientists continue to explore settlement of animals and plants on this completely bare area. After 3 years, the first living organisms were observed on the island: the blue-green algae inhabited areas of cooled volcanic ash. Along came mosses, they are the pioneer species that first colonists of a habitat.

Pioneer species created the basis for the settlement of other organisms. After 25 years, the tropical forest appeared and with it the characteristic fauna of birds, snails, butterflies, bats, etc. In 1983, scientists counted a total of 420 taxa on Inland Rakata (Figure 2.12). During the colonisation of the island Rakata, several communities replaced one after another. In succession, plants are always preceded the animals, creating a habitat for them.

2.2.5.1 Climax

According to the classical ecological theory, succession ceases when species of late successional stages become dominant. At this point, the community has achieved stability, and any disturbance cannot significantly change the system. This final point of succession is called the climax.

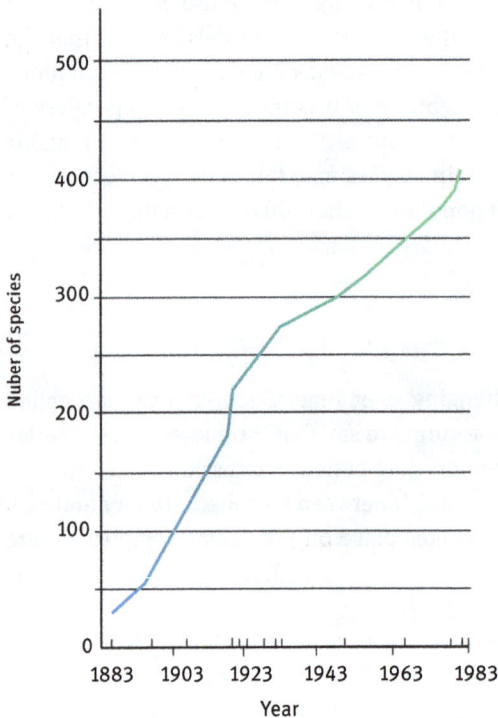

Figure 2.12: The number of species on Rakata in time, after Krakatau explosion; inspired by [28].

Community climax theoretically has certain characteristics: community achieved the balance between total primary production and total respiration; energy obtained from sunlight and decomposition; between the consumption of nutrients and nutrients returned by a process of decomposition. Each individual in the stage of climax is replaced after the death with another individual of the same species; species composition is actually in a state of equilibrium.

Many communities that seemingly appear balanced actually undergo successive changes a lot more often than we perceive. Tropical forests in Central America are exposed to frequent natural disasters. Tropical storms and hurricanes occur very often in this area and have a great impact on the communities. According to some calculations, tropical storms affect the same area (community) every 60 years, leaving behind devastation: broken trees or full or partial defoliation of vegetation. Destruction of vegetation cover, of course, affects the wildlife causing also a different composition and distribution of animal habitats. Therefore, climax should be considered as a dynamic equilibrium of the community. Even "old" and apparently stable community is constantly in the process of dynamic change.

2.2.6 Biomes

Climate largely determines plant formations and life in them. These formations are called biomes. There are several major terrestrial biomes: tundra, grassland, desert, taiga, forest temperate zone, Mediterranean vegetation and tropical rainforest. They are further divided into types depending on the climate and altitude (Figure 2.13).

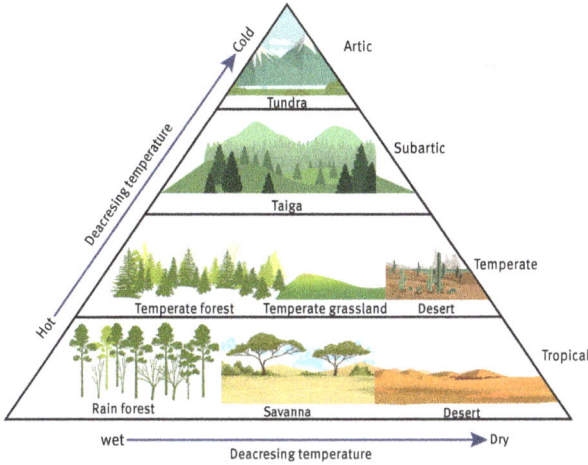

Figure 2.13: Biomes; inspired by [29].

Although inland water (freshwaters) account just a little bit over 0.5 % of all water on Earth, its importance is very high. It is used as drinking water, in industry and agriculture. Inland waters are groundwaters and surface waters. Surface waters are running waters like rivers and streams and standing waters like pools, ponds and lakes.

Seas and oceans cover more than 70 % of the Earth's surface and are inhabited by living organisms from the surface to bottom. The chemical composition of sea water is very stable, more than 85 % of the total amount consists NaCl. Temperature fluctuations of the surface layer depend on the geographical location. In addition to temperature, pressure has a significant impact on life in the sea. Pressure increases with depth. Deep layers have a constant temperature, while in the surface layers (up to 200 m) temperature varies. There are two main habitats in the sea: bottom and water column. The sea bottom is very diverse habitat it is inhabited by benthic organisms. Plankton and nekton organisms live in the water column.

2.2.7 Global changes

Today, human activities cause environmental changes that disturb the normal functioning of the biosphere. Major changes are greenhouse effect that causes the global warming, ozone depletion, acid rains and the destruction of habitats.

2.2.7.1 Greenhouse effect

In the last 100 years, the concentration of CO_2 in the atmosphere has increased about 25 % particularly due to a greater exploitation and use of fossil fuels. Excessive amounts of CO_2 in the Earth's atmosphere are causing the greenhouse effect. The earth's surface absorbs solar radiation and then transmits it as longwave radiation or thermal energy. CO_2 traps longwave, thus heating up the Earth's atmosphere. This effect is called the greenhouse effect and causes global warming. According to some predictions, by the year 2100, the average temperature on Earth will increase in about 2 °C. The consequences could be climate change that can lead to drastic changes in the distribution of individual plant and animal species, which in turn can lead to problems in food production, and transmission of various diseases (malaria, influenza and yellow fever).

2.2.7.2 Acid rains

Precipitation such as rain and snow have naturally slightly acidic pH: between 5 and 5.6 (due to a certain amount of CO_2 in the air). Acid rain is any precipitation with pH lower than 5. Such acidic precipitation is caused by higher amounts of various air pollutants particularly sulfur dioxide (SO_2) and nitrogen oxides (NOx). Precipitation with low pH damages plants, causes acidification of lakes and rivers and directly influences the living world in them. Acid rain may also cause acidification of soil. The consequences of this are heavy metal leeching in terrestrial and aquatic ecosystems. Heavy metals are toxic to living organisms: in vertebrates, they are deposited in the liver and brain and can cause serious damage to these organs.

2.2.7.3 Ozone depletion

Ozone, O_3, is a form of elemental oxygen that is important for the absorption of ultraviolet radiation in the stratosphere. It absorbs 99 % of UV radiation which can cause severe damage to DNA. The depletion of the ozone layer can be caused by freons (a number of halocarbon products which have typically been used as refrigerants and as aerosol propellants), the detonation of nuclear weapons and the exhaust gases over large urban centres.

2.2.7.4 Habitat fragmentation

We already at the beginning of this chapter emphasised that the human's impact on nature is great. One of the global environmental problems is the destruction of habitats. Construction of roads, railways, power lines, expansion of settlements and agricultural land use caused the destruction of natural habitats and/or their fragmentation. Such fragments of natural habitat are often insufficient to maintain populations. One of the major problems of habitat destruction is certainly deforestation, particularly in the tropics. In addition to degradation of the forest community, it also causes a drastic climate change.

2.2.8 Applied ecology

2.2.8.1 Constructed wetlands

Constructed wetlands are treatment systems designed for the purpose of improvement of water quality. They use natural processes involving wetland vegetation, soils and associated microbial assemblages in treating anthropogenic discharge such as municipal or industrial wastewater or storm water runoff.

Constructed wetlands can be designed to emulate the features of natural wetlands, such as acting as biofilters for removing sediments and pollutants from the water. Some constructed wetlands may also serve as a habitat for native and migratory wildlife, although that is usually not their main purpose.

Vegetation in a constructed wetland provides a substrate (roots, stems and leaves) upon which microorganisms can grow as they break down organic materials (Figure 2.14 and Figure 2.15). The plants remove about 7–10% of pollutants and act as a carbon source for the microbes which are responsible for approximately 90 % of pollutant removal.

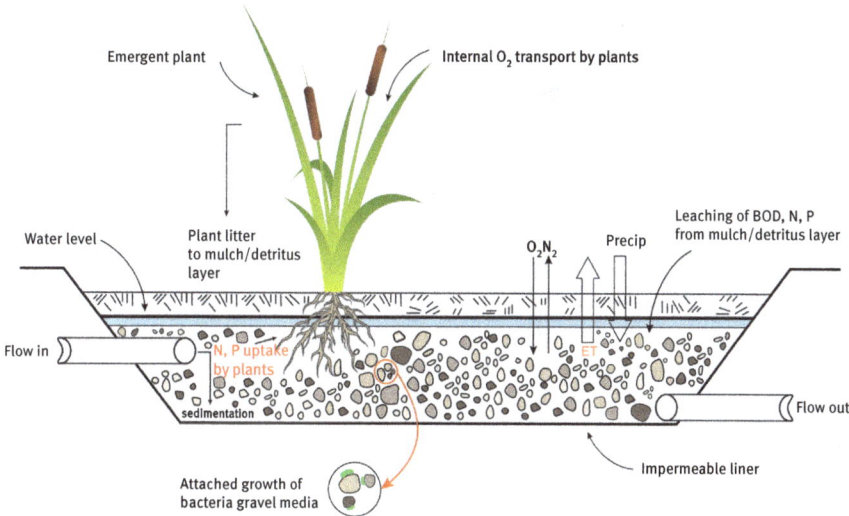

Figure 2.14: Constructed wetland; inspired by [30].

2.2.8.2 Water framework directive

Freshwater ecosystems are among the most degraded on the planet. Numerous human activities impact water quality, including agriculture, industry, urbanisation, disposal of human waste, mining and climate change. In 2000, the European Union established its framework for the protection, improvement and sustainable use of Europe's water resources by launching the Water Framework Directive (WFD) [31]. The WFD is considered to be pioneering because it champions the

Figure 2.15: Constructed wetland in location Hrušćica near Zagreb; photo by Renata Horvat.

ecosystem approach, that is, the integrated consideration of chemical and ecological status in defining water quality [32]. Although the Directive requires the assessment of hydromorphological and physico-chemical elements, a biological assessment is given priority. The assessment of the ecological status establishes the analysis of the structure of biological communities of macroinvertebrates, fish, macrophytes, phytobenthos and phytoplankton. Figure 2.16 illustrates how these various elements are combined and how the "one-out, all-out" principle is applied.

Measuring ecological status makes it possible to evaluate the effect of human activity on water ecosystems, and it will be an essential tool for the sustainable management of water resources. The legislation requires the periodic assessment of all water bodies, including rivers, lakes, estuaries, coastal waters and groundwaters. The first and basic step in the implementation of the WFD in the segment of surface waters is their typification, so that a type-specific approach to estimation water quality could be applied [33]. Water types are then assigned to a classification system that grades their deviation from reference state (high, good, moderate, poor and bad), with no, or very minor, disturbance from human activities. In Croatia, 19 river types were identified and large river types were intercalibrated at EU level. The WFD requires the EU Member States to develop classification systems to describe the ecological status of a given water body at a given time. The main objective of the WFD is to implement measures to achieve "good ecological status" of all natural water bodies. The emphasis given by the WFD to ecological elements has been widely welcomed by scientists and environmental managers as it focuses risk management efforts on restoring those impacted ecosystems that are classified as being of less than "good ecological status."

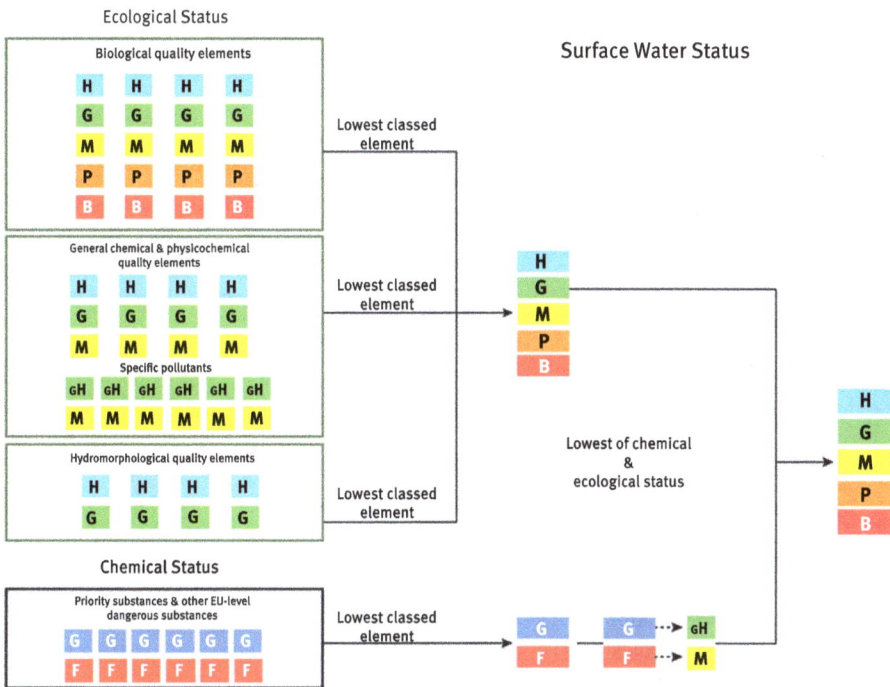

Figure 2.16: Classification scheme of surface water status using "one-out, all-out" principle; H – high status, G – good status, M – moderate status, P – poor status, B – bad status.

References

[1]   Darwin C. On the origin of species by means of natural selection or the preservation of favoured races in the struggle for life. London: John Murray, 1859.

[2]   Haeckel EH. Generelle Morphologie der Organismen. Allgemeine Grundzüge der organischen Formen-Wissenschaft, mechanische Begründet durch die von Charles Darwin reformirte Descendenz-Theorie. Volume I: Allgemeine Anatomie der Organismen. 32 + 574 pages; volume II: Alllgemeine Entwickelungsgeschichte der Organismen. 140 + 462 pages. Berlin, Germany: Georg Reimer, 1866.

[3]   Shelford VE. Animal communities in temperate America. Chicago: University of Chicago Press, 1913.

[4]   Krohne DT. General ecology, 2nd ed. Pacific Grove, CA: Brooks/Cole, 2001.

[5]   Postlethwait JH, Hopson J. The nature of life. New York: McGraw-Hill, 1992.

[6]   MacArthur RC, Wilson EO. The theory of Island biogeography. Princeton: Princeton University Press, 1967.

[7]   Connell JH. The influence of intra-specific competition and other factors on the distribution of the barnacle *Chthamalus stellatus*. Ecology 1961;42:710–723.

[8]   Jelić D, Duplić A, Ćaleta M, Žutinić P. Endemske vrste riba jadranskog sliva. Zagreb: Agencij a za zaštitu okoliša, 2008.

[9]   UNEP-WCMC. Review of the grey squirrel *Sciurus carolinensis*. Cambridge: NEP- WCMC, 2010.

[10]   Bukvić I. Trofička struktura makrozooplanktona u krškim jezerima Visovac i Vrana (Cres). Disertation, Zagreb: Faculty of Science, University of Zagreb, 1998.

[11]   Grant PR. Convergent and divergent character displacement. Biol J Linnean Soc 1972;4:39–68.

[12]   Rasmussen DI. Biotic communities of Kaibab Plateau, Arizona. Ecol Monogr 1941;11:229–275.

[13]   U.S. Geological Survey, Nonindigenous aquatic species, *Petromyzon marinus* Available at: https://nas.er.usgs.gov/queries/FactSheet.aspx?speciesID=836). Accessed: 6 Feb 2017.

[14]   Warner RE. The role of introduced diseases in the extinction of the endemic Hawaiian avifauna. Condor 1969;70:101–120.

[15]   Janzen DH. Coevolution of mutualism between ants and acacias in Central America. Evolution 1966;20:249–275.

[16]   Clements FE Plant succession: Analysis of the development of vegetation. 1916, Carnegie Institute of Washington Publications 242, 1–512.

[17]   Clements FE. Nature and structure of the climax. J Ecol 1936;24:252–284.

[18]   Gleason HA. The individualistic concept of the plant association. Torrey Bot Club Bull 1926;53:7–26.

[19]   Gleason HA. The individualistic concept of the plant association. American Midland Naturalist 1939;21:92–110.

[20]   Gaston KJ. Global patterns in biodiversity. Nature 2000;405(6783):220–227.

[21]   Cox B, Moore PD. Biogeography: an ecological and evolutionary approach, 7th ed. Oxford: Wiley-Blackwell, 2009.

[22]   Rosing M, Bird D, Sleep N, Bjerrum C. No climate paradox under the faint early Sun. Nature 2010;464(7289):744–747.

[23]   Krohne DT. General ecology. Pacific Grove: Brooks/Cole,Thomson Learning Inc., 2001.

[24]   Elton C. Animal ecology. London: Didgwick & Jackson, 1927.

[25]   Lawton JH. What do species do in ecosystem? Oikos 1994;71:367–374.

[26]   Naeem S, Thompson LJ, Lawler SP, Lawton JH, Woodfin RM. Empirical evidence that declining species diversity may alter the performance of terrestrial ecosystems. Phil Trans R Soc London B 1995;347:249–262.

[27]   Symstad AJ, Tilman D, Wilson J, Knops MH. Species loss and ecosystem functioning: Effect of species identity and community composition. Oikos 1998;81:389–397.

[28]   Whitaker RH. Plant recolonization and vegetation succession on the Krakatau Island. Ecol Monogr 1989;59:59–123.

[29]   Whitaker RH. 1989 communities and ecosystem, 2nd ed. New York: Macmillan, 1975.

[30]   Tilley E, Ulrich L, Lüthi C, Reymond P, Zurbrügg C. Compendium of sanitation systems and technologies, (2nd Revised Edition). Duebendorf, Switzerland: Swiss Federal Institute of Aquatic Science and Technology (Eawag), 2014.

[31]   EU. Directive 2000/60/EC of the European Parliament and of the Council establishing a framework for the Community action in the field of water policy – EU Water Framework Directive (WFD), 2000.

[32]   Vincent C, Heinrich H, Edwards A, Nygaard K, Haythornthwaite K. Guidance on typology, reference conditions and classification systems for transitional and coastal waters. CIS Working Group 2.4 (COAST), Common Implementation Strategy of the Water Framework Directive, European Commission, 2002.

[33]   Mihaljevic Z. Typology of non-stagnant waters in Croatia based on macrozoobenthos community (In Croatian with English summary). Hrvatske Vode 2011;76:93–180.

Study questions

1. What do several different populations living together make?
   A) a biosphere; B) an organism; C) a community; D) an ecosystem
2. When there is a lot of pollution, rain can be acidic, harming plants and animals. What is this an example of?
   A) competition between a population and a community; B) a mutualistic type of symbiosis; C) an abiotic factor affecting an ecosystem; D) an individual affecting a community
3. Bears eat fruits such as berries and animals such as fish. They hibernate in the winter. Which of these terms applies to bears?
   A) they have a mutualistic relationship with berries; B) they are at the bottom of the energy pyramid; C) they are decomposers; D) they are consumers.

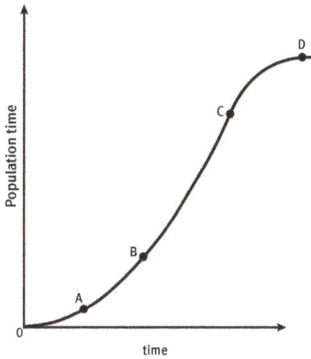

4. The carrying capacity of the environment, for the population shown in the diagram, is the best represented by a point on the curve marked with a letter:
   A) A; B) B; C) C; D) D.

Table 1: Population density.

Level	Population
1	53
2	361
3	823
4	6283

5. In a simple ecosystem, a census of the populations in four successive trophic levels was taken as shown in the Table 1. If Level 1 is composed of photosynthetic autotrophs, then the trophic level with 823 individuals will most likely represent?
   A) primary consumers; B) secondary consumers; C) decomposers; D) producers.

6. What effect does biodiversity have on a community?
A) it makes destruction by insects impossible; B) it makes primary succession more likely; C) it enables species to survive in a desert; D) it enables the community to withstand changes.

7. Precipitation and evaporation are important components of the
A) nitrogen cycle; B) water cycle; C) carbon cycle; D) phosphorus cycle.

8. Match each item with the correct statement:
A) the populations in an ecosystem 1) food web
B) made up of interconnected food chains 2) abiotic
C) the model that shows feeding relationships 3) food chain
D) living things in an ecosystem 4) biotic
E) non-living things in an ecosystem 5) community

9. A certain area in north-west Europe (peatlands) store 30 million tonnes of carbon. In pristine condition, these peatlands can store an additional 15,000 tonnes of carbon per year.
A) Given this rate of productivity, how long would it take for the peatlands to sequester this much carbon?
B) One prediction of climate change models is a reduction in rainfall in this part of Europe. This would transform the peatlands into a carbon source, emitting up to 462,000 tonnes of carbon per year. How many years would it take to respire all the carbon stored in the peatland?
C) What is the effective residence time of the carbon that makes it into long-term storage in the peatland if there is no disturbance in the ecosystem? (Note: it is not a calculating problem!)

10. Base your answer to the following question on the diagram below. A pair of wild goat, A, lives on 1 square kilometre of land. A second pair, B, lives on 4 square kilometres of land, in which 0.25 of the 4 square kilometres overlaps the territory of the first pair.

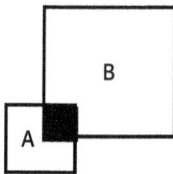

Both pairs A and B are members of the same species. Which of the following explains why area B is larger than area A?
A) Pair B is bigger; B) Pair B has less food in its area; C) Pair A has fewer offspring D) Pair B has more available resources.

Solutions

1. C; 2. C; 3. D; 4. B; 5. B; 6. D; 7. B; 10. B;

8. A) the populations in an ecosystem 5) community

B) made up of interconnected food chains 1) food web

C) model that shows feeding relationships 3) food chain

D) living things in an ecosystem 4) biotic

E) non-living things in an ecosystem 2) abiotic

9. A) 30,000,000 tonnes C / 15,000 tonnes C yr-1 = 2000 years;

B) 30,000,000 tonnes C / 462,000 tonnes C yr-1 = 64,9 years;

C) Theoretically, infinite: the net community productivity is stored in peat indefinitely.

Tomislav Bolanča, Šime Ukić and Mirjana Novak Stankov

3 Environmental Chemistry

Abstract: Environmental engineers need to be good in chemistry because pollutants are chemicals too and, in order to fight the pollution, they should know the properties of these chemicals. Environmental chemistry is a scientific discipline that studies chemical principles, analyses the environment and uses the obtained results in understanding of the processes occurring in environment. The discipline has arisen due to the recognition that chemical compounds, whether anthropogenic or natural, have important effects on naturally occurring processes and organisms. One of the goals of the discipline is to quantify these effects through controlled experiments and measurements. Using the above mentioned approach, a fundamental understanding of the chemical processes occurring in a given system can be achieved. This chapter is dealing with fundaments of Environmental chemistry.

Keywords: environment, chemistry, chemical engineering

3.1 Introduction

Most of environmental problems and processes have chemical nature. Therefore, being familiar with basics of environmental chemistry is crucial for understanding of processes that directly affect the environment. The basics of environmental chemistry are strictly related with chemical equilibrium phenomena that are explained in this chapter.

This article is focused on physical-chemical phenomena that occur in water, soil, and atmosphere. Phenomena related to biosphere will be discussed elsewhere.

Chemistry of water, soil, and atmosphere environment can be discussed separately. However, one should be aware that each of these three environments: water, soil or atmosphere, actually incorporates all three phases. This emphasizes importance of understanding chemical equilibrium as a foundation for any further upgrade that could help in explaining more complex environmental engineering problems.

3.2 Electrolytes and non-electrolytes

Some substances dissociate to ions when dissolving in polar solvents. They are called electrolytes since, due to the occurrence of ionic species, this solutions conduct electricity. Further, those substances that do not show this ability are called non-electrolytes. Electrolytes that dissociate completely (100 %) are called strong electrolytes, while those that show partial dissociation are weak ones. The degree of

https://doi.org/10.1515/9783110468038-003

dissociation, i. e. degree of ionization, is commonly marked with Greek letter α and is defined as:

$$\alpha = \frac{n_{\text{DISSOCIATED MOLECULES}}}{n_{\text{TOTAL AMOUNT OF MOLECULES}}} \tag{3.1}$$

Letter n represents the amount of molecules of monitored substance expressed in moles.

3.3 Concentration and activity

Concentration of electrolyte has significant influence on majority of equilibria that occur in solutions. Many experiments have shown that even presence of foreign electrolytes, i. e. the electrolytes that actually have no part in considered chemical reactions, can affect the related equilibrium. Moreover, electrolytes concentration is not the only factor that affects the equilibrium but the charge of the ions also. Scientists related these phenomena with a quantity called *ionic strength* of solution, I, which can be calculated according to eq. (3.2).

$$I = \frac{1}{2} \cdot \sum_i c_i \cdot z_i^2 \tag{3.2}$$

Symbol c_i represents molar concentration of electrolyte's ion i, while z_i is the charge of the same ion. For low values of ionic strength ($I < 0.1\,\text{M}$), electrolyte effect depends only on the strength value and has no relation with the kind of present ions [1].

There is lot of strong interactions in electrolytic environment and, accordingly, behavior of any ion in such environment is subjected by influence of the surrounding ions. Due to the interactions, many ions behave less active then they should be, so it seems like they are less concentrated than they really are. Such active or "effective concentration" is called the activity, a. Activities are related with actual molar concentration, c, through *activity coefficient*, γ:

$$a = \gamma \cdot c \tag{3.3}$$

For ideal solutions (i. e. highly diluted solutions) activity coefficient has value 1 and the activity is considered equal to the molar concentration. Activity coefficient decreases as the solution becomes more concentrated. Accordingly, it is lower than 1 in moderate concentrated solution so in these solutions activity differs from the molar concentration. Activity coefficient is related with the ionic strength of the solution and can be calculated according to Debye–Hückel equation:

$$- \log \gamma_i = \frac{A \cdot z_i^2 \sqrt{I}}{1 + B \cdot \alpha_i \sqrt{I}} \tag{3.4}$$

Symbol γ_i is activity coefficient of ion i (*single-ion activity coefficient*), A and B are model constants ($A = 0.512$ and $B = 3.3$ for aqueous solution at 25 °C) [1], and α is effective diameter of hydrated ion i, expressed in nanometers. Debye–Hückel equation is usually used in simplified form for aqueous solutions at 25 °C with values of the ionic strength less than 0.01:

$$- \log \gamma_i = 0.512 \cdot z_i^2 \sqrt{I} \tag{3.5}$$

Mean activity coefficient, $\gamma_\pm$, is also applied in many cases. This mean-value can be calculated as geometric mean of single-ion activity-coefficients of cation (+) and anion (−):

$$\gamma_\pm = \sqrt{\gamma_+ \cdot \gamma_-} \tag{3.6}$$

New form of Debye–Hückel eq. (3.5) can be obtained by incorporation of mean activity coefficient:

$$- \log \gamma_\pm = 0.512 \cdot |z_+ \cdot z_-| \sqrt{I} \tag{3.7}$$

Charge numbers, z, of cation and anion are marked with + and −, respectively.

Obviously, it is more appropriate to use activity instead of molar concentration when describing some chemical system. Yet, in order to simplify the calculations and due to the fact that most chemical reactions occur in relatively diluted solutions, the chemists knowingly use concentration instead of activity. The authors applied same practice in this article. The exceptions are cases where difference between activity and real concentration has significant role in understanding of system behavior.

3.4 Ionization of water

Johannes Nicolaus Brønsted and Thomas Martin Lowry gave probably the most common definition of acids and bases in 1923 [2, 3]. They defined acids as substances that can donate proton (*proton donors*) and, accordingly, all substances with the ability to accept proton (*proton acceptors*) were classified as bases. The substances that can behave either as acids or as bases are called amphoteric substances, i. e. *ampholytes*.

Water is one of the amphoteric substances. Also, it is one of the most commonly used polar solvents. Due to its amphoteric properties, it shows effect of autoionization (eq. (3.8)).

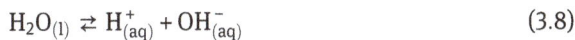

$$H_2O_{(l)} \rightleftarrows H^+_{(aq)} + OH^-_{(aq)} \tag{3.8}$$

The obtained hydrogen ion is bare nucleus of hydrogen atom, i. e. it is a proton. Due to its positive charge, proton is attracted to any part of a nearby molecule in which an excess of negative charge exists. In the case of water solution, proton is attracted by the unshared pair of oxygen electrons from water molecule. So it is more correct to write H_3O^+ instead H^+. Yet, for simplicity, the using of H^+ symbol will remain, knowing that it actually represents hydrated proton, i. e. hydronium ion, H_3O^+ [4].

The equilibrium constant for chemical reaction (8) can be written as:

$$K = \frac{[H^+] \cdot [OH^-]}{[H_2O]} \tag{3.9}$$

The square parentheses in eq. (3.8) represent equilibrium concentrations. The concentration of water molecules in water solutions is practically constant and as such it can be incorporated into equilibrium constant. Let's call this new constant the *ionization constant of water* and mark it as K_W:

$$K_W = [H^+] \cdot [OH^-] \tag{3.10}$$

Ionization constant of water depends on temperature, just like all other constants of chemical equilibrium. At room temperature (25 °C) it has value of $1.01 \cdot 10^{-14}$ M^2 [5]. This determines water as very weak electrolyte.

Concentration of ions in solutions might vary over large range of concentrations, meaning many orders of magnitude. Therefore, it is practical to use logarithmic scale while expressing the concentrations. Danish chemist Søren Peder Lauritz Sørensen introduced concept of "*p*-values" for this purpose [6]. The concept defines *p*-value as negative logarithm of an observed quantity. For example:

$$pH = - \log[H^+] \tag{3.11}$$

$$pOH = - \log[OH^-] \tag{3.12}$$

$$pK_W = - \log K_W \tag{3.13}$$

The value inside logarithm cannot have dimension. Therefore, it is important to emphasize that values of concentration used in the calculation must match mol/dm^3 (shortly written as M) and values of equilibrium constant must be derived from equally expressed concentrations. For example, ionization constant of water must be expressed as 10^{-14} since this value relates M^2 unit. This gives $\log K_W$ value of 14. Accordingly, eq. (3.10) can be transformed into:

$$pH + pOH = pK_W = 14 \tag{3.14}$$

3.5 Strong and weak acids

Let's observe a hypothetic weak acid HA, consisted of proton and a hypothetic anion A^-. In water solution, this acid reacts with the solvent and releases proton.

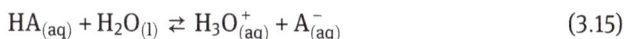

$$HA_{(aq)} + H_2O_{(l)} \rightleftarrows H_3O^+_{(aq)} + A^-_{(aq)} \tag{3.15}$$

The constant that defines this equilibrium is called *acid ionization constant*, K_A:

$$K_A = \frac{[H_3O^+] \cdot [A^-]}{[HA]} \tag{3.16}$$

According to Brønsted–Lowry definition, HA is acid, while at the same time the obtained product A^- is base since it can accept proton:

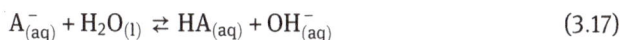

$$A^-_{(aq)} + H_2O_{(l)} \rightleftarrows HA_{(aq)} + OH^-_{(aq)} \tag{3.17}$$

Base A^- is considered conjugate base of acid HA. Equilibrium for chemical reaction represented by eq. (3.17) is defined by *base ionization constant*, K_B:

$$K_B = \frac{[HA] \cdot [OH^-]}{[A^-]} \tag{3.18}$$

Equilibrium concentration of conjugate base A^- can be expressed from eq. (3.18) and transferred into eq. (3.16) bringing a new equation:

$$K_A = \frac{[H_3O^+] \cdot \dfrac{[HA] \cdot [OH^-]}{K_B}}{[HA]} \tag{3.19}$$

Equation (3.19) can be additionally simplified to:

$$K_A \cdot K_B = [H_3O^+] \cdot [OH^-] \tag{3.20}$$

The obtained equation is the general relation that connects acids with their conjugate bases:

$$K_A \cdot K_B = K_W \tag{3.21}$$

$$pK_A + pK_B = pK_W = 14 \tag{3.22}$$

Acids that act like strong electrolytes are called strong acids. The equilibrium constant for ionization of such acids, K_A, tends to infinity, since the equilibrium concentration of non-ionized form of acid practically equals 0 M. The example of strong acid is hydrochloric acid, HCl.

Concentration of protons that originate from autoionization of water can be ignored during calculation of solution's pH-value for cases where initial concentration of strong acid is significantly higher than 10^{-7} M. Since strong acids ionized completely, all protons at equilibrium state practically originate from the acid and the influence of water-originated protons is negligible.

Problem 3.1.

Calculate pH-value of 0.1 M HCl solution.

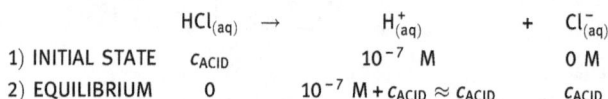

	$HCl_{(aq)}$	$\rightarrow$	$H^+_{(aq)}$	$+$	$Cl^-_{(aq)}$
1) INITIAL STATE	c_{ACID}		10^{-7} M		0 M
2) EQUILIBRIUM	0		10^{-7} M $+ c_{ACID} \approx c_{ACID}$		c_{ACID}

$[H^+] = c_{ACID}$

$pH = -\log[H^+] = -\log c_0 = -\log 0.1 = 1$

According to above performed calculation, general expression for pH-value of strong acids is:

$$pH = -\log c_{ACID} \qquad (3.23)$$

Protons that originate from autoionization of water must be taken in consideration in cases when concentration of strong acid is not significantly higher or is even lower than 10^{-7} M.

Problem 3.2.

Calculate pH-value of 10^{-8} M HCl solution.

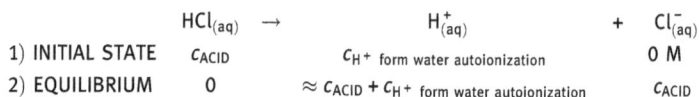

	$HCl_{(aq)}$	$\rightarrow$	$H^+_{(aq)}$	$+$	$Cl^-_{(aq)}$
1) INITIAL STATE	c_{ACID}		c_{H^+} form water autoionization		0 M
2) EQUILIBRIUM	0		$\approx c_{ACID} + c_{H^+}$ form water autoionization		c_{ACID}

HCl is a strong electrolyte and it ionizes completely. This generates an amount of H^+ ion equal to the concentration of HCl, i.e. 10^{-8} M. At the same time, 10 times higher concentration of protons is already present in water due to water autoionization. Therefore, concentration of protons from the autoionization must be taken in calculations.

$$[H^+] \approx c_{ACID} + c_{H^+} \text{ form water ionization} = 10^{-8} \text{ M} + 10^{-7} \text{ M} = 1.1 \cdot 10^{-7} \text{ M}$$
$$pH = -\log[H^+] = -\log(1.1 \cdot 10^{-7}) = 6.96$$

Weak acids are weak electrolytes since they ionize partially. Calculation of pH-value for weak acids is somewhat different then for strong ones. Let's observe weak acetic acid, CH_3COOH, in water. At equilibrium state, the concentration of ionized molecules of acid, x, is negligible related to the non-ionized ones. Yet, if initial concentration of acid, c_{ACID}, is large enough, concentration of protons obtained from the ionization of the acid (also x) will be significantly higher than the one obtained from ionization of water. Accordingly, protons obtained from water ionization can be ignored during the calculation.

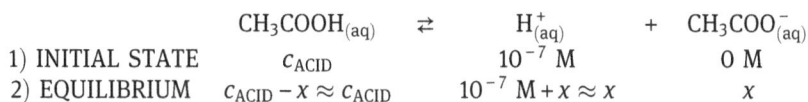

$$CH_3COOH_{(aq)} \quad \rightleftarrows \quad H^+_{(aq)} \quad + \quad CH_3COO^-_{(aq)}$$

1) INITIAL STATE $\quad c_{ACID} \qquad\qquad 10^{-7}\ M \qquad\qquad 0\ M$

2) EQUILIBRIUM $\quad c_{ACID} - x \approx c_{ACID} \quad 10^{-7}\ M + x \approx x \qquad x$

It is important to notice that equal concentration of proton and the conjugate base originates from ionization of the acid. Therefore ionization constant can be written as:

$$K_A = \frac{[H^+] \cdot [CH_3COO^-]}{[CH_3COOH]} \approx \frac{[H^+]^2}{c_{ACID}} \tag{3.24}$$

Equation (3.24) can be rearranged as follows:

$$[H^+] = \sqrt{K_A \cdot c_{ACID}} \tag{3.25}$$

$$-\log[H^+] = -\log\left(K_A \cdot c_{ACID}\right)^{\frac{1}{2}} \tag{3.26}$$

This provides, finally, the general expression for calculation of pH-value of weak acid:

$$pH = \frac{1}{2} \cdot pK_A - \frac{1}{2} \cdot \log c_{ACID} \tag{3.27}$$

Sometimes, in cases of not so weak or much diluted acids, the concentration of ionized molecules of acid cannot be neglected. For such cases, quadratic equation instead of eq. (3.24) is recommended:

$$K_A = \frac{x^2}{c_{ACID} - x} \tag{3.28}$$

$$x^2 + K_A \cdot x - K_A \cdot c_{ACID} = 0 \tag{3.29}$$

Every quadratic equation, from mathematical point of view, has two solutions: one positive and the other negative. Positive solutions should be taken in consideration for this specific case since concentration cannot have negative value:

$$x = \frac{-K_A + \sqrt{K_A^2 - 4 \cdot 1 \cdot (-K_A \cdot c_{ACID})}}{2 \cdot 1} \tag{3.30}$$

i. e.:

$$[H^+] = \frac{-K_A + \sqrt{K_A^2 + 4 \cdot K_A \cdot c_{ACID}}}{2} \tag{3.31}$$

Many acids have more than one proton bounded. Such acids are called polyprotic acids. One of the well-known polyprotic acids is phosphoric acid, H_3PO_4. This acid is ionizing in three steps and each step is characterized by its own equilibrium constant (Table 3.1).

Table 3.1: Ionization steps of phosphoric acid [1].

Reaction of ionization	Ionization constant	p$K_A$value
$H_3PO_{4(aq)} \rightleftarrows H^+_{(l)} + H_2PO^-_{4(aq)}$	$K_{A1} = \dfrac{[H^+] \cdot [H_2PO_4^-]}{[H_3PO_4]} = 7.1 \cdot 10^{-3}$ M	2.1
$H_2PO^-_{4(aq)} \rightleftarrows H^+_{(l)} + HPO^{2-}_{4(aq)}$	$K_{A2} = \dfrac{[H^+] \cdot [HPO_4^{2-}]}{[H_2PO_4^-]} = 6.3 \cdot 10^{-8}$ M	7.2
$HPO^{2-}_{4(aq)} \rightleftarrows H^+_{(l)} + PO^{3-}_{4(aq)}$	$K_{A3} = \dfrac{[H^+] \cdot [PO_4^{3-}]}{[HPO_4^{2-}]} = 4.2 \cdot 10^{-13}$ M	12.4

The first ionization step generates significantly higher concentration of protons then the remaining two steps, according to ionization constants. Therefore pH-value of solution that contains phosphoric acid can be calculated according to eq. (3.27) or (3.32), with appliance of appropriate ionization constant. Ions $H_2PO_4^-$ and HPO_4^{2-} are amphoteric substances, while PO_4^{3-} is pure base.

Examples for exercise
Example 3.1.
What is the pH value of 10 mM water solution of nitric acid?
Nitric acid is a strong acid which ionizes completely in water (100 %).

$$HNO_{3(aq)} \rightarrow H^+_{(aq)} + NO^-_{3(aq)}$$

Therefore, in 10 mM solution of nitric acid, concentration of H^+ ions that originate from HNO_3 is 10 mM, i. e. 10^{-2} M. (It was said before that concentration values used in calculation of pH must be expressed as mol/L.) This concentration is much higher than concentration of protons that originate from autoionization of water (10^{-2} M $\ll 10^{-7}$ M). Thus, concentration of protons from the autoionization can be neglected while calculating pH. The calculation is performed according the eq. (3.23):

$$pH = -\log[H^+] = -\log c(HNO_3) = -\log 10^{-2} = 2$$

Example 3.2.
What is the pH of 0.01 M water solution of hydrofluoric acid?

$$K_A = 6.6 \cdot 10^{-4} \text{ M}$$

Hydrofluoric acid ionizes in water according the equation:

$$HF_{(aq)} \rightleftarrows H^+_{(aq)} + F^-_{(aq)}$$

It is a weak acid that ionizes only partially. This can be concluded from value of the acid ionization constant. Yet, the concentration of acid is high enough and allows us to neglect protons from water while calculating pH. Therefore, pH value can be calculated according to eq. (3.27):

$$pH = \frac{1}{2} \cdot pK_A - \frac{1}{2} \cdot \log c(HF)$$

$$pK_A = - \log K_A = - \log(6.6 \cdot 10^{-4}) = 3.18$$

$$pH = \frac{1}{2} \cdot 3.18 - \frac{1}{2} \cdot \log 0.01 = 2.59$$

Example 3.3.
What is the pH of 0.1 M water solution of carbonic acid?

$$K_{A1} = 4.2 \cdot 10^{-7} \text{ M}, K_{A2} = 4.8 \cdot 10^{-11} \text{ M}$$

Carbonic acid is a weak polyprotic acid that ionizes in water solutions in two further steps:

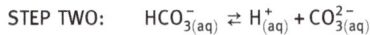

STEP ONE: $H_2CO_{3(aq)} \rightleftarrows H^+_{(aq)} + HCO^-_{3(aq)}$

STEP TWO: $HCO^-_{3(aq)} \rightleftarrows H^+_{(aq)} + CO^{2-}_{3(aq)}$

Acid ionization constant for the first step is approximately 10^4 times higher than the one for the second step. Accordingly, the first step will generate significantly higher amount of protons so the second step can be omitted from the pH calculation. Thus, pH can be calculated according to eq. (3.27) but counting only first ionization step:

$$pH = \frac{1}{2} \cdot pK_{A1} - \frac{1}{2} \cdot \log c(H_2CO_3)$$

$$pK_{A1} = - \log K_{A1} = - \log(4.2 \cdot 10^{-7}) = 6.38$$

$$pH = \frac{1}{2} \cdot 6.38 - \frac{1}{2} \cdot \log 0.1 = 3.69$$

3.6 Strong and weak bases

Definition of strong and weak bases is analogous to those of strong and weak acids. Accordingly, the expression for pH-value for solution of a strong base, for example NaOH, can be derived as follows:

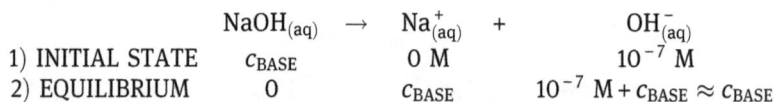

$$NaOH_{(aq)} \rightarrow Na^+_{(aq)} + OH^-_{(aq)}$$

1) INITIAL STATE $\quad c_{BASE} \qquad$ 0 M $\qquad 10^{-7}$ M

2) EQUILIBRIUM $\qquad$ 0 $\qquad c_{BASE} \qquad 10^{-7}$ M $+ c_{BASE} \approx c_{BASE}$

$[OH^-] = c_{BASE}$
$pOH = - \log[OH^-] = - \log c_{BASE}$
$pH = pK_W - pOH$

This leads to a general pH expression for solution of a strong base:

$$pH = 14 + \log c_{BASE} \tag{3.32}$$

Calculation is more complicated in case of a weak base, like ammonium hydroxide: NH_4OH.

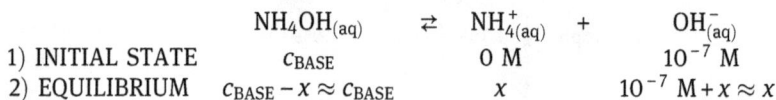

$$NH_4OH_{(aq)} \rightleftharpoons NH_4^+_{(aq)} + OH^-_{(aq)}$$

1) INITIAL STATE $\qquad c_{BASE} \qquad$ 0 M $\qquad 10^{-7}$ M

2) EQUILIBRIUM $\quad c_{BASE} - x \approx c_{BASE} \qquad x \qquad 10^{-7}$ M $+ x \approx x$

Mark x represents concentration of ionized molecules of base after equilibrium is reached. A general form for pH calculation can be obtained in a similar way as for weak acids:

$$K_B = \frac{[NH_4^+] \cdot [OH^-]}{[NH_4OH]} = \frac{[OH^-]^2}{c_{BASE}} \tag{3.33}$$

$$\frac{K_W}{K_A} = \frac{\left(\frac{K_W}{[H^+]}\right)^2}{c_{BASE}} \tag{3.34}$$

$$[H^+] = \sqrt{K_W \cdot K_A \cdot \frac{1}{c_{BASE}}} \tag{3.35}$$

$$- \log[H^+] = - \log \left(K_W \cdot K_A \cdot \frac{1}{c_{BASE}}\right)^{\frac{1}{2}} \tag{3.36}$$

$$pH = 7 + \frac{1}{2} \cdot pK_A + \frac{1}{2} \cdot \log c_{BASE} \tag{3.37}$$

K_A represents, in this specific case, ionization constant of NH_4^+, which is conjugate acid of base NH_4OH.

Of course, quadratic equation has to be solved for cases when x is not negligible in comparison with the c_{BASE}:

$$K_B = \frac{[NH_4^+] \cdot [OH^-]}{[NH_4OH]} = \frac{x^2}{c_{BASE} - x} \tag{3.38}$$

$$[OH^-] = \frac{-K_B + \sqrt{K_B^2 + 4 \cdot K_B \cdot c_{BASE}}}{2} \tag{3.39}$$

Examples for exercise
Example 3.4.
What is the pH value of 0.02 M water solution of ammonium hydroxide?

$$K_B = 1.8 \cdot 10^{-5} \text{ M}$$

Ammonium hydroxide is a weak base which ionizes partially in water:

$$NH_4OH_{(aq)} \rightleftarrows NH_{4(aq)}^+ + OH_{(aq)}^-$$

Thus, pH is calculated according the eq. (3.37):

$$pH = 7 + \frac{1}{2} \cdot pK_A + \frac{1}{2} \cdot \log c(NH_4OH)$$

Ionization constant of conjugate acid NH_4^+ is required for solving the previous equation. Therefore, eq. (3.22) will be applied:

$$pK_B = -\log K_B = -\log(1.8 \cdot 10^{-5}) = 4.74$$
$$pK_A = pK_W - pK_B = 14.00 - pK_B = 14.00 - 4.74 = 9.26$$

And the final step is:

$$pH = 7 + \frac{1}{2} \cdot 9.26 + \frac{1}{2} \cdot \log 0.02 = 10.78$$

Example 3.5.
What is the pH value of 0.15 M water solution of methylamine?

$$K_B = 5.0 \cdot 10^{-4} \text{ M}$$

Methylamine is also a weak base. When dissolved in water, it reacts with the solvent's molecules according the reaction:

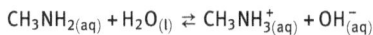

$$CH_3NH_{2(aq)} + H_2O_{(l)} \rightleftarrows CH_3NH_{3(aq)}^+ + OH_{(aq)}^-$$

Calculation is similar as in previous example:

$$pK_B = -\log K_B = -\log(5.0 \cdot 10^{-4}) = 3.30$$

$$pK_A = pK_W - pK_B = 14 - pK_B = 14 - 3.30 = 10.70$$

$$pH = 7 + \frac{1}{2} \cdot pK_A + \frac{1}{2} \cdot \log c(CH_3NH_2) = 7 + \frac{1}{2} \cdot 10.70 + \frac{1}{2} \cdot \log 0.15 = 11.94$$

3.7 Hydrolysis of salts

Sometimes water solutions become acidic or basic after dissolution of a salt. Yet, in some cases the solutions remain neutral. How can salts influence pH-value of the solutions? The answer lies in a phenomenon called *hydrolysis*. The term is derived from Greek words *hydro*, meaning water, and *lysis*, meaning breaking or solving.

It is well known that chemical reaction between acidic and basic reactants generates salt and water. This reaction is called neutralization. Salt is consisted of conjugate base of the acidic reactant, and conjugate acid of the basic reactant. During dissolution process, salt completely or partially dissociates, releasing thus these conjugate components into the solvent. For example:

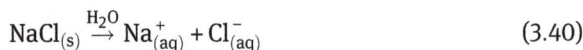

$$NaCl_{(s)} \overset{H_2O}{\rightarrow} Na^+_{(aq)} + Cl^-_{(aq)} \tag{3.40}$$

Hydrolysis is a reaction somehow opposite to neutralization: some of the ions released by salt dissociation can react with water, producing thus acidic compounds, basic compounds or both. It was said in chapter 5 that conjugate base can react with water, producing hydroxide ion and related weak acid (see eq. (3.17)). The same situation is with conjugate acids which in reaction with water produce weak base and proton. Yet, it is important to say that hydrolysis will not occur in cases when conjugate base or conjugate acid originate from strong acid or strong base, respectively. So the salt NaCl from eq. (3.40) originates from strong acid HCl and strong base NaOH. Dissolving of this salt in water will not initiate hydrolysis and pH-value of the solution remains neutral.

Let's discuss salt ammonium chloride, NH_4Cl. This salt dissociates completely while dissolved in water, producing ammonium and chloride ions:

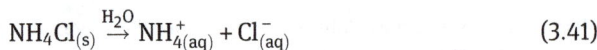

$$NH_4Cl_{(s)} \overset{H_2O}{\rightarrow} NH^+_{4(aq)} + Cl^-_{(aq)} \tag{3.41}$$

Ammonium ion originates from weak ammonium hydroxide and therefore will partially react with water. Chloride originates from strong hydrochloric acid and will not hydrolyze.

$$NH_4^+{}_{(aq)} \quad + \quad H_2O_{(l)} \quad \rightleftarrows \quad NH_4OH_{(aq)} \quad + \quad H^+_{(aq)}$$

1) INITIAL STATE $\quad c_{SALT}$ $\qquad\qquad\qquad\qquad$ 0 M $\qquad$ 10^{-7} M

2) EQUILIBRIUM $\quad c_{SALT} - x \approx c_{SALT}$ $\qquad\qquad\quad$ x $\qquad$ 10^{-7} M $+ x \approx x$

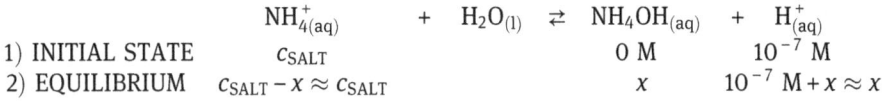

Concentration of ammonium ion in initial state equals concentration of dissolved salt what is consequence of full dissociation of salt in water. Concentration of ammonium ions that will react with water in equilibrium state is defined by the size of constant of hydrolysis, K_H:

$$K_H = \frac{[H^+] \cdot [NH_4OH]}{[NH_4^+]} \tag{3.42}$$

Let's expand the fraction from eq. (3.42) by multiplying and dividing it with the equilibrium concentration of hydroxide ion.

$$K_H = \frac{[H^+] \cdot [NH_4OH]}{[NH_4^+]} \cdot \frac{[OH^-]}{[OH^-]} \tag{3.43}$$

Incorporation of eqs (3.10), (3.20) and (3.33) in eq. (3.43) confirms that constant of hydrolysis for ammonium chloride is actually ionization constant of acid NH_4^+.

$$K_H = \frac{K_W}{K_B} = K_A \tag{3.44}$$

Ammonium chloride in water solution acts like a weak acid. Therefore, the expression for pH-value of weak acids (3.28) can be applied for such solution. A short derivation from eqs (3.42) and (3.44) will be presented in order to prove this statement. It is assumed that equilibrium concentration of ammonium equals initial concentration of the salt, and that equilibrium concentration of hydronium ions is equal to equilibrium concentration of ammonium hydroxide.

$$K_A = \frac{[H^+]^2}{c_{SALT}} \tag{3.45}$$

$$[H^+] = \sqrt{K_A \cdot c_{SALT}} \tag{3.46}$$

$$pH = \frac{1}{2} \cdot pK_A - \frac{1}{2} \cdot \log c_{SALT} \tag{3.47}$$

The analogy with general expression for calculation of pH-value of weak acids (eq. (3.27)) is clearly evident.

The salt sodium acetate dissociates completely in water solution:

$$CH_3COONa_{(s)} \xrightarrow{H_2O} Na^+_{(aq)} + CH_3COO^-_{(aq)} \tag{3.48}$$

The salt consists of sodium cation, which originates from strong NaOH base, and acetate anion from weak acetic acid. Therefore, only acetate anion does hydrolysis in water solutions.

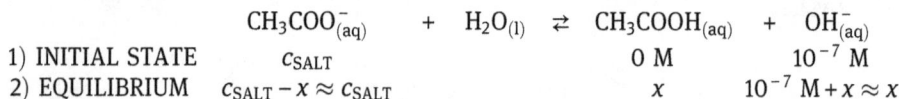

	$CH_3COO^-_{(aq)}$	$+$	$H_2O_{(l)}$	$\rightleftarrows$	$CH_3COOH_{(aq)}$	$+$	$OH^-_{(aq)}$
1) INITIAL STATE	c_{SALT}				0 M		10^{-7} M
2) EQUILIBRIUM	$c_{SALT} - x \approx c_{SALT}$				x		10^{-7} M $+ x \approx x$

Concentration of acetate ion in initial state equals concentration of dissolved salt. Since hydrolysis occurs on anion only, the salt acts like weak base:

$$K_B = \frac{[HAc] \cdot [OH^-]}{[Ac^-]} \approx \frac{[OH^-]^2}{c_{SALT}} \tag{3.49}$$

$$\frac{K_W}{K_A} = \frac{(K_W[H^+])^2}{c_{SALT}} \tag{3.50}$$

$$[H^+] = \sqrt{K_W \cdot K_A \cdot \frac{1}{c_{SALT}}} \tag{3.51}$$

$$pH = 7 + \frac{1}{2} \cdot pK_A + \frac{1}{2} \cdot \log c_{SALT} \tag{3.52}$$

Once again, the analogy with expression for calculation of pH value of weak base (eq. (3.37)) is present.

Ammonium acetate is an example for amphoteric salt.

$$CH_3COONH_{4(s)} \xrightarrow{H_2O} NH^+_{4(aq)} + CH_3COO^-_{(aq)} \tag{3.53}$$

$$NH^+_{4(aq)} + CH_3COO^-_{(aq)} + H_2O_{(l)} \rightleftarrows NH_4OH_{(aq)} + CH_3COOH_{(aq)} \tag{3.54}$$

The salt is comprised from ammonium and acetate ion. The cation is a weak acid, while the anion originates from weak acetic acid. Ionization constants that relate these two acids are required for calculation of pH-value of the solution. The general expression is:

$$pH = \frac{1}{2} \cdot (pK_{A1} + pK_{A2}) \tag{3.55}$$

Examples for exercise
Example 3.6.
What is the pH value of 0.17 M solution of ammonium nitrate?

$$K(NH_4^+) = 5.6 \cdot 10^{-10} \text{ M}$$

Ammonium nitrate dissociates completely in water solutions:

$$NH_4NO_{3(s)} \xrightarrow{H_2O} NH_{4(aq)}^+ + NO_{3(aq)}^-$$

Only one of the ions obtained by dissociation has ability to react with water, i. e. ability to do hydrolysis. It is ammonium ion since its origin is the weak base NH_4OH.

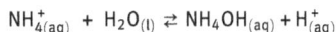

$$NH_{4(aq)}^+ + H_2O_{(l)} \rightleftarrows NH_4OH_{(aq)} + H_{(aq)}^+$$

Obviously, this salt acts like weak acid so eq. (3.47) can be used for calculation of pH value of the solution:

$$pK_A = pK(NH_4^+) = -\log K(NH_4^+) = -\log(5.6 \cdot 10^{-10}) = 9.25$$

$$pH = \frac{1}{2} \cdot pK_A - \frac{1}{2} \cdot \log c(NH_4NO_3) = \frac{1}{2} \cdot 9.25 - \frac{1}{2} \cdot \log 0.17 = 4.24$$

Example 3.7.
0.3100 g of sodium nitrite, $NaNO_2$, was dissolved in 1 L of distilled water. What is the pH of the solution?

$$K(HNO_2) = 7.2 \cdot 10^{-4} \text{ M}$$

Primary, it is necessary to determine molar concentration, c, of $NaNO_2$ in the solution. The amount of dissolved sodium nitrate expressed in mols, n, can be obtained from a ratio of the salt's dissolved mass, m, and its molar mass, M:

$$n(NaNO_2) = \frac{m(NaNO_2)}{M(NaNO_2)} = \frac{0.3100 \text{ g}}{69.00 \frac{\text{g}}{\text{mol}}} = 4.45 \cdot 10^{-3} \text{ mol}$$

Now it is easy to calculate the molar concentration knowing that $NaNO_2$ is dissolved in 1 L, i. e. 1 dm³, of water.

$$c(NaNO_2) = \frac{n(NaNO_2)}{V_{WATER}} = \frac{4.45 \cdot 10^{-3} \text{ mol}}{1 \text{ dm}^3} = 4.45 \cdot 10^{-3} \text{ M}$$

Sodium nitrite is a salt that also dissociate completely in water:

$$NaNO_{2(s)} \xrightarrow{H_2O} Na_{(aq)}^+ + NO_{2(aq)}^-$$

The obtained nitrite ion comes from a weak acid HNO_2 and accordingly will react with water.

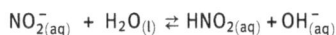

$$NO_{2(aq)}^- + H_2O_{(l)} \rightleftarrows HNO_{2(aq)} + OH_{(aq)}^-$$

Sodium ion comes from a strong base NaOH and accordingly will not hydrolyze. Therefore, sodium nitrite in water media acts like weak base, and eq. (3.52) can be applied for calculation of pH value:

$$pK_A = -\log K(HNO_2) = -\log(7.2 \cdot 10^{-4}) = 3.14$$

$$pH = 7 + \frac{1}{2} \cdot pK_A + \frac{1}{2} \cdot \log c(NaNO_2) = 7 + \frac{1}{2} \cdot 3.14 + \frac{1}{2} \cdot \log(4.45 \cdot 10^{-3}) = 7.40$$

Example 3.8.
100 mL of 0.300 M solution of HF was added into 300 mL of 0.100 M ammonium hydroxide. What is the pH value of such obtained solution?

$$K(HF) = 6.6 \cdot 10^{-4} \text{ M}, K(NH_4^+) = 5.6 \cdot 10^{-10} \text{ M}$$

Total volume of the mixed solutions, V_{TOTAL}, is sum of volumes of the separate solutions, V_{HF} and V_{NH_4OH}:

$$V_{TOTAL} = V_{HF} + V_{NH_4OH} = 100 \text{ mL} + 300 \text{ mL} = 400 \text{ mL} = 0.400 \text{ dm}^3$$

Addition of 100 mL of hydrofluoric acid into 300 mL of ammonium hydroxide solution is causing dilution of both solutions and, accordingly, concentrations of both chemicals: HF and NH_4OH, are changed. New molar concentrations are:

$$c(HF)_{NEW} = c(HF)_{OLD} \cdot \frac{V_{HF}}{V_{TOTAL}} = 0.300 \text{ M} \cdot \frac{100 \text{ mL}}{400 \text{ mL}} = 0.075 \text{ M}$$

$$c(NH_4OH)_{NEW} = c(NH_4OH)_{OLD} \cdot \frac{V_{NH_4OH}}{V_{TOTAL}} = 0.100 \text{ M} \cdot \frac{300 \text{ mL}}{400 \text{ mL}} = 0.075 \text{ M}$$

Addition of hydrofluoric acid into ammonium hydroxide solution is also causing the neutralization:

	$HF_{(aq)}$	+	$NH_4OH_{(aq)}$	→	$NH_{4(aq)}^+$	+	$F_{(aq)}^-$	+	$H_2O_{(l)}$
1) INITIAL STATE	0.075 M		0.075 M		0 M		0 M		
2) REACTION END	0 M		0 M		0.075 M		0.075 M		

Stoichiometrically equivalent amounts of acid and base are joined together in this specific case. So the reactants react completely and only the resulting NH_4F salt and water remain at the reaction end. NH_4F salt is completely dissociated in water solutions and exists only as separated ions NH_4^+ and F^-. Both of these ions have the ability to do hydrolysis, so this is an amphoteric salt. pH value of the salt solution is calculating from the eq. (3.55):

$$pH = \frac{1}{2} \cdot (pK_{A1} + pK_{A2})$$

$$pK_{A1} = pK(HF) = -\log K(HF) = -\log(6.6 \cdot 10^{-4}) = 3.18$$

$$pK_{A2} = pK(NH_4^+) = -\log K(NH_4^+) = -\log(5.6 \cdot 10^{-10}) = 9.25$$

$$pH = \frac{1}{2}(3.18 + 9.25) = 6.21$$

3.8 Buffers

Some solutions show significant ability to resist the change of pH-value upon the addition of acidic or basic components. These solutions are called buffers. Buffer solutions can be combination of a weak acid and its salt (e. g. CH_3COOH/CH_3COONa), a weak base and its salt (e. g. NH_4OH/NH_4Cl), or even two salts of which first salt must act like weak base or weak acid while the other salt must be derived from the first one (e. g. $NaHCO_3/Na_2CO_3$). How is buffer capable to resist the change of pH-value? Let's take CH_3COOH/CH_3COONa buffer for example. Sodium acetate is completely dissociated in water solution according to eq. (3.48). Therefore, equilibrium involves further species:

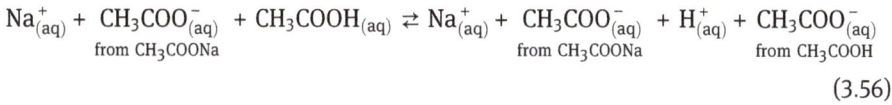

$$\underset{\text{from } CH_3COONa}{Na^+_{(aq)} + CH_3COO^-_{(aq)}} + CH_3COOH_{(aq)} \rightleftarrows Na^+_{(aq)} + \underset{\text{from } CH_3COONa}{CH_3COO^-_{(aq)}} + H^+_{(aq)} + \underset{\text{from } CH_3COOH}{CH_3COO^-_{(aq)}}$$

$$(3.56)$$

The acetate ions in solution mostly originate from the salt.

$$c(CH_3COO^-)_{\text{from } CH_3COONa} \gg c(CH_3COO^-)_{\text{from } CH_3COOH} \tag{3.57}$$

Adding an additional amount of protons into buffer solution will not cause significant change of the pH-value since added protons will instantly react with acetate ions, already present in the solution, thus forming poorly dissociated acetic acid:

$$H^+_{(aq)} + CH_3COO^-_{(aq)} \rightleftarrows CH_3COOH_{(aq)} \tag{3.58}$$

Addition of hydroxide ions will not significantly change pH-value of buffer solutions, also. Added hydroxide ions are instantly reacting with present protons, forming poorly dissociated water molecules (eq. (3.8)). This reaction removes protons from equilibrium described by eq. (3.56). Yet, there is a large reserve of acetic-acid molecules that are not ionized (left side of the eq. (3.57)). These molecules are instantly compensating lack of protons by additional ionization.

The similar situation is happening in case of basic buffer NH_4OH/NH_4Cl. Equilibrium in the buffer solution can be described as follows:

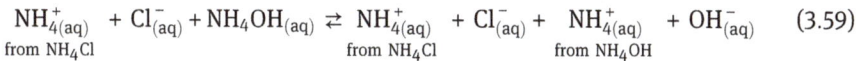

$$\underset{\text{from } NH_4Cl}{NH_4^+_{(aq)} + Cl^-_{(aq)}} + NH_4OH_{(aq)} \rightleftarrows \underset{\text{from } NH_4Cl}{NH_4^+_{(aq)} + Cl^-_{(aq)}} + \underset{\text{from } NH_4OH}{NH_4^+_{(aq)}} + OH^-_{(aq)} \tag{3.59}$$

There is a large amount of ammonium ions in solution that mostly originates from the salt.

$$c(NH_4^+)_{\text{from } NH_4Cl} \gg c(NH_4^+)_{\text{from } NH_4OH} \tag{3.60}$$

Potentially added hydroxide ions instantly react with large amount of available ammonium ions, forming poorly dissociated molecules of ammonium hydroxide and thus pH-value of the solution remains preserved.

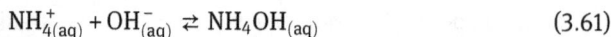

$$NH_{4(aq)}^+ + OH_{(aq)}^- \rightleftarrows NH_4OH_{(aq)} \tag{3.61}$$

Added protons will react with hydroxide ions thus forming water (eq. (3.8)) as a weak electrolyte. Removal of hydroxide ions from equilibrium (3.60) will provoke further ionization of large amount of still non-ionized molecules of ammonium hydroxide and the pH-value of solution will maintain unchanged.

Equilibrium in the discussed NH_4OH/NH_4Cl buffer system is described by ionization constant of ammonium hydroxide:

$$K_B = \frac{[NH_4^+] \cdot [OH^-]}{[NH_4OH]} \tag{3.62}$$

This equation can be easily transformed into general equation for buffer solutions, known as the Henderson–Hasselbalch equation (eq. (3.66)).

$$\frac{K_w}{K_A} = \frac{[NH_4^+] \cdot \frac{K_w}{[H^+]}}{[NH_4OH]} \tag{3.63}$$

$$[H^+] = K_A \cdot \frac{[NH_4^+]}{[NH_4OH]} \tag{3.64}$$

$$pH = pK_A + \log\frac{[NH_4OH]}{[NH_4^+]} \tag{3.65}$$

$$pH = pK_A + \log\frac{c_{BASIC\ FORM}}{c_{ACIDIC\ FORM}} \tag{3.66}$$

According to the Henderson–Hasselbalch equation, pH of a buffer is determined by two factors: acid ionization constant of related acid and the ratio of equilibrium concentrations of basic and acidic component from the buffer solution.

Be aware that buffer solutions have their buffer capacity and cannot resist indefinite addition of acid or base substances.

Examples for exercise
Example 3.9.
500 mg of sodium nitrite, $NaNO_2$, was added in 500 mL of 0.1 M nitrous acid, HNO_2. What is the pH of the solution?

$$K(HNO_2) = 7.2 \cdot 10^{-4} \text{ M}$$

Molar concentration of added sodium nitrite, c, can be calculated in following way:

$$n(NaNO_2) = \frac{m(NaNO_2)}{M(NaNO_2)} = \frac{0.500 \text{ g}}{69.00 \frac{\text{g}}{\text{mol}}} = 7.25 \cdot 10^{-3} \text{ mol}$$

$$c(NaNO_2) = \frac{n(NaNO_2)}{V_{\text{NITROUS ACID SOLUTION}}} = \frac{7.25 \cdot 10^{-3} \text{ mol}}{0.5 \text{ dm}^3} = 1.45 \cdot 10^{-2} \text{ M}$$

Calculation of pH is done using eq. (3.66):

$$pH = pK_A + \log\frac{c_{\text{BASIC FORM}}}{c_{\text{ACIDIC FORM}}}$$

$$pK_A = -\log K(HNO_2) = -\log(7.2 \cdot 10^{-4}) = 3.14$$

$$pH = pK_A + \log\frac{c(NaNO_2)}{c(HNO_2)} = 3.14 + \log\frac{1.45 \cdot 10^{-2}}{0.1} = 2.30$$

Example 3.10.
700 mg of sodium phosphate (Na_3PO_4) was dissolved in 1 L of $1.0 \cdot 10^{-3M}$ phosphoric acid, H_3PO_4. What is the pH value of such obtained solution?

$$K_{A1} = 7.1 \cdot 10^{-3} \text{ M}, K_{A2} = 6.3 \cdot 10^{-8} \text{ M}, K_{A3} = 4.2 \cdot 10^{-11} \text{ M}$$

The first step is to calculate molar concentration of added sodium phosphate:

$$n(Na_3PO_4) = \frac{m(Na_3PO_4)}{M(Na_3PO_4)} = \frac{0.700 \text{ g}}{163.94 \frac{\text{g}}{\text{mol}}} = 4.27 \cdot 10^{-3} \text{ mol}$$

$$c(Na_3PO_4) = \frac{n(Na_3PO_4)}{V_{\text{PHOSPHORIC ACID SOLUTION}}} = \frac{4.27 \cdot 10^{-3} \text{ mol}}{1 \text{ dm}^3} = 4.27 \cdot 10^{-3} \text{ M}$$

Sodium phosphate dissociates completely in water solutions. Thus, concentration of phosphate anion equals concentration of the salt:

$$c(PO_4^{3-}) = c(Na_3PO_4) = 4.27 \cdot 10^{-3} \text{ M}$$

Na_3PO_4 acts like base in water solutions due to the hydrolysis of phosphate anion:

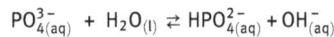

$$PO_{4(aq)}^{3-} + H_2O_{(l)} \rightleftarrows HPO_{4(aq)}^{2-} + OH_{(aq)}^{-}$$

Therefore, if Na_3PO_4 is added into acidic solution, like water solution of phosphoric acid is, the neutralization will happen:

	$H_3PO_{4(aq)}$	$+$	$PO_{4(aq)}^{3-}$	$\rightarrow$	$H_2PO_{4(aq)}^-$	$+$	$HPO_{4(aq)}^{2-}$
1) INITIAL STATE	$1.0 \cdot 10^{-3}$ M		$4.3 \cdot 10^{-3}$ M		0 M		0 M
2) END OF FIRST NEUTRALIZATION STEP	0 M		$3.3 \cdot 10^{-3}$ M		$1.0 \cdot 10^{-3}$ M		$1.0 \cdot 10^{-3}$ M

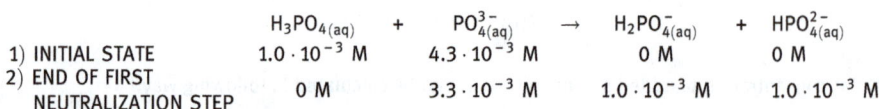

There is more than one neutralization step in this specific solution. The first neutralization step includes reaction between phosphoric acid and phosphate ion. Phosphoric acid is the limiting reactant in this case and therefore the reaction is happening until practically all phosphoric acid reacts with phosphate ions. The reaction results with two products: $H_2PO_4^-$ and HPO_4^{2-}, both in concentrations equal to initial concentration of the limiting reactant. Concentration of the remaining phosphate ions is the difference between initial concentration of phosphate ions and initial concentration of phosphoric acid.

So, three ionization forms of phosphoric acid are present in the solution at the end of the first neutralization step: $H_2PO_4^-$, HPO_4^{2-}, and PO_4^{3-}. An additional neutralization will happen between the remaining phosphate ions and dihydrogen phosphate ion:

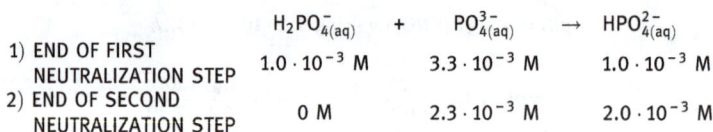

	$H_2PO_{4(aq)}^-$	$+$	$PO_{4(aq)}^{3-}$	$\rightarrow$	$HPO_{4(aq)}^{2-}$
1) END OF FIRST NEUTRALIZATION STEP	$1.0 \cdot 10^{-3}$ M		$3.3 \cdot 10^{-3}$ M		$1.0 \cdot 10^{-3}$ M
2) END OF SECOND NEUTRALIZATION STEP	0 M		$2.3 \cdot 10^{-3}$ M		$2.0 \cdot 10^{-3}$ M

The limiting reactant in this reaction is $H_2PO_4^-$. Only two ionic forms of the phosphoric acid remain in the solution at the end of the reaction: HPO_4^{2-}, and PO_4^{3-}. Further neutralization is impossible, since these two ions form the equilibrium according the reaction:

$$HPO_{4(aq)}^{2-} \rightleftarrows PO_{4(aq)}^{3-} + H_{(aq)}^+$$

This presents a buffer solution so pH can be calculated according the eq. (3.66):

$$pH = pK_A + \log \frac{c_{BASIC\ FORM}}{c_{ACIDIC\ FORM}}$$

Acidic form of the buffer is HPO_4^{2-}, while PO_4^{3-} is the basic one. K_{A3} will be used for calculation of pK_A value, since it relates the existing equilibria. Concentration of phosphate ions, at the end of the reaction, is reduced for the initial concentration of dihydrogen phosphate ions, while the concentration of hydrogen phosphate ions in enlarged for the same amount.

$$pK_{A3} = -\log pK_{A3} = -\log 4.2 \cdot 10^{-11} = 10.38$$

$$pH = pK_{A3} + \log \frac{c\left(PO_4^{3-}\right)}{c\left(HPO_4^{2-}\right)} = 10.38 + \log \frac{2.3 \cdot 10^{-3}}{2.0 \cdot 10^{-3}} = 10.44$$

3.9 Reactions of complexation

Chemical complex consists of a central atom surrounded by a certain number of ligands [7]. Ligands are anions or neutral molecules that contain one or more electron-pairs which can be shared with the central atom. The sharing of the electrons is one-sided: the ligands are exclusively donors of the electrons and provide whole electron pair for the incurring bond with the central atom. In compliance, the central atom acts exclusively as acceptor of the electrons. The formed bond has both covalent and ionic character [8]. According to *ligand field theory* [9], the electrostatic field of present ligands causes splitting in the energy levels of the d-orbitals of central ion what results with stabilization of the complex compound.

Let's take $Cd(CN)_4^{2-}$ complex for further consideration. This complex is formed by bonding four cyanide ligands on Cd^{2+} central ion:

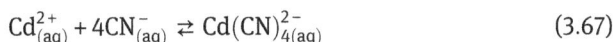

$$Cd^{2+}_{(aq)} + 4CN^-_{(aq)} \rightleftarrows Cd(CN)^{2-}_{4(aq)} \tag{3.67}$$

The equilibrium can be described using the *stability constant* of complex compound, K_{ST} (sometimes also called the *complex formation constant, K_F*):

$$K_{ST} = \frac{\left[Cd(CN)_4^{2-}\right]}{\left[Cd^{2+}\right] \cdot \left[CN^-\right]^4} \tag{3.68}$$

or the *instability constant, K_{INST}*, that relates the opposite reaction, i. e. disintegration of the complex:

$$K_{INST} = \frac{\left[Cd^{2+}\right] \cdot \left[CN^-\right]^4}{\left[Cd(CN)_4^{2-}\right]} \tag{3.69}$$

It is obvious that stability and instability constants of complex compounds are reciprocal:

$$K_{ST} = \frac{1}{K_{INST}} \tag{3.70}$$

$$pK_{ST} = -pK_{INST} \tag{3.71}$$

Equation (3.67) is just a simplified way to present real path of formation of $Cd(CN)_4^{2-}$ complex. In real systems, like water solution, Cd^{2+} ions tend to form stable complexes. Water molecules are the only available ligands. Therefore, in water solutions, Cd^{2+} ions are present in $Cd(H_2O)_4^{2+}$ form and formation of $Cd(CN)_4^{2-}$ complex is practically reaction of ligand-exchange. This exchange performs stepwise as presented in Table 3.2.

The constant for cumulative equation from Table 3.2 is obtained by multiplying equilibrium constants of each exchange step:

Table 3.2: Stepwise replacement of H_2O ligands by CN^- ions in formation of $Cd(CN)_4^{2-}$ complex.

Replacement steps	Equilibrium constant
$Cd(H_2O)_4^{2+} + CN^- \rightleftarrows Cd(H_2O)_3(CN)^+ + H_2O$	$K_{ST1} = \dfrac{[Cd(H_2O)_3(CN)^+]}{[Cd(H_2O)_4^{2+}] \cdot [CN^-]}$
$Cd(H_2O)_3(CN)^+ + CN^- \rightleftarrows Cd(H_2O)_2(CN)_2 + H_2O$	$K_{ST2} = \dfrac{[Cd(H_2O)_2(CN)_2]}{[Cd(H_2O)_3(CN)^+] \cdot [CN^-]}$
$Cd(H_2O)_2(CN)_2 + CN^- \rightleftarrows Cd(H_2O)(CN)_3^- + H_2O$	$K_{ST3} = \dfrac{[Cd(H_2O)(CN)_3^-]}{[Cd(H_2O)_2(CN)_2] \cdot [CN^-]}$
$Cd(H_2O)(CN)_3^- + CN^- \rightleftarrows Cd(CN)_4^{2-} + H_2O$	$K_{ST4} = \dfrac{[Cd(CN)_4^{2-}]}{[Cd(H_2O)(CN)_3^-] \cdot [CN^-]}$

$$K_{ST} = K_{ST1} \cdot K_{ST2} \cdot K_{ST3} \cdot K_{ST4} = \frac{\left[Cd(CN)_4^{2-}\right]}{\left[Cd(H_2O)_4^{2+}\right] \cdot [CN^-]^4} \tag{3.72}$$

Complexes have different applications in environmental chemistry practice. Those of high stability can be used to improve solubility of highly insoluble minerals or to immobilize (mask) the trouble-making compounds in analytical matrices while performing chemical analysis. Also, transition metals form highly colorable complexes what is commonly used in qualitative or quantitative chemical analysis.

3.10 Redox reactions

Chemical reactions that include change of oxidation state of reacting compounds, as a direct consequence of electron exchange, are called redox reactions. For example:

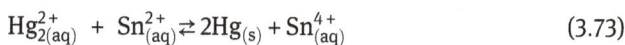

$$Hg_{2(aq)}^{2+} + Sn_{(aq)}^{2+} \rightleftarrows 2Hg_{(s)} + Sn_{(aq)}^{4+} \tag{3.73}$$

In the considered redox reaction, mercury changes its oxidation state from +1 into 0, and tin from +2 into +4.

Each redox reaction is comprised of two half-reactions: the reaction of oxidation and the reduction. During oxidation process, reactant releases one or more electrons and accordingly increases its oxidation state. For considered example, it is further half-reaction:

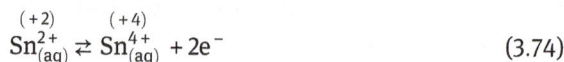

$$\overset{(+2)}{Sn_{(aq)}^{2+}} \rightleftarrows \overset{(+4)}{Sn_{(aq)}^{4+}} + 2e^- \tag{3.74}$$

The reactant that is oxidized is called *reducer* or *reducing agent*.

The opposite half-reaction in which reactant accepts one or more electrons and reduces its own oxidation state is called the reduction:

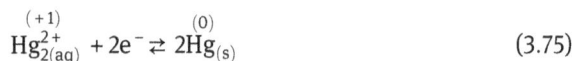

$$\overset{(+1)}{Hg^{2+}_{2(aq)}} + 2e^- \rightleftarrows 2\overset{(0)}{Hg_{(s)}} \tag{3.75}$$

The reactant that reduces its oxidation number and therefore oxidizes reactant from other half-reaction is called *oxidant* or *oxidizing agent*.

Each of these half-reactions has its *relative standard electrode potential*, E^0. The value of this physical-chemical property is commonly expressed considering reaction of the chemical reduction, and that will be the rule in this book also. Relative standard electrode potential is an electrode potential measured against reference electrode [10]. Commonly used reference electrode is *standard hydrogen electrode* which has for every temperature and under standard conditions nominated value of 0 V. Therefore, relative standard electrode potential for system described by eq. (3.75) is actually the potential difference between this system and system described by reaction:

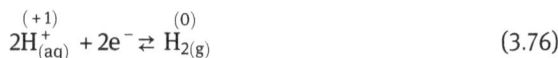

$$\overset{(+1)}{2H^+_{(aq)}} + 2e^- \rightleftarrows \overset{(0)}{H_{2(g)}} \tag{3.76}$$

Driving force for each redox system is redox potential of the system. Redox potential is difference between relative electrode-potentials of the reduction reaction and the oxidation reaction. Relative electrode potential of some half-system depends on relative standard electrode potential, number of exchanged electrons (n) and concentration of oxidized and reduced form of the agent. The related mathematical expression is known as Nernst equation:

$$E = E^0 + \frac{R \cdot T}{n \cdot F} \cdot \ln \frac{c_{\text{OXIDIZED FORM}}}{c_{\text{REDUCED FORM}}} \tag{3.77}$$

Fraction that multiplies logarithm in eq. (3.77) has two constants: constant R is universal gas constant with value 8.314 J/K mol, and F presents Faraday constant of 96485 C/mol. T is absolute temperature. Nernst equation can be simplified for room temperature condition (273.15 K):

$$E = E^0 + \frac{0.059 \text{ V}}{n} \cdot \log \frac{c_{\text{OXIDIZED FORM}}}{c_{\text{REDUCED FORM}}} \tag{3.78}$$

The concentration in Nernst equation will be replaced by partial pressure of the component (expressed in atmospheres; 1 atm = 101325 Pa) if the component is present in gas form. Also, concentration of pure solids is by default equal to 1 [1].

Nernst equation applied on system presented by reaction (3.75) is:

$$E_{\text{Sn}^{4+}/\text{Sn}^{2+}} = E^0_{\text{Sn}^{4+}/\text{Sn}^{2+}} + \frac{0.059\ \text{V}}{2} \cdot \log\frac{c(\text{Sn}^{4+})}{c(\text{Sn}^{2+})} \tag{3.79}$$

For some redox systems, relative electrode potential can be function of concentration of hydronium ion, i. e. pH-value of the solution. Namely, Nernst equation for system $\text{MnO}_4^-/\text{Mn}^{2+}$ has following form:

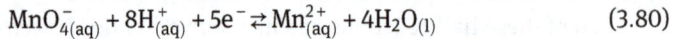

$$\text{MnO}_{4(\text{aq})}^- + 8\text{H}_{(\text{aq})}^+ + 5e^- \rightleftarrows \text{Mn}_{(\text{aq})}^{2+} + 4\text{H}_2\text{O}_{(1)} \tag{3.80}$$

$$E_{\text{MnO}_4^-/\text{Mn}^{2+}} = E^0_{\text{MnO}_4^-/\text{Mn}^{2+}} + \frac{0.059\ \text{V}}{5} \cdot \log\frac{c(\text{MnO}_4^-) \cdot c^8(\text{H}^+)}{c(\text{Mn}^{2+})} \tag{3.81}$$

The potential of this and similar systems can be, obviously, additionally altered by changing acidity of the solution.

Which reactant will act like oxidant, and which one will be the reducer in some redox system? For the answer, one needs to compare relative electrode-potentials of both redox half-systems. Half-system with higher value of the relative electrode potential will be the oxidant.

Problem 3.3.

Compare values of relative standard electrode potentials and give answers on two questions:
1. Which system can oxidize all other systems from the Table 3.3?
2. Which system can reduce all other systems from the Table 3.3?

Among the presented systems, system $\text{H}_2\text{O}_2/2\text{H}_2\text{O}$ has highest value of relative standard electrode potential and therefore can oxidize all other systems from Table 3.3.

Table 3.3: Relative standard electrode potentials of several reduction systems [1].

Reaction of reduction	E^0/V
$\text{Cr}^{3+} + e^- \rightleftarrows \text{Cr}^{2+}$	−0.41
$\text{V}^{3+} + e^- \rightleftarrows \text{V}^{2+}$	−0.26
$2\text{H}^+ + 2e^- \rightleftarrows \text{H}_2$	0.00
$\text{I}_3^- + 2e^- \rightleftarrows 3\text{I}^-$	0.54
$\text{Fe}^{3+} + e^- \rightleftarrows \text{Fe}^{2+}$	0.77
$2\text{Hg}^{2+} + 2e^- \rightleftarrows \text{Hg}_2^{2+}$	0.91
$\text{Cr}_2\text{O}_7^{2-} + 14\text{H}^+ + 6e^- \rightleftarrows 2\text{Cr}^{3+} + 7\text{H}_2\text{O}$	1.33
$\text{MnO}_4^- + 8\text{H}^+ + 5e^- \rightleftarrows \text{Mn}^{2+} + 4\text{H}_2\text{O}$	1.51
$\text{H}_2\text{O}_2 + 2\text{H}^+ + 2e^- \rightleftarrows 2\text{H}_2\text{O}$	1.78

Among the presented systems, system Cr^{3+}/Cr^{2+} has lowest value of relative standard electrode potential and therefore it is capable to reduce all other systems from Table 3.3.

Let's discuss redox system composed of MnO_4^-/Mn^{2+} and Fe^{3+}/Fe^{2+} half-systems. Suppose that concentrations of ionic forms are 1 M. According to data presented in Table 3.3, the oxidant will be permanganate ion, MnO_4^-, and the system will behave as follows:

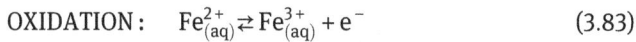

$$\text{REDUCTION}: \quad MnO_{4(aq)}^- + 8H_{(aq)}^+ + 5e^- \rightleftarrows Mn_{(aq)}^{2+} + 4H_2O_{(l)} \tag{3.82}$$

$$\text{OXIDATION}: \quad Fe_{(aq)}^{2+} \rightleftarrows Fe_{(aq)}^{3+} + e^- \tag{3.83}$$

Five electrons are exchanged between these two half-systems. Therefore, eq. (3.83) has to be multiplied by 5 and then sum with eq. (3.82) to obtain cumulative chemical equation:

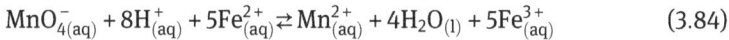

$$MnO_{4(aq)}^- + 8H_{(aq)}^+ + 5Fe_{(aq)}^{2+} \rightleftarrows Mn_{(aq)}^{2+} + 4H_2O_{(l)} + 5Fe_{(aq)}^{3+} \tag{3.84}$$

Driving force of the system, ΔE, is defined as:

$$\Delta E = E_{MnO_4^-/Mn^{2+}} - E_{Fe^{3+}/Fe^{2+}} = 1.51 \text{ V} - 0.77 \text{ V} = 0.74 \text{ V} \tag{3.85}$$

ΔE is biggest at initial state since the initial state is characterized by highest difference of the potentials. Concentration of permanganate ion will be dwindling during the process of electron exchange as well as the concentration of Fe^{2+} ion. At the same time concentrations of Mn^{2+} and Fe^{3+} ions will rise.

Due to the concentration changes, and following the Nernst equation (eq. (3.78)), potential of system MnO_4^-/Mn^{2+} will decreasing while the one of Fe^{3+}/Fe^{2+} system will increasing, causing diminishing of the driving force. Potentials of both half-systems will become equal by reaching the equilibrium:

$$E_{MnO_4^-/Mn^{2+}} = E_{Fe^{3+}/Fe^{2+}} \tag{3.86}$$

$$E_{MnO_4^-/Mn^{2+}}^0 + \frac{0.059 \text{ V}}{5} \cdot \log \frac{[MnO_4^-] \cdot [H^+]^8}{[Mn^{2+}]} = E_{Fe^{3+}/Fe^{2+}}^0 + \frac{0.059 \text{ V}}{5} \cdot \log \frac{[Fe^{3+}]^5}{[Fe^{2+}]^5}$$

$$\tag{3.87}$$

Let's transfer the standard potentials on one side of the equation, and the logarithmic expressions on the other one.

$$E^0_{MnO_4^-/Mn^{2+}} - E^0_{Fe^{3+}/Fe^{2+}} = \frac{0.059 \text{ V}}{5} \cdot \log \frac{[Mn^{2+}] \cdot [Fe^{3+}]^5}{[MnO_4^-] \cdot [H^+]^8 \cdot [Fe^{2+}]^5} \qquad (3.88)$$

It is easy to observe that expression inside logarithm in eq. (3.88) presents equilibrium constant of redox system from eq. (3.84):

$$K_{RO} = \frac{[Mn^{2+}] \cdot [Fe^{3+}]^5}{[MnO_4^-] \cdot [H^+]^8 \cdot [Fe^{2+}]^5} \qquad (3.89)$$

Therefore, general equation that relates equilibrium constant of some redox system with relative standard electrode potentials is:

$$E^0_1 - E^0_2 = \frac{0.059 \text{ V}}{n_1 \cdot n_2} \cdot \log K_{RO} \qquad (3.90)$$

System 1 from eq. (3.90) is system with higher value of relative electrode potential.

Examples for exercise
Example 3.11.
What is the potential of Pb^{2+}/Pb system in solution containing 0.100 M of Pb^{2+} ions?

$$E^0 (Pb^{2+}/Pb) = -0.126 \text{ V}$$

The related reaction is:

$$Pb^{2+}_{(aq)} + 2e^- \rightleftarrows Pb_{(s)}$$

Nernst equation for this reaction is:

$$E_{Pb^{2+}/Pb} = E^0_{Pb^{2+}/Pb} + \frac{0.059 \text{ V}}{2} \cdot \log \frac{c(Pb^{2+})}{c(Pb)}$$

Concentration of pure solid phase, like solid lead in this example, is by default equal to 1 [1]. This simplifies the equation:

$$E_{Pb^{2+}/Pb} = E^0_{Pb^{2+}/Pb} + \frac{0.059 \text{ V}}{2} \cdot \log c(Pb^{2+})$$

Finally, the potential is calculated as follows:

$$E_{Pb^{2+}/Pb} = -0.126 \text{ V} + \frac{0.059 \text{ V}}{2} \log 0.100 = -0.156 \text{ V}$$

Example 3.12.
What is the value of equilibrium constant for the following redox system?

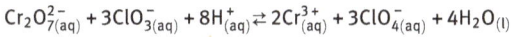

$$Cr_2O_{7(aq)}^{2-} + 3ClO_{3(aq)}^{-} + 8H_{(aq)}^{+} \rightleftarrows 2Cr_{(aq)}^{3+} + 3ClO_{4(aq)}^{-} + 4H_2O_{(l)}$$

$$E^0\left(Cr_2O_7^{2-}/Cr^{3+}\right) = 1.333 \text{ V}, E^0\left(ClO_4^{-}/ClO_3^{-}\right) = 1.189 \text{ V}$$

The equilibrium constant for the given system is defined as follows:

$$K_{RO} = \frac{[Cr^{3+}]^2 \cdot [ClO_4^{-}]^3}{[Cr_2O_7^{2-}] \cdot [ClO_3^{-}]^3 \cdot [H^+]^8}$$

Equation (3.90) will provide the value of the constant:

$$E_1^0 - E_2^0 = \frac{0.059 \text{ V}}{n \text{ EXCHANGED ELECTRONS}} \cdot \log K_{RO}$$

$$E^0\left(Cr_2O_7^{2-}/Cr^{3+}\right) - E^0\left(ClO_4^{-}/ClO_3^{-}\right) = \frac{0.059 \text{ V}}{6} \cdot \log K_{RO}$$

$$1.333 \text{ V} - 1.189 \text{ V} = \frac{0.059 \text{ V}}{6} \cdot \log K_{RO}$$

$$\log K_{RO} = 14.64$$

$$K_{RO} = 4.4 \cdot 10^{14} \text{ M}^{-7}$$

3.11 Solubility of minerals

The emergence of solid phase from water solution, i. e. the precipitation, is important phenomena in environmental chemistry since it is commonly used in many processes relating water treatment. Dissolution of minerals is the opposite process that requires contact between mineral and water, i. e. the existence of boundary between solid and liquid phase. Minerals dissolved in water can change chemical characteristics of the water. Just remember the occurrence of hydrolysis (chapter 7). Maximal amount of solid that can be dissolved in certain volume of solvent at specified temperature is called *solubility*.

Chemical reactions of dissolving/precipitation also tend to reach equilibrium state. The equilibrium is described by *solubility product constant*, K_{SP}, sometimes simply called *solubility product*. The K_{SP} is the product of the concentrations of ions obtained by dissolution of the mineral, with each ion concentration raised to a power equal to the coefficient of that ion in balanced equation for the solubility equilibrium. Concentration of solid phase is not taken in consideration when defining K_{SP}. Taking $FeCl_3$ salt as an example, the solubility product will be:

$$K_{SP} = \left[Fe^{3+}\right] \cdot \left[Cl^-\right]^3 \qquad (3.91)$$

since this salt dissolves according to further chemical equation:

$$FeCl_{3(s)} \rightleftarrows Fe^{3+}_{(aq)} + 3Cl^-_{(aq)} \qquad (3.92)$$

There are three possible states of solution according to solubility of some precipitate: unsaturated, saturated and supersaturated solution. Unsaturated solution is the one in which the product of concentrations of ions that constitute precipitate (each ion concentration should be raised to a power equal to the ion's stoichiometric coefficient from the solubility equation) is lower than related solubility product:

$$c\left(Fe^{3+}\right) \cdot c^3(Cl^-) < K_{SP} \qquad (3.93)$$

Any addition of new solid particles of $FeCl_3$ into unsaturated solution will cause further dissolution of the salt, in order to reach the equilibrium. Solution becomes saturated in cases when the product of concentrations of ions reaches the value of solubility product. The equilibrium is established at this point, and the speeds of dissolving and precipitation became equal. Supersaturated solution is highly unstable since the product of concentrations of ions is higher than the solubility product. Precipitation occurs only in supersaturated solutions.

Several effects, like nature of solvent and solute, temperature, common-ion effect, effect of complex formation, electrolyte effect, and pH-value of solution, have influence on solubility of minerals. These effects will be discussed in the following text.

3.11.1 Nature of solvent and solute

Effect that nature of solvent has on solids solubility follows general rule: "like dissolves like". This means that a solute will be dissolved best in a solvent that has chemical properties similar to the solute. The most important property is polarity. Therefore polar solvents are required to dissolve compounds with ionic characteristics, like salts [11].

For example, water is a polar solvent with partially negative charge on oxygen atom and partially positive charge on hydrogen atoms. Thus, ionic compounds like NaCl dissolve easily in water due to the attractive forces between positive Na^+ ions and negatively charged part of water molecule, and, in the equivalent way, the attractions between Cl^- and positively charged part of the water. On the other side, non-polar compound naphthalene is practically insoluble in water, but it is highly soluble in non-polar solvent like benzene.

3.11.2 Temperature effect

Temperature has dual effect on solubility of minerals: it affects equilibrium of the precipitation and the reaction rate. The solubility generally increases with rising of the temperature. Yet, it is important to say that for some very common environmental minerals, such as $CaCO_3$, $Ca_3(PO_4)_2$, $CaSO_4$ and $FePO_4$, the dissolution is exothermic process and accordingly, the exceptions were noticed [4]. These salts are practically insoluble in water and only very small concentrations of the ions that constitute these salts are present in water solution at room temperature. Yet, at higher temperature, due to the inverse solubility, even these small concentrations will be sufficient for further precipitation of the salt. This causes severe problems in processes using heat exchangers. Deposition of salt on the exchanger's surface decreases the efficiency of the heat exchange. This is reason why water with high content of calcium ions usually needs a treatment. The treatment includes ion-exchange of calcium with sodium ions; the process is called water softening.

The temperature effect on equilibrium is described by Van't Hoff's equation:

$$\ln\frac{K_1}{K_2} = \frac{\Delta H^0}{R} \cdot \left(\frac{1}{T_2} - \frac{1}{T_1}\right) \tag{3.94}$$

T_1 and T_2 are absolute temperatures, K_1 and K_2 equilibrium constants at the related temperatures, R ideal gas constant, and ΔH^0 standard enthalpy change.

3.11.3 Common-ion effect

Let's take barium chromate to explain common-ion effect. Interactions between water molecules and the salt crystal are strong due to water polarity and ionic nature of bond between barium and chromate. Water molecules are pulling apart the ions from salt crystal and thus dissolving the salt. This can be described by simplified chemical reaction:

$$BaCrO_{4(s)} \rightleftarrows Ba^{2+}_{(aq)} + CrO^{2-}_{4(aq)} \tag{3.95}$$

All dissolved $BaCrO_4$ units are present in solution in form of solvated ions. The amounts of barium and chromate ions equal the amount of dissolved $BaCrO_4$ units and for saturated solution these amounts are equal to the solubility of the salt.

$$s(BaCrO_4) = \left[Ba^{2+}\right] = \left[CrO_4^{2-}\right] \tag{3.96}$$

Solubility can be now easy calculated from known value of the solubility product:

$$K_{SP}(BaCrO_4) = [Ba^{2+}] \cdot [CrO_4^{2-}] \tag{3.97}$$

$$s(BaCrO_4) = \sqrt{K_{SP}(BaCrO_4)} \tag{3.98}$$

But, what happens with solubility of a mineral if one of the ions that form the mineral originates from some other chemical? Let's imagine that someone disrupted the existed solubility equilibrium by adding extra amount of one of the precipitate ions, i. e. the common ion, for example chromate. Consequently, the reaction between barium and chromate ion will occur, according to Le Chatelier's [12] principle. The reaction will occur in order to lower concentration of added chromate ions and thus reestablish the equilibrium (eq. (3.97)). So, the presence of chromate ions from some other source will cause inevitable decrease of concentration of barium ions. Since barium ions come exclusively from the salt, their concentration determines solubility of the salt.

$$s(BaCrO_4) = \frac{K_{SP}(BaCrO_4)}{[CrO_4^{2-}]} \tag{3.99}$$

The conclusion is that addition of common ion reduces solubility of salt.

3.11.4 Effect of complex formation

Solubility of the minerals can be improved by addition of agent that form stable chemical complex with one of precipitate ions. AgCl is heavily insoluble salt which dissociates according to reaction:

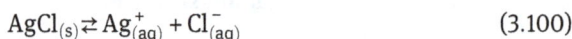

$$AgCl_{(s)} \rightleftarrows Ag^+_{(aq)} + Cl^-_{(aq)} \tag{3.100}$$

Yet, in presence of ammonia, silver forms highly stable silver ammine complex ($K_{ST} = 1.6 \cdot 10^7$ M):

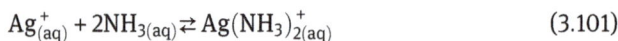

$$Ag^+_{(aq)} + 2NH_{3(aq)} \rightleftarrows Ag(NH_3)^+_{2(aq)} \tag{3.101}$$

This reaction practically is pulling silver ion out of reaction (100), causing further dissolution of the salt. More accurate statement is that there are two systems in solution that highly affect each other. Chemical reaction that describes integrated behavior of both systems can be obtained as sum of eqs. (3.100) and (3.101):

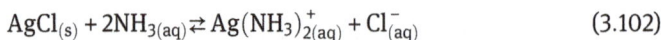

$$AgCl_{(s)} + 2NH_{3(aq)} \rightleftarrows Ag(NH_3)^+_{2(aq)} + Cl^-_{(aq)} \tag{3.102}$$

$$K_R = \frac{\left[Ag(NH_3)_2^+\right] \cdot \left[Cl^-\right]}{\left[NH_3\right]^2} \qquad (3.103)$$

Analysis of expression for constant of the summary reaction, K_R (eq. (3.103)), leads to general conclusion that addition of complexing agent increases solubility of the salt.

The constant of the summary reaction, K_R, can be calculated as a combination of the solubility product constant and the complex stability constant. K_R for selected example is:

$$K_R = \frac{\left[Ag(NH_3)_2^+\right] \cdot \left[Cl^-\right]}{\left[NH_3\right]^2} \cdot \frac{\left[Ag^+\right]}{\left[Ag^+\right]} = K_{ST} \cdot K_{SP} \qquad (3.104)$$

It is important to know that some common ions show different effects on mineral solubility, depending on their concentration. At low concentrations common-ion effect causes reduction of metal solubility while at higher concentrations common ions form complexes and consequently increase the solubility.

3.11.5 Electrolyte effect

Solubility can be, like other chemical reactions, also affected with electrolyte effect. Ag_2CrO_4 dissolves according to reaction:

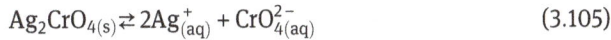

$$Ag_2CrO_{4(s)} \rightleftarrows 2Ag^+_{(aq)} + CrO^{2-}_{4(aq)} \qquad (3.105)$$

The activity, as more accurate approach, will be used instead molar concentration for calculation of solubility product during this specific discussion.

$$K_{SP}(Ag_2CrO_4) = a^2(Ag^+) \cdot a\left(CrO_4^{2-}\right) \qquad (3.106)$$

This equation can be modified taking in consideration the relation between activity and molar concentration (eq. (3.3)):

$$K_{SP}(Ag_2CrO_4) = \gamma^2(Ag^+) \cdot \left[Ag^+\right]^2 \cdot \gamma\left(CrO_4^{2-}\right) \cdot \left[CrO_4^{2-}\right] \qquad (3.107)$$

$$\frac{K_{SP}(Ag_2CrO_4)}{\gamma^2(Ag^+) \cdot \gamma\left(CrO_4^{2-}\right)} = \left[Ag^+\right]^2 \cdot \left[CrO_4^{2-}\right] \qquad (3.108)$$

Left side of eq. (3.108) has value higher than solubility product since activity coefficients are lower than 1 in concentrated electrolyte. This indicates that solubility of minerals increases with increase of the ionic strength (see chapter 3) [1].

3.11.6 Effect of pH of the solution

Solubility of minerals can be affected by change of solution's pH-value. For example, in case of minerals which perform hydrolysis, like AgCN. This salt dissolves according to reaction:

$$AgCN_{(s)} \rightleftarrows Ag^+_{(aq)} + CN^-_{(aq)} \tag{3.109}$$

$$K_{SP}(AgCN) = 7.2 \cdot 10^{-11} \ M^2 \tag{3.110}$$

One of the products of dissolving process is cyanide ion which, in reaction with hydronium ion, forms hydrogen cyanide, a very weak acid:

$$H^+_{(aq)} + CN^-_{(aq)} \rightleftarrows HCN_{(aq)} \tag{3.111}$$

$$K_A = 2 \cdot 10^{-9} \ M \tag{3.112}$$

Therefore, in acid environment, the summary reaction must be considered. The reaction can be obtained by adding together the eqs. (3.109) and (3.111):

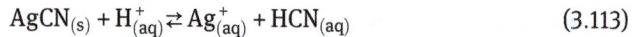

$$AgCN_{(s)} + H^+_{(aq)} \rightleftarrows Ag^+_{(aq)} + HCN_{(aq)} \tag{3.113}$$

$$K_R = \frac{[Ag^+] \cdot [HCN]}{[H^+]} = \frac{s^2(AgCl)}{[H^+]} \tag{3.114}$$

$$s(AgCl_{(s)}) = \sqrt{K_R \cdot [H^+]} \tag{3.115}$$

The equilibrium constant, K_R, can be obtained as combination of the related solubility product and ionization constant of the acid:

$$K_R = \frac{K_{SP}(AgCN)}{K_A} = \frac{7.2 \cdot 10^{-11} \ M^2}{2 \cdot 10^{-9} \ M} = 3.6 \cdot 10^{-2} \ M \tag{3.116}$$

An interesting example is aluminum hydroxide due to its amphoteric properties.

Same effects like those presented for silver cyanide occurs for aluminum hydroxide in highly acid solutions. Aluminum hydroxide is dissolved in such environment and aluminum exists in forms of Al^{3+}, $AlOH^{2+}$ or $Al(OH)_2^+$ ions. Precipitation of aluminum hydroxide occurs at medium pH-values where addition of hydroxide ion decreases the solubility due to common-ion effect:

$$Al(OH)_{3(s)} \rightleftarrows Al^{3+}_{(aq)} + 3OH^-_{(aq)} \qquad (3.117)$$

$$K_{SP}(Al(OH)_3) = [Al^{3+}] \cdot [OH^-]^3 \qquad (3.118)$$

The concentration of Al^{3+} is in fact solubility of $Al(OH)_3$:

$$K_{SP}(Al(OH)_3) = s(Al(OH)_3) \cdot [OH^-]^3 \qquad (3.119)$$

To express the solubility, the eq. (3.119) can be rearranged as:

$$s(Al(OH)_3) = \frac{K_{SP}(Al(OH)_3)}{[OH^-]^3} \qquad (3.120)$$

Since K_{SP} is a constant, it is obvious from eq. (3.120) that the solubility of $Al(OH)_3$ decreases with the increase of OH^- concentration. But, with the further increase of OH^- concentration (in basic environment) dissolving of aluminum hydroxide occurs as a result of complex formation:

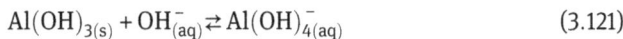

$$Al(OH)_{3(s)} + OH^-_{(aq)} \rightleftarrows Al(OH)^-_{4(aq)} \qquad (3.121)$$

$$K_R = \frac{[Al(OH)_4^-]}{[OH^-]} = \frac{s(Al(OH)_3)}{[OH^-]} \qquad (3.122)$$

$$s(Al(OH)_3) = K_R \cdot [OH^-] \qquad (3.123)$$

Examples for exercise
Example 3.13.
What is the solubility of Hg_2Cl_2 in water?

$$K_{SP}(Hg_2Cl_2) = 1.3 \cdot 10^{-18} \ M^3$$

Solubility of Hg_2Cl_2 in water is only affected by polarity of water molecules. Therefore, it will be calculated from the expression for solubility product.

$$Hg_2Cl_{2(s)} \rightleftarrows Hg^{2+}_{2(aq)} + 2Cl^-_{(aq)}$$

$$K_{SP}(Hg_2Cl_2) = [Hg_2^{2+}] \cdot [Cl^-]^2$$

One diatomic mercury(I) ion (Hg_2^{2+}) and two chloride ions arise from one particle of Hg_2Cl_2. Therefore solubility of the salt, s, equals the concentration of mercury(I) ions, i. e. the one half of the concentration of chloride ions.

$$K_{SP}(Hg_2Cl_2) = s \cdot (2s)^2 = 4s^3$$

This leads to the final solubility calculation:

$$s = \sqrt[3]{\frac{K_{SP}(Hg_2Cl_2)}{4}} = \sqrt[3]{\frac{1.3 \cdot 10^{-18}\ M^3}{4}} = 6.9 \cdot 10^{-7}\ M$$

Example 3.14.
What is the mass of silver bromide that can be dissolved in 1 L of 1.00 M solution of sodium thiosulfate?

$$K_{SP}(AgBr) = 5.2 \cdot 10^{-13}\ M^2,\ K_{ST}\left(Ag(S_2O_3)_2{}^{3-}\right) = 2.9 \cdot 10^{13}\ M^{-2}$$

Silver bromide salt has very low value of the solubility product. Presence of thiosulfate ions in the solution will significantly increase its solubility due to the formation of quite stabile complex. Silver bromide dissociates according the reaction:

$$AgBr_{(s)} \rightleftarrows Ag^+_{(aq)} + Br^-_{(aq)}$$

$$K_{SP}(AgBr) = [Ag^+] \cdot [Br^-]$$

The second important reaction that occurs in the solution is reaction of formation of $Ag(S_2O_3)_2{}^{3-}$ complex:

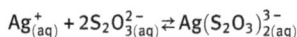

$$Ag^+_{(aq)} + 2S_2O_3^{2-}_{(aq)} \rightleftarrows Ag(S_2O_3)_2^{3-}_{(aq)}$$

$$K_{ST}\left([Ag(S_2O_3)_2]^{3-}\right) = \frac{\left[Ag(S_2O_3)_2^{3-}\right]}{[Ag^+] \cdot [S_2O_3^{2-}]^2}$$

Sum of these two reactions gives the final reaction that controls the solubility of AgBr:

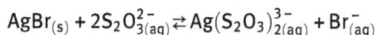

$$AgBr_{(s)} + 2S_2O_3^{2-}_{(aq)} \rightleftarrows Ag(S_2O_3)_2^{3-}_{(aq)} + Br^-_{(aq)}$$

$$K_R = \frac{\left[Ag(S_2O_3)_2^{3-}\right] \cdot [Br^-]}{[S_2O_3^{2-}]^2}$$

Value of the equilibrium constant of final reaction, K_R, is currently unknown, but it is a combination of $K_{SP}(AgBr)$ and $K_{ST}(Ag(S_2O_3)_2{}^{3-})$:

$$K_R = \frac{\left[Ag(S_2O_3)_2^{3-}\right]}{[Ag^+] \cdot [S_2O_3^{2-}]^2} \cdot [Ag^+] \cdot [Br^-] = K_{ST}\left(\left[Ag(S_2O_3)_2\right]^{3-}\right) \cdot K_{SP}(AgBr)$$

$$K_R = 2.9 \cdot 10^{13} \text{ M}^{-2} \cdot 5.2 \cdot 10^{-13} \text{ M}^2 = 15.08$$

According the final reaction, dissolving of silver bromide in thiosulfate solution produces equal concentrations of bromide and thiosulfate complex. These concentrations match the concentration of the dissolved AgBr particles, i. e. they correspond to the solubility, s, of the AgBr:

$$K_R = \frac{s^2(AgBr)}{[S_2O_3^{2-}]^2}$$

Concentration of thiosulfate ions equals the concentration of sodium thiosulfate, since sodium thiosulfate dissociates completely in water solutions. This information leads to the final calculation for solubility:

$$s(AgBr) = \sqrt{K_R \cdot [S_2O_3^{2-}]^2} = \sqrt{15.08 \cdot (1.00 \text{ M})^2} = 3.88 \text{ M}$$

Example 3.15.
What is the solubility of $Ni(OH)_2$ in water solution that has pH = 8.5?

$$K_{SP}\left(Ni(OH)_2\right) = 2.0 \cdot 10^{-15} \text{ M}^3$$

Nickel hydroxide forms the following equilibrium in water solutions:

$$Ni(OH)_{2(s)} \rightleftarrows Ni^{2+}_{(aq)} + 2OH^-_{(aq)}$$

$$K_{SP}\left(Ni(OH)_2\right) = [Ni^{2+}] \cdot [OH^-]^2$$

The presented water solution is slightly basic, meaning that this solution has an elevated concentration of hydroxide ions. The hydroxide concentration is known from the solution's pH value.

$$pOH = 14 - pH = 14 - 8.5 = 5.5$$

$$pOH = -\log[OH^-]$$
$$[OH^-] = 10^{-pOH} = 10^{-5.5} = 3.2 \cdot 10^{-6} \text{ M}$$

Dissolution of nickel hydroxide in basic solutions is influenced by the common ion effect. Therefore, concentration of the dissolved $Ni(OH)_2$ is equal to the concentration of nickel ions.

$$K_{SP}\left(Ni(OH)_2\right) = s\left(Ni(OH)_2\right) \cdot [OH^-]^2$$

$$s\left(Ni(OH)_2\right) = \frac{K_{SP}\left(Ni(OH)_2\right)}{[OH^-]^2} = \frac{2.0 \cdot 10^{-15} \text{ M}^3}{(3.2 \cdot 10^{-6} \text{ M})^2} = 1.9 \cdot 10^{-4} \text{ M}$$

Example 3.16.

30 mL of 0.200 M water solution of $Na_2C_2O_4$ was joined with 30 mL of saturated $CaCO_3$ solution. Please calculate if CaC_2O_4 precipitate is present in such obtained solution.

$$K_{SP}(CaCO_3) = 3.36 \cdot 10^{-9} \ M^2, K_{ST}(CaC_2O_4) = 2.32 \cdot 10^{-9} \ M^2$$

Total volume of the joined solution is 60 mL. Monitored precipitate in solution is CaC_2O_4, which incurs according the reaction:

$$Ca^{2+}_{(aq)} + C_2O^{2-}_{4(aq)} \rightleftarrows CaC_2O_{4(s)}$$

$$K_{SP}(CaC_2O_4) = [Ca^{2+}] \cdot [C_2O_4^{2-}]$$

Precipitation will occur if product of concentrations of ions that constitute CaC_2O_4 raised to a power equal to the ions' stoichiometric coefficient from solubility reaction is higher than solubility product. Therefore, it is important to calculate these concentrations.

Concentration of oxalate ion, $C_2O_4^{2-}$, equals the concentration of $Na_2C_2O_4$ from the joined solution, since $Na_2C_2O_4$ is totally dissociated in water solutions:

$$c(Na_2C_2O_4)_{NEW} = c(Na_2C_2O_4)_{OLD} \cdot \frac{V_{Na_2C_2O_4}}{V_{TOTAL}} = 0.200 \ M \cdot \frac{30 \ mL}{60 \ mL} = 0.100 \ M$$

$$c\left(C_2O^{2-}_{4(aq)}\right) = c(Na_2C_2O_4)_{NEW} = 0.100 \ M$$

Concentration of calcium ion in saturated $CaCO_3$ solution is determined from the related solubility product.

$$CaCO_{3(s)} \rightleftarrows Ca^{2+}_{(aq)} + CO^{2-}_{3(aq)}$$

$$K_{SP}(CaCO_3) = [Ca^{2+}] \cdot [CO_3^{2-}]$$

Also, concentration of calcium ions in this solution equals the concentration of carbonate ions, what provides easy calculation of the calcium concentration:

$$[Ca^{2+}] = [CO_3^{2-}]$$

$$K_{SP}(CaCO_3) = [Ca^{2+}]^2$$

$$[Ca^{2+}] = \sqrt{K_{SP}(CaCO_3)} = \sqrt{3.36 \cdot 10^{-9} \ M^2} = 5.80 \cdot 10^{-5} \ M$$

This concentration has been changed by dilution:

$$c(Ca^{2+})_{NEW} = c(Ca^{2+})_{OLD} \cdot \frac{V_{CaCO_3}}{V_{TOTAL}} = 5.80 \cdot 10^{-5} \ M \cdot \frac{30 \ mL}{60 \ mL} = 2.90 \cdot 10^{-5} \ M$$

Finally, the product of concentrations of calcium and oxalate ions is:

$$c(Ca^{2+}) \cdot c(CO_3^{2-}) = 2.90 \cdot 10^{-5} \text{ M} \cdot 0.100 \text{ M} = 2.90 \cdot 10^{-6} \text{ M}$$

The obtained value is higher than solubility product of calcium oxalate. This indicates that calcium oxalate is precipitating from the solution.

3.12 Solubility of gases

The solubility of gases in water decreases with increasing temperature. This is one of reasons why ecologists are concerned about increasing of thermal pollution of water. Among other bad effects, high temperatures of water cause hypoxia. This phenomenon refers to concentration of dissolved oxygen that is below acceptable level and consequently has devastating effect on water-ecosystems [11].

Pressure is the other important factor that has significant influence on solubility of gases. In cases with no chemical reaction, only partial pressure of the gas in surrounding atmosphere affects its solubility. This can be expressed by Henry's law:

$$s_i = K_{H,i} \cdot p_i \tag{3.124}$$

Symbol s_i marks solubility of gas i, p_i represents related partial pressure, and $K_{H,i}$ is Henry's law constant applicable for particular gas i at specified temperature [13].

The existence of additional chemical reactions, like in case of ammonia or carbon dioxide which are reacting with water, may increase the solubility of gases.

$$NH_{3(g)} \rightleftarrows NH_{3(aq)} \tag{3.125}$$

$$NH_{3(aq)} + H_2O_{(l)} \rightleftarrows NH_{4(aq)}^+ + OH_{(aq)}^- \tag{3.126}$$

The summary reaction can be obtained from eqs (3.125) and (3.126):

$$NH_{3(g)} + H_2O_{(l)} \rightleftarrows NH_{4(aq)}^+ + OH_{(aq)}^- \tag{3.127}$$

and is defined by equilibrium constant, K_R:

$$K_R = \frac{[NH_4^+] \cdot [OH^-]}{p(NH_3)} = K_H \cdot K_B \tag{3.128}$$

References

[1] Skoog DA, West DM, Holler FJ, Crouch SR. Fundamentals of analytical chemistry. 9th ed. Belmont, CA: Brooks/Cole, 2014.

[2] Brönsted JN. EinigeBemerkungenüber den Begriff der Säuren und Basen. Recl Trav Chim Pays-B. 1923;42:718–28.

[3]   Lowry TM. The uniqueness of hydrogen. J Soc Chem Ind-L. 1923;42:43–47.

[4]   Snoeyink VL, Jenkins D. Water chemistry. New York, NY: John Wiley & Sons, 1980.

[5]   Evangelou VP. Environmental soil and water chemistry: principles and applications. New York, NY: John Wiley & Sons, 1998.

[6]   Myers RJ. One-hundred years of pH. J Chem Educ. 2010;87:30–32.

[7]   Vogel Al. Vogel's textbook of macro and semimicro qualitative inorganic analysis. 5th ed. London, UK: Longman Group Limited, 1979.

[8]   Frenking G. Understanding the nature of the bonding in transition metal complexes: from Dewar's molecular orbital model to an energy partitioning analysis of the metal–ligand bond. J Organomet Chem. 2001;635:9–23.

[9]   Ballhausen CJ. Introduction to Ligand field theory. New York, NY: McGraw-Hill, 1962.

[10]   Trassati S. The absolute electrode potential: an explanatory note (Recommendations 1986). Pure Appl Chem. 1986;58:955–66.

[11]   Chang R. Chemistry. 10th ed. New York, NY: McGraw Hill, 2000.

[12]   Harvey D. Modern analytical chemistry. Boston, MA: McGraw-Hill, 2000.

[13]   Manahan SE. Environmental chemistry. 7th ed. Boca Raton, FL: CRC Press LLC, 2000.

Felicita Briški and Marija Vuković Domanovac

4 Environmental microbiology

Abstract: For most people, microorganisms are out of sight and therefore out of mind but they are large, extremely diverse group of organisms, they are everywhere and are the dominant form of life on planet Earth. Almost every surface is colonized by microorganisms, including our skin; however most of them are harmless to humans. Some microorganisms can live in boiling hot springs, whereas others form microbial communities in frozen sea ice. Among their many roles, microorganisms are necessary for biogeochemical cycling, soil fertility, decomposition of dead plants and animals and biodegradation of many complex organic compounds present in the environment. Environmental microbiology is concerned with the study of microorganisms in the soil, water and air and their application in bioremediation to reduce environmental pollution through the biological degradation of pollutants into nontoxic or less toxic substances. Field of environmental microbiology also covers the topics such as microbially induced biocorrosion, biodeterioration of constructing materials and microbiological quality of outdoor and indoor air.

Keywords: microorganisms, environment, indicator microorganisms, biodegradation, bioremediation

> *Gentlemen, it is the microbes who will have the last word.*
>
> (Louis Pasteur)

4.1 Evolution of microorganisms

Earth is about 4.5 billion years old and scientists estimate that life first emerged at least 3.8 billion years ago after the surface of crust had cooled enough to allow liquid water to form. Early Earth was inhospitable from time to time because space rocks crushed into the Earth's surface. Some impacts were powerful enough to vaporize oceans and create clouds of steam which sterilized the Earth's surface. Nonetheless, some microorganisms were able to survive this period deep underground while some may have had the capacity of modern microorganism to produce survival forms called endospores. Early in the planet's history conditions were harsh. The Earth's surface was exposed to strong ultraviolet (UV) radiation because the ozone layer was not yet formed in the atmosphere. Nevertheless, the prokaryotic microorganisms began to

This article has previously been published in the journal Physical Sciences Reviews. Please cite as: Briški, F., Vuković Domanovac, M. Environmental Microbiology. *Physical Sciences Reviews* [Online] **2017**, *2* (11). DOI: 10.1515/psr-2016-0118

https://doi.org/10.1515/9783110468038-004

develop. The first prokaryotic microorganisms lived in anaerobic environment because atmosphere was a mixture of CO_2, N_2, and H_2O vapour and in traces H_2. O_2 gas began to appear in significant amount in the Earth's atmosphere between 2.5 and 2 billion years ago as a result of microbial metabolic process called oxygenic photosynthesis. Oxygenic photosynthesis which started around 3 billion years ago differed from earlier forms of photosynthesis and the bacteria responsible for this type of photosynthesis are called **cyanobacteria**. Cyanobacteria brought the O_2 level of the Earth's atmosphere up to 10 % of today's level and due to it the formation of ozone layer started. O_2 level was high enough to enable evolution of oxygen-utilizing organisms [1, 2]. An approximate timeline of the development of life on Earth is presented in Figure 4.1. Since many eukaryotes are O_2 dependent, researchers had theorized that protists first appeared around 2 billion years ago but according to recent evidence the first protists appeared about 3 billion years ago. Although bacteria and archaea are older than protists, an early appearance of the first eukaryotes is evidenced with a high degree of diversity in this group. Algae appeared after cyanobacteria within the last 2 billion years, because their chloroplasts were derived from cyanobacteria. Fungi appeared during the last several hundred million years.

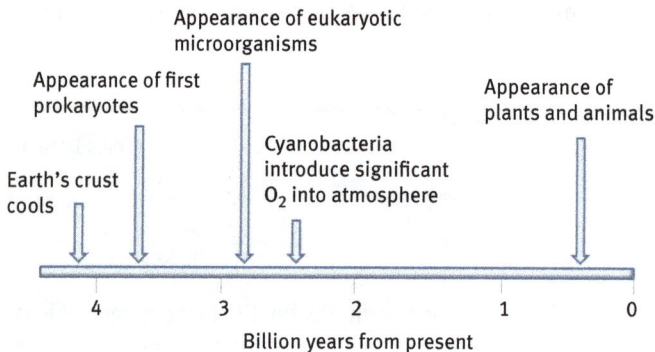

Appearance of eukaryotic
microorganisms

Appearance of first
prokaryotes

Appearance of
plants and animals

Earth's crust
cools

Cyanobacteria
introduce significant
O_2 into atmosphere

4   3   2   1   0

Billion years from present

Figure 4.1: Approximate timeline in the history of life on Earth [1].

4.2 The microbial world: classification, metabolism and growth

Microorganisms are the foundation for all life on Earth. They vary in their appearance, ability to carry out different biochemical transformations, ability to grow in wide variety of environments and in their interactions with other organisms. Due to a great variety of different organisms on Earth, a systematic approach to classifying these organisms is necessary. The science of classifying organisms is called **taxonomy** and the groups making up the classification hierarchy are called taxa. **Nomenclature** refers to the actual naming of organisms. The binominal system of nomenclature is used for microorganisms. The names are given in Latin or are Latinized. The first word in the name is the **genus**, with the first letter always capitalized; the second is the **species** name, which is not capitalized. Both words are always italicized. Classification of well-known bacterium *Escherichia coli* is presented in Example 4.1.

Example 4.1. Classification of bacterium *Escherichia coli*

   E. coli is a member of genus *Escherichia*. It was named after the person who first isolated and described it, Theodor Escherich. The species name comes from the location where it was found, in this case in human intestinal tract. Its classification is as follows:

   Domain: Bacteria, **Phylum**: Proteobacteria, **Class**: Gammaproteobacteria, **Order**: Enterobacteriales, **Family**: Enterobacteriaceae, **Genus**: *Escherichia*, **Species**: *E. coli*.

Traditional classification of living organisms was based on morphological differences like shape, colouring or appendages (flagella) extending from the cells. In the 1970s, a research by Woese and others suggested that life on Earth evolved along three evolutionary lineages. Analysis of 16S rRNA (ribosomal ribonucleic acid) has led to the modern **phylogenetic classification** of living organisms in **three domains: Archaea, Bacteria** and **Eukarya**. Phylogenetic information along with other taxonomic information has been used to construct the phylogenetic tree (Figure 4.2).

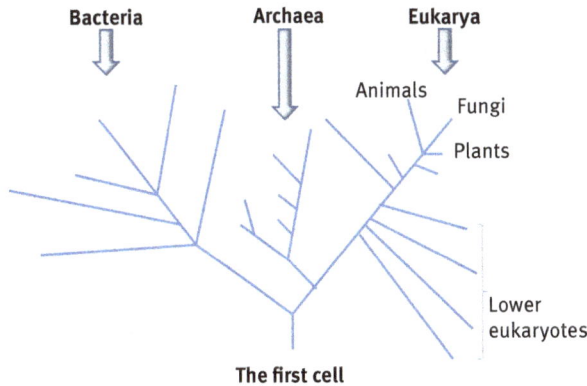

Figure 4.2: Schematics of universal phylogenetic tree evolved from a common ancestor.

Using the rRNA gene sequences of many different organisms resulted in formation of a tree with longer and shorter branches; the longer the branches lengths, the greater the diversity in the group. The branches lengths are shorter for newly evolved organisms such as fungi, plants and animals in comparison with the branches lengths for prokaryotic and lower eukaryotic microorganisms [1, 3].

   Members of microbial world consist of two major cell types: the simple **prokaryotic** and the complex **eukaryotic cells**. Microscopically, the cells in domains Bacteria and

Archaea look similar. They do not contain a membrane-bound nucleus or other intracellular lipid-bound organelles. Their genetic information is stored in a circular strand of deoxyribonucleic acid (DNA) in the nucleotide and cytoplasm is surrounded by a rigid cell wall. Due to structural similarities bacteria and archaea are **prokaryotes** but they differ in their chemical composition and therefore they are in different domains. The domain Eukarya is comprised of four kingdoms: Animalia, Fungi, Plantae and Protista. All members in this domain contain eukaryotic cells and this microbial world is composed of single-celled and multicellular organisms. The cells contain a membrane-bound nucleus and other internal organelles (mitochondria, chloroplast and cytoskeleton) that make **eukaryotes** more complex than the simple prokaryotes.

Term microorganism is used to describe an organism that is so small that it cannot be seen without the use of microscope. Microscopy plays an integral role in the study of microorganisms and can provide extremely useful information about them. Even today, one of the most important tools for studying microorganisms is the **bright-field light microscope** that can magnify images 1000×. Most living microorganisms are nearly transparent and move rapidly about the slide. Consequently, cells must be immobilised and stained with dyes. Simple staining employs one dye to stain the cells. Differential staining is used to distinguish one group of bacteria from another. One of most frequently used staining procedures is the Gram stain (Example 4.2).

Example 4.2. Gram staining is the procedure (Figure 4.3) by which bacteria can be separated in two major groups: Gram positive and Gram negative. Difference in the staining properties of these two groups is due to the difference in the chemical structure of their cell walls. This procedure was developed over a century ago by Dr. Hans Christian Gram.

PROCEDURE

Thin film of specimen spread over slide

Drying at room temperature

Passing slide through flame

Flooding with stain

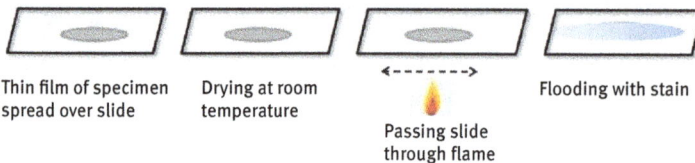

Steps in Gram staining

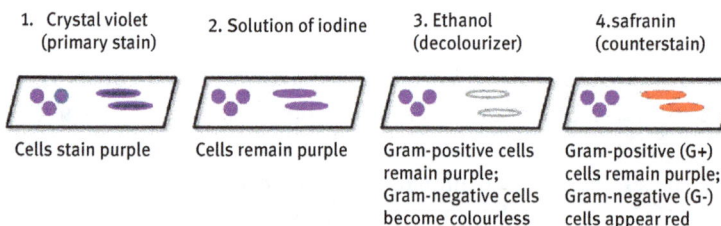

1. Crystal violet (primary stain)

2. Solution of iodine

3. Ethanol (decolourizer)

4. safranin (counterstain)

Cells stain purple

Cells remain purple

Gram-positive cells remain purple; Gram-negative cells become colourless

Gram-positive (G+) cells remain purple; Gram-negative (G-) cells appear red

Figure 4.3: Staining bacteria for microscopic observation.

The **bacteria** are the smallest prokaryotic cells and the least structurally complex unicellular microorganisms. They possess the greatest metabolic flexibility. Typical shapes of common bacteria are: (a) sphere (coccus), (b) rod (bacillus) and (c) spiral (spirillum). Diameter of a typical cell is 1 µm. A representative of rod-shaped bacteria is presented in Figure 4.4(a). The cell wall contains chemical compound peptidoglycan, which is not found in organisms in other domains. They multiply by binary fission in which one cell divides in two cells, and many can move using flagella. The **archaea** are a group of ancient organisms and subdivisions include methanogens (methane producing), halobacteria (live in high-salt environments) and thermoacidophiles (grow best under high temperature and high acidity). The cells of archaea are somewhat similar to bacteria in size and shape; however, they are genetically and biochemically quite different and may be the oldest form of life on Earth. They also multiply by binary fission and move by means of flagella, but archaea do not have peptidoglycan in their cell walls. Members in the domain *Eukarya* consist of eukaryotic cells and diameter of a typical cell is 10–100 µm. Many **fungi** are microscopic, while others, like mushrooms, are macroscopic. All fungi have chitin in their cell walls. Yeasts are single-celled fungi. **Yeasts** can be spherical, oval or cylindrical, and are usually 3–5 µm in diameter (Figure 4.4(b)). Yeasts can reproduce by asexual or sexual means.

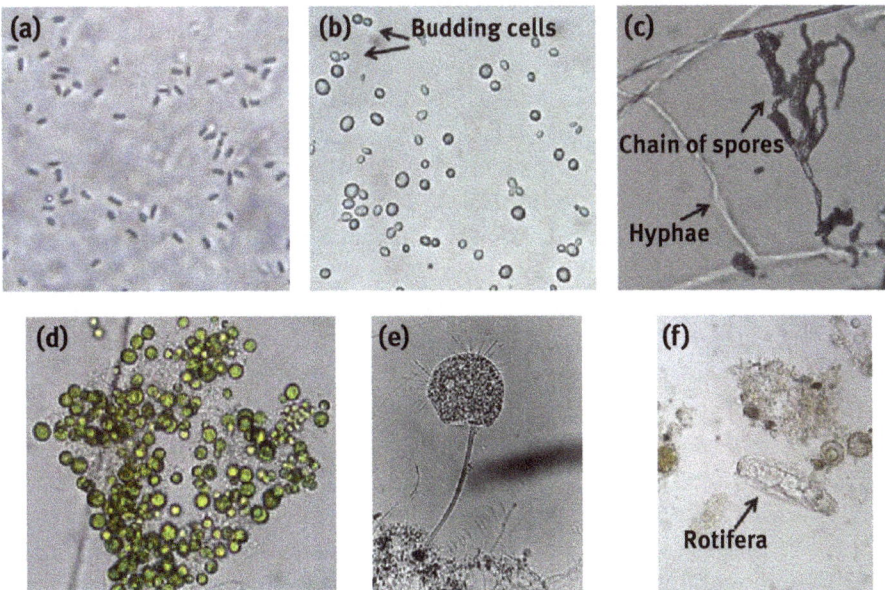

Figure 4.4: Bright-field photomicrograph of cells: (a) rod-shaped bacterium *Bacillus* sp., (1000×), (b) yeast *Saccharomyces* sp., (400×), (c) mould *Penicillium* sp., (400×), (d) green algae *Chlorella*, (400×), (e) single-celled attached protist (stalked ciliates), (400×) and (f) multicellular Rotifera (100×).

Asexual reproduction is by budding or fission. **Moulds** are filamentous fungi and have a mycelial structure. Long thin filaments on the mycelium are called hyphae. Some branches of mycelium may grow in the air and asexual spores (conidia) are formed on these aerial branches. Fungal spores are typically a single cell of about 3–30 μm in diameter, depending on the species (Figure 4.4(c)). Protozoa, slime moulds and unicellular algae as a single-celled organisms belong to the taxonomic kingdom of Protista. The **algae** are a diverse group of photosynthetic protists. Some of them are single-celled while others are multicellular, and they come in a great range of shapes and sizes, from spherical cells with 0.5 μm diameter (*Prochlorococcus*) to 60 m long multicellular thalli (*Macrocystis*). They all contain green pigment chlorophyll, but some of them also contain other pigments (Figure 4.4(d)). They reproduce asexually or sexually. The role of pigment is to absorb sunlight, which algae use as a source of energy. The cell walls of algae are rigid and are composed of cellulose. Diatoms are algae that have silicon dioxide incorporated into their cell walls. Some multicellular species contain other compounds such as carrageenan and agar in their cell walls. Protozoa are diverse groups of microscopic single-cell organisms and they are much larger (5–500 μm) than prokaryotes (Figure 4.4(e)). Protozoa are classified into three groups based on shape: ciliates, flagellates and amoebae. They do not have a rigid cell wall and many of them have a specific shape due to cytoskeleton just beneath the outer membrane of the cell. Metazoa – invertebrates are microscopic multicellular organisms (0.1–0.5 mm long) divided in several subgroups and among them are rotifers (Rotifera, Figure 4.4(f)). The body of a rotifer is divided into a head, trunk and foot, and is typically somewhat cylindrical. The organisms discussed above are all living members of the microbial world. **Viruses** are not living organisms and they consist of a piece of nucleic acid surrounded by protective protein coat. Viruses can only multiply inside living cells and therefore they are considered obligate intracellular parasites [1–4].

Environmental factors that influence microbial growth are temperature, oxygen requirements and pH and water availability. The right temperature, pH and moisture levels vary from one organism to another. Some of them can grow at –20 °C in a brine to prevent freezing, while others can grow at 120 °C where water is under high pressure to prevent boiling. Microorganisms that grow optimally below 20 °C are called **psychrophiles**, those with temperatures optima in the range of 20 °C to 50 °C are **mesophiles**, while those that grow best at temperatures higher than 50 °C are **thermophiles. Hyperthermophiles** have an optimum growth temperature between 70 °C and 110 °C and these are usually members of the Archaea. Cells that require O_2 for growth and metabolism are obligate **aerobes; obligate anaerobes** cannot multiply in the presence of O_2, while **facultative anaerobes** grow best if O_2 is present, but can also grow without it. **Microaerophiles** require small amounts of O_2 and **aerotolerant anaerobes** (obligate fermenter) are indifferent to O_2. Each bacterial species can survive within a range of pH values, but within this range it has a pH optimum. **Neutrophiles** multiply in the range of pH 5 to 8, **acidophiles** grow optimally at a pH below 5.5 and **alkalophiles** above 8.5. Finally, all microorganisms require water for growth. Growth of any microorganisms depends not only on a suitable physical environment, but also on available source of chemicals to use as **nutrients**. All cells contain the macromolecules as follows: carbohydrates, lipids, proteins, nucleotides. Nucleotides as monomers are needed for DNA and RNA synthesis and they play important role in cellular energetics. Elements that make up cell constituents are called macronutrients or **major elements** (C, O, H, N, S, P, K, and Mg). **Trace elements** are also essential to microbial nutrition. The most widely needed trace elements are Fe, Zn and Mn. The elements needed under specific growth condition are Cu, Co, Mo, Ca, Na, Cl, Ni and Se. Some bacteria, especially methanogens, need Na for ion balance. Element Na is important in the transport of charged species in eukaryotic cells. Some halobacteria and marine microorganisms need Cl^-. Microorganisms derive energy either from sunlight or by metabolising chemical compounds. Prokaryotes can live in many environmental habitats because they are able to use diverse sources of energy and carbon. **Photoautotrophs** use the energy of sunlight and CO_2 in the atmosphere to synthesize organic compounds required by many other organisms. Cyanobacteria are important photoautotrophs that inhabit fresh and saltwater habitats. **Chemoautotrophs** or chemolithotrophs use inorganic compounds (H_2, NH_3, NO_2^-, Fe^{2+}, and H_2S) for energy and derive their carbon from CO_2. **Photoheterotrophs** use the energy of sunlight and derive their carbon from organic compounds. **Chemoheterotrophs** or chemoorganotrophs use organic compounds for energy and as a carbon source.

Cells synthesize only the **enzymes** they require for growth. Under the changed environmental conditions, additional or different enzymes may be important. Enzymes are proteins that act as biological catalysts, facilitating the conversion of substrate into product. They usually act on only one, or a very limited number of substrates. More than a 2000 different enzymes are known. They are named by adding the suffix **–ase** to the end of the substrate name, like amyl**ase**, or reaction catalysed, such as alcohol dehydrogen**ase**. Some enzymes have a simple structure, while others have more than one subunit. Some protein enzymes require a non-protein group for their activity, such as cofactor (metal ions Mg, Zn, Cu), or coenzyme that are non-protein organic compounds (e.g. nicotinamide adenine dinucleotide – NAD^+), or some vitamins.

Differences in **microbial metabolisms** can be attributed partly to genetic differences and partly to differences in their responses to changes in their environment. Even the same species may produce different products when grown under different nutritional and environmental conditions. Living cells require energy for biosynthesis, transport of nutrients, motility and maintenance. Two primary functions of cellular metabolism are **catabolism** and **anabolism**. Catabolism is the degradation of a substrate to more highly oxidized end products for the purpose of generating energy and reducing power. Anabolism is the biosynthesis of complex compounds from simpler compounds, usually with the consumption of energy and reducing power. Energy in biological systems is stored and transferred via **adenosine triphosphate (ATP)**, which contains high energy phosphate bonds. Chemoheterotrophs use two processes to provide energy to form ATP: substrate-level phosphorylation and oxidative phosphorylation. Photosynthetic organisms can generate ATP using the process of photophosphorylation, utilizing radiant energy of the sun. There are other compounds analogue to ATP, which store and transfer high-energy phosphate bonds, but not to the extent of ATP. Glucose is a major carbon and energy source for many heterotrophic organisms. Several different metabolic pathways are used by different organisms for catabolism of glucose. Three of the most important metabolic pathways are:

a.   Embden-Meyerhof-Parnas (EMP) pathway or glycolysis – converts glucose to pyruvate and produces initial amount of ATP.
b.   Krebs, tricarboxylic acid cycle (TCA), or citric acid cycle – oxidizes pyruvate through acetyl-CoA into CO_2 and H_2O, and produces a large amount of ATP.
c.   The pentose-phosphate or hexose-monophosphate (HMP) pathway – converts glucose-6-phosphate into variety of carbon skeletons (C3, C4, C5), with glyceraldehyde-3-phosphate as the end product.

Although all three pathways can have catabolic and anabolic roles, the EMP and TCA cycles are the primary means for **energy generation**, while HMP plays a key role in supplying carbon skeletons and reducing powers for use in biosynthesis. In prokaryotes a single glucose molecule will yield up to 24 mol of ATP, while in eukaryotes up to 36 mol of ATP. **Reducing power** is stored in coenzyme nicotinamide adenine dinucleotide NADH (oxidized form is NAD^+) and nicotinamide adenine dinucleotide phosphate NADPH (oxidized form is $NADP^+$). Reducing power can be used to generate ATP through electron transport chain. When O_2 is the final electron acceptor for this reducing power, the process is called **aerobic respiration**. If another electron acceptor such as NO_3^- or SO_4^{2-} is used in conjunction with electron transport chain, then the process is called **anaerobic respiration**. Anaerobic respiration is a less efficient form of energy transformation than aerobic respiration. Cells that obtain energy without using the electron transport chain use **fermentation**. Substrate-level phosphorylation supplies ATP. The end products of fermentative metabolism such as ethanol or lactic acid are formed in response to the cells' need to balance consumption and the production of reducing power. **Autotrophic** organisms use CO_2 as their carbon source to incorporate carbon from CO_2 into cellular material. Energy is obtained either through sunlight (photoautotroph) or oxidation of reduced inorganic chemicals such as H_2S, NH_3 or Fe^{2+} (chemoautotroph).

By knowing environmental and nutritional factors that influence growth of specific prokaryotes, it is possible to provide appropriate conditions for their cultivation in laboratory. These include a medium on which to grow the organism and a suitable atmosphere. For routine purposes many types of complex media are used. Commonly used **complex medium**, nutrient broth, consists of only peptone (protein hydrolysed to amino acids and short peptides), beef extract and water. Nutrient agar results if agar (polysaccharide from algae) is added to nutrient broth. **Chemically defined media** are composed of mixtures of pure chemicals. An example is glucose-salts medium, which supports the growth of bacterium *E. coli*, and contains ingredients as follows: glucose, K_2HPO_4, KH_2PO_4, $MgSO_4$, $(NH_4)_2SO_4$, $CaCl_2$, $FeSO_4$ and water. **Selective** and **differential media** are used for the purpose of detecting or isolating an organism that is a part of a mixed bacterial population. Selective medium types are formulated to support the growth of one group of organisms such as Gram-negative bacteria, but inhibit the growth of other like Gram-positive bacteria. Differential media are widely used for differentiating closely related organisms or groups of organisms. Some media may possess both, selective and differential properties, allowing them to grow certain groups of organisms of interest and give the investigator a way to discern differences in the group based on a visible reaction with the media. These media can be powerful diagnostic tools in environmental settings. Furthermore, an appropriate **atmospheric condition** for microbial growth has to be obtained. Obligate aerobes grow best when tubes or flasks containing media are shaken, providing maximum aeration. Some other microbial species requires atmospheric condition with increased CO_2, microaerophilic or anaerobic conditions.

The most widely used methods [1–3, 5] in the study of microorganisms present in different samples is their growth in a liquid medium, followed by dilution of the sample and plating on a solid agar medium. The theory is that one colony arises on a solid agar medium from one microbial cell. Each colony is then referred to as a **colony forming unit** (**CFU**) and the result of the analysis expressed in terms of CFU/mL or CFU/g of sample. Except for providing an estimate of bacterial numbers, this procedure allows obtaining pure culture isolates. Bacterial growth can be measured by **direct cell counts** and **viable cell counts**. Direct microscopic count is one of the most rapid methods of determining the number of cells in a suspension, but generally does not distinguish living and dead cells. Viable cell counts are used to quantify the number of cells capable of multiplying. Plate counts and membrane filtration both measure the concentration of cells by determining the number of colonies that arise from a sample added to an agar plate. The two different plating methods **pour plate** and **spread plate**, differ in how the suspension of microorganisms is applied to the agar plate. A simple count of the colonies determines how many cells were in the initial sample. The ideal number of colonies to count is between 30 and 300. However, samples frequently contain many more microbial cells than this number and therefore it is necessary to dilute the samples before plating out the cells. The sample is diluted in 10-fold increments and a sterile physiological saline (0.9% NaCl in water) is used to make the dilutions (1 mL sample in 9 mL saline solution). In the pour plate method, 1.0 mL of the final dilution is transferred into sterile Petri dish and then overlaid with melted nutrient agar that has been cooled to 45 °C. At this temperature, agar is still liquid. Petri dish is then gently swirled to mix microbial cells with liquid agar. When the agar hardens the individual cells are fixed in place and, after incubation, distinguishable colonies appear on the agar surface (Figure 4.5).

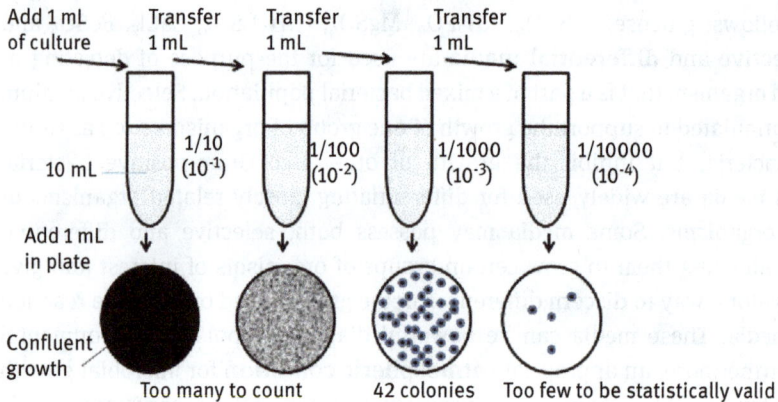

Number of colonies × dilution factor = cells (CFU)/mL of original culture
$42 \times 10^3 \times 1$ (1 mL plated out) = 4.2×10^4 CFU/mL

Figure 4.5: Pour plate method – serial dilutions for the estimation of viable cell count.

In the spread plate method 0.1 mL of the final dilution is transferred directly onto the plate already containing a solidified nutrient agar medium. This solution is then spread over the surface of agar with previously sterilized bent glass rod. **Membrane filtration** is used when the numbers of organisms in a sample are relatively low, as might occur in dilute environment such as natural waters. This method concentrates the bacteria by filtration before they are plated. A known amount of liquid is passed through a sterile membrane filter (pore sizes 0.2–0.45 µm), the filter is placed on an appropriate agar medium and then incubated. The number of colonies that grow on the filter indicates the number of microbial cells that were in the volume filtered and the result is reported per 100 mL of original sample. The **most probable number** (MPN) method is a statistical assay of cell numbers based on the theory of probability. To determine the MPN, three sets of three or five tubes containing the same growth medium are prepared. Each set receives a measured amount of sample such as water or soil. What is important is that the second set receives 10-fold less than the first and the third set 100-fold less. After incubation, the presence (e.g. gas production or precipitate) or absence of growth in each set is noted. The results are then compared against an MPN table, which gives a statistical estimate of the cell concentration, and expressed as MPN/100 mL of original sample. Instead of measuring the number of cells the **cell mass** can be determined.

This can be done by measuring the **turbidity** or **dry weight**. A spectrophotometer is used to measure cloudiness or turbidity of a bacterial suspension as a broth culture. To measure the wet weight, cells growing liquid mediums are centrifuged down, liquid removed and cell mass weighted. The dry weight of the cell mass is determined by drying the centrifuged cells at 105 °C for several hours before weighing them. In the laboratory, bacteria are typically grown in broth contained in a tube or flask. When grown under batch conditions a population of bacteria goes through four distinct stages: in **lag phase** cells synthesize macromolecules prior to active multiplication; in **exponential** or **log phase**, cells divide at a constant rate and their numbers increase by the same percentage during each time interval (generation time is measured during this active multiplication); cells enter the **stationary phase** when they have exhausted their supply of energy and nutrients and the total number of viable cells in the population remains relatively constant; **death phase** is the period when total number of viable cells in the population decreases as cells die off at a constant rate. Theoretically, the time taken for cell division to occur is the **mean generation time** or **doubling time**. This varies depending on the species of microorganisms and conditions in which it is grown. For example, one of the most intensively studied prokaryotic and eukaryotic model single-celled organisms bacterium *E. coli* can double in 20 minutes under the laboratory conditions and yeast *Saccharomyces cerevisiae* in 90 to 120 minutes. Most bacteria reproduce by binary fission, which results in doubling of the number of viable bacterial cells. The relationship of cell numbers and generations can be expressed in a series of equations. Starting with an initial population N_0, the total population N at the end of a given time period would be expressed as eq. (4.1):

$$N = N_0 \cdot 2^n \tag{4.1}$$

where 2^n is the bacterial population after n generations.
    Solving the equation for n, eq. (4.2):

$$\log N = \log N_0 \cdot n \log 2 \quad \rightarrow \quad n = \frac{\log N - \log N_0}{\log 2} \tag{4.2}$$

The number of generations can be calculated if the initial population N_0 and the population N after time t are known. The generation time t_g is equal to t (the time which elapsed between N_0 and N) divided by the number of generations n, eq. (4.3):

$$t_g = \frac{t}{n} \quad \rightarrow \quad t_g = \frac{t \cdot \log 2}{\log \dfrac{N}{N_0}} \tag{4.3}$$

Calculation of mean generation time of bacterial population is presented in Example 4.3.

Example 4.3. What is a generation time of bacterial population that increases from 1×10^4 cells per 1 mL to 1×10^7 cells per 1 mL in four hours of growth?

$$t_g = \frac{t \cdot \log 2}{N} \rightarrow t_g = \frac{240 \cdot \log 2}{\dfrac{10^7}{10^4}} \rightarrow t_g = \frac{240 \cdot 0.301}{3}$$

Answer: t_g = 24 minutes

The selection of methods used for **microbial identification** depends on the type and nature of the microorganism. The methods commonly used in identification and substantiation of taxonomic classification of microorganisms are **phenotypic analysis** and **molecular methods**. Phenotypic analysis often involves growth of microorganisms as pure culture on an artificial media, followed by visual observation of colony morphology such as shape, size, surface characteristics and pigmentation and cell morphology (cell wall characteristics – Gram-staining, sporulation characteristics, mechanisms of motility and other cellular inclusions) using microscopy. For biochemical test specific growth media, nutrients, chemicals or growth conditions are used. These tests include: utilization of carbon and nitrogen sources, growth requirements (anaerobic or aerobic; temperature-optimum and range, pH optimum and range), generation of fermentation products, production of extracellular enzymes (e.g. amylase, protease and cellulase, which hydrolyses starch, protein and cellulose, respectively; Figure 4.6) or other compounds.

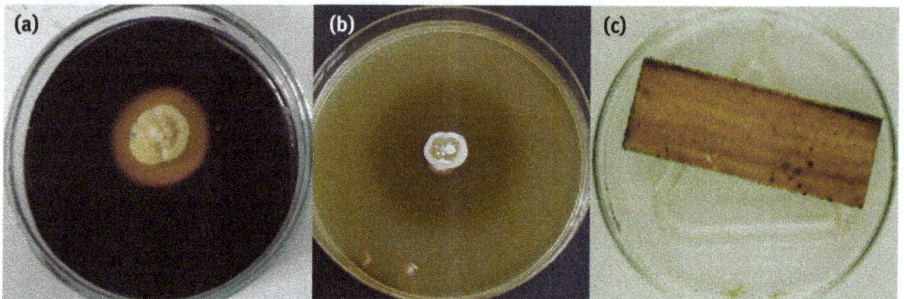

Figure 4.6: Amylolytic activity (a) and proteolytic activity (b) indicated by haloes around a colonies of mould *Aspergillus niger* and filamentous bacterium *Streptomyces* sp. and (c) cellulolytic activity indicated by growth of mould *Trichoderma viride* on wood chip as a sole source of carbon.

Identification, using molecular methods, relies on the comparison of the nucleic acid sequences (DNA, RNA) or protein profiles of a microorganism with documented data on known organisms. The molecular methods are considered sensitive enough to allow detection of low concentrations of viable or non-viable microorganisms in both pure cultures and complex samples (e.g. soil, water and other samples). One of the methods is polymerase chain reaction (PCR), which consists of sequence comparisons of conserved genomic regions such as 16S or 18S rRNA. Reliable genotypic identification requires databases with accurate and complete sequence information from a large number of taxa. Commonly used gene sequence database is GenBank provided by the National Centre for Biotechnology Information (NCBI, http://www.ncbi.nih.gov). Microbial identification of bacterium *Pseudomonas aeruginosa* is described in Example 4.4.

Example 4.4. A nicotine degrading bacterium was isolated from the composting mass [6]. To identify isolated culture phenotypic and molecular method were conducted. According to phenotypic analysis colonies of isolated pure culture were circular, raised and produced water soluble yellow-green pigment. Bacterial cells observed with a bright field microscope were rod-shaped and motile, and after Gram staining cell were Gram negative. A series of biochemical tests were carried out applying API 20 NE test. The biochemical reactions were read according to the Reading Table and by referring to the Analytical Profile Index, the bacterial culture was identified as *Pseudomonas aeruginosa*. Although it is known that the maximum temperature for growth of *P. aeruginosa* is 42 °C, this isolated species was quite actively degrading nicotine at 50 °C during composting tobacco waste. The molecular characterisation that uses 16S rRNA gene sequence to identify bacterium, confirmed that isolated species is *P. aeruginosa* FN (FN-fast nicotine degrading).

4.3 Microbial diversity in the environment

Microorganisms inhabit terrestrial and aquatic ecosystems, including geographical locations considered to be extreme to life [2, 5, 7]. Among their many roles, they are necessary for cycling of carbon, nitrogen, phosphorus and sulphur (biogeochemical cycles), all essential components of living organisms. Since the Earth is a closed system, in order for life to continue, dead organisms must be degraded to their component parts to provide the ingredients needed to grow new living organisms. Microorganisms are recyclers of biomass on Earth and without their recycling activities life on Earth would quickly come to a halt. Microorganisms are also food for some higher organisms and they are at the base of all the Earth's food chains. Apart from recycling dead biological material, some microorganisms also utilize metals and thus recycle non-biological materials. The biodeterioration of man-made compounds by some microorganisms can be a problem (e.g. in the case of paints and plastics). Some fungi attack and degrade man-made materials such as paper, leather, jet fuel, etc.

Terrestrial environments are the richest and the most complex and vary from a rain forest to a desert, but all soils contain a rich diversity of microorganisms. **Soil** is three-phase system consisting of: (a) a solid or mineral inorganic phase that is often associated with organic matter; (b) a liquid or solution phase that contains almost all essential minerals (e.g. NO_3^-, Cl^-, CO_3^{2-}, SO_4^{2-}, Ca^{2+}, Mg^{2+}, Na^+, K^+), which are primary source of inorganic nutrients for plant roots and (c) a gas phase or atmosphere. The density and composition of microbial species of the soil is affected by environmental conditions. For example, wet soils are unfavourable for aerobic microorganisms because the free air spaces are filled up with water. When the water content of soil drops to a very low level (drought or desert) the metabolic activity and number of soil microorganisms decrease. Acidity, temperature and nutrient supply also have influence on microbial activity and diversity. Humification – humus as regulator of soil structure, source of plant nutrients, carbon pool and carbon – CO_2 sequestration. Representative **sampling of soils** is crucial for determination of microbial diversity, interactions between them and shifts in microbial population in soil. Soil samples are obtained with a shovel or a soil auger. Sampling strategy requires taking into consideration the objective of the study, a level of precision of the data needed and sampling requirements (sampling sites, numbers and types of samples, size of sample units, etc.). Methods of soil sampling can be random, transect, two-stage or systematic grid [5, 7, 8]. The traditional methods of microbiological **analysis** of soil samples include cultural assays utilizing dilution and plating methodology (Figure 4.7 and Figure 4.8) or direct count

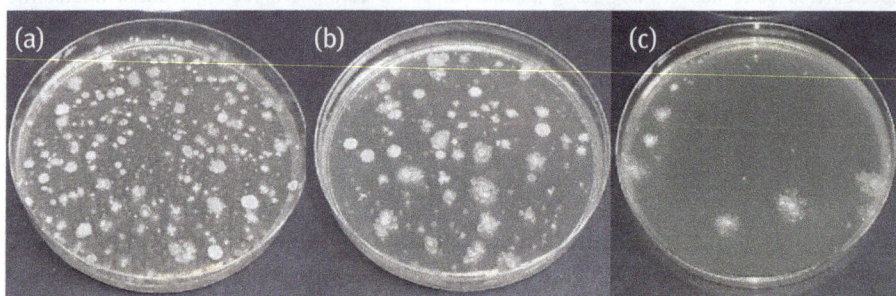

Figure 4.7: Bacterial colonies that result from incubation following dilution and plating of soil sample. The plate on the left (a) has the most bacterial colonies (lowest soil dilution). The other two plates (b and c) result from subsequent dilutions. From these three plates, plate B should be counted since it has between 30 and 300 colonies.

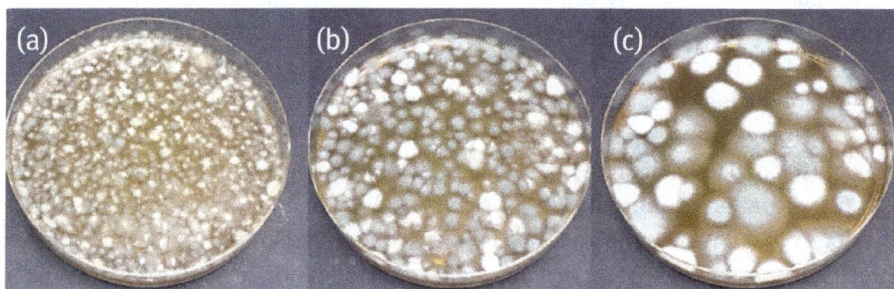

Figure 4.8: Fungal colonies that result from incubation following dilution and plating of soil sample. The plate on the left (a) has the most fungal colonies (lowest soil dilution). The other two plates (b and c) result from subsequent dilutions. From these three plates, plate C should be counted since it has between 30 and 300 colonies.

assays. The number of bacteria in sample is expressed in terms of CFU/g soil (dry weight). As moist soil is originally weighed out, therefore actual dry weight of soil depends on soil moisture at the time of diluting. Molecular methods such as DNA or RNA sequencing are applied for identification of nonculturable microorganisms.

Surface soils are occupied by indigenous populations of archaea, bacteria, fungi, algae, and protozoa, and the microbial distribution depends on soil texture and structure. The microbial communities within the subsurface are lower in number and less diverse than in surface soils. Since nutrients are much more limited in the subsurface, a greater proportion of the population may be in a nonculturable state. In deep saturated zones, the numbers are even lower with limited microbial diversity and include a range of aerobic and facultative anaerobic chemoheterotrophs such as denitrifiers, methanogens, sulphate-reducers, sulphur-oxidizers, nitrifiers, and nitrogen-fixing bacteria.

Prokaryotes are the most numerous soil inhabitants. Gram-positive (G+) bacteria are more abundant in soils than Gram-negative (G-) bacteria. Among the most common are G (+) spore-forming, rod-shaped bacteria of genera *Bacillus* and *Clostridium*. These species form endospores inside vegetative cells, enabling them to survive adverse periods such as drought or extreme heat. Filamentous G (+) bacteria of the genus *Streptomyces* produce desiccation-resistant spores unrelated to endospores. They also produce metabolites called geosmin, which give soil its characteristic musty odour. Some G (-) chemoheterotrophic bacteria adapted to live in terrestrial environments are species of *Azotobacter, Agrobacterium, Rhizobium* and *Pseudomonas*, and among chemoautotrophs nitrifying bacteria such as species of *Nitrosomonas* (transform ammonium to nitrite) and *Nitrobacter* (oxidize nitrite to nitrate). Importance and practical use of nitrogen-fixing bacteria is for bacterisation of legume seeds (symbiotic – *Rhisobium* sp. and non-symbiotic *Azotobacter chroococum*). Prokaryotes are most numerous in soil, but biomass of fungi (eukaryotes) is much greater. Most fungi are aerobes and they are usually found in the top 10 cm of soil. Fungi derive energy from organic material and they are found wherever organic materials are present. The soil fungi are crucial in decomposing plant matter, degrading and using complex macromolecules such as lignin and cellulose. Many soil bacteria and fungi are important in nutrient cycling and removal of organic contaminants by the processes of biodegradation and bioremediation. Various protists are also found in most soils. Algae generally depend on sunlight for energy and most protozoans require O_2, so they are found near the soil surface [9].

Aquatic environments occupy more than 70 % of the Earth's surface and main types are: surface waters (springs, streams, rivers and lakes), marine environments (seas, oceans) and ground waters. There are many types of springs (cold, thermal, hot and specialized – iron, sulphur, radioactive). Cold springs are habitat for bacteria and algae, thermal or hot springs for many different species of archaea and thermophilic bacteria, iron springs for bacterial species *Gallionella* and *Sphaerotilus* and sulphur springs for photosynthetic and non-photosynthetic sulphur bacteria. Water from a source like a spring, rain or snow melt begins to flow down, and small streams are formed. These streams may slowly join together to form a larger stream or river. Streams and rivers accumulate organic matter and heterotrophic population from surrounding terrestrial environment thus a profile of microorganisms in streams and rivers resembles terrestrial. The types and relative numbers of microorganisms inhabiting lakes and marine environments depend on concentration of dissolved O_2 and nutrients, temperature and sunlight. The upper layer is oxygen-rich due to the activities of many photosynthetic organisms such as cyanobacteria. In contrast, the bottom layer of lakes may be anaerobic due to the consumption of O_2 by heterotrophs. In deep sea, water is O_2-saturated due to mixing associated with tides, currents, and wind action. Ground waters are found inland in the subsurface zone and include shallow or deep aquifers. Microorganisms are the sole inhabitants and bacteria are predominant. In general, here microbial activity is low due to the poor concentration of nutrients. Ground water, rivers and lakes are important sources of fresh water. Man-made aquatic environment comprises of stagnant ponds, swimming pools, drainage ditches and wastewaters treatment plants in industrial and urbanized regions [9].

Water and wastewater **sampling** is generally performed by one of two methods; grab sampling or composite sampling. Grab samples are single samples collected from a specific spot at a site over a short period of time (typically seconds or minutes) for: protected groundwater supplies, water supplies receiving conventional treatment, some well-mixed surface waters, wastewater streams, rivers, large lakes, shorelines or estuaries. Composite samples can be obtained by combining portions of multiple grab samples or by using specially designed automatic sampling devices. Membrane filtration (Figure 4.9(a)) and multiple-tube MPN method (Figure 4.10(a))

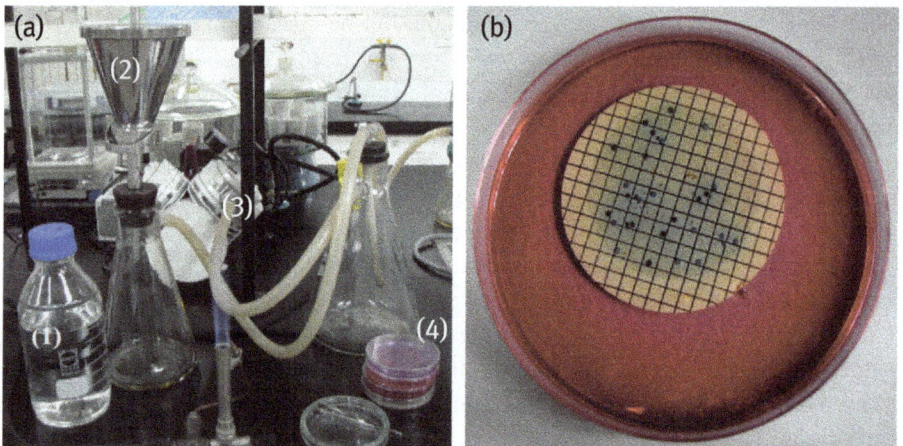

Figure 4.9: (a) Membrane filtration apparatus: 1. water sample, filtration funnel, 3. vacuum pump, 4. selective agar media; and (b) appearance of blue colonies of faecal coliform bacteria on membrane filter after incubation on mFC agar for 24 hours at 44.5 °C

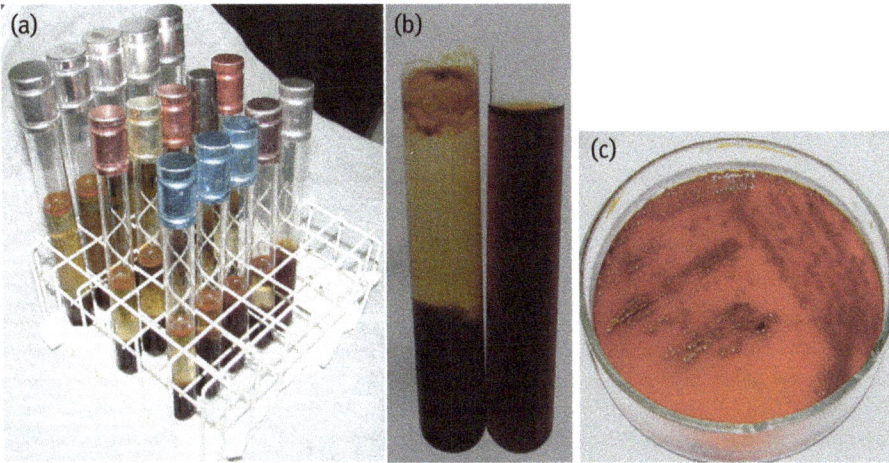

Figure 4.10: MPN dilution series: (a) test tubes after incubation - three series with five replicate tubes, (b) positive test tube in comparison with negative one and (c) dark colour colonies of iron bacteria with metallic shine on Winogradsky agar.

are applied for microbiological **analysis** of natural waters and wastewater streams; however more polluted water will likely require a greater dilution range [10]. Total and faecal coliform bacteria (TC and FC) have been used to evaluate the general quality of water. **Faecal coliform bacteria** and their representative E. coli originate specifically from the intestinal tract of warm-blooded species (e.g. humans, beavers, racoons, etc.) and are cultured by increasing the incubation temperature to 44.5 C and using selective growth medium such as mFC broth or mFC agar (Figure 4.9(b)).

Two other groups of bacteria that are present in faeces are: faecal streptococci (FS) and *Clostridium*. Clostridia spores can survive a long time during adverse conditions. These genera occur naturally in soils and polluted waters, but also indicate a faecal contamination by warm-blooded animals. The ratio of faecal coliforms to faecal streptococci (FC:FS) can provide information on the source of contamination. Human faeces have much more FC than FS, while FS are more common in animal faeces. Comparing the numbers of FC to FS is handy for getting an idea as to whether the water has been polluted with human faecal wastes (>4:1 ratio) or animal wastes (<1:1). For example FC:FS ratio from human faecal wastes is 4.4:1, while from duck and cow 0.6:1 and 0.2:1, respectively.

While faecal coliform bacteria and their representative E. coli are a health problem, **iron bacteria** are not hazardous to health, but in some cases they cause problems in wells, which are a part of water distribution system. Iron bacteria are a natural part of the soil and aqueous environments (ocean, river, or groundwater) in most parts of the world. Iron bacteria use iron as an energy source for their life functions and form chains of cells encased within a tube, or sheath. They are also characterized by the deposition of ferric hydroxide in their sheaths. This clustering within a sheath enables iron bacteria to attach to a solid surface such as rocks or wood in flowing water. The genera *Gallionella*, *Leptothrix* and *Crenothrix*, which are Gram-negative rods, belong to the iron bacteria group. If surface water or soil enters a well, the iron bacteria may be introduced and begin to thrive if conditions are favourable. A major problem associated with iron in wells is production of reddish-brown slime as a metabolic by-product from the oxidation of iron. The slime generated by iron bacteria can corrode pipes and plumbing equipment, and clog pipes, screens, and other components of the well system. To estimate the number of iron bacteria in water well the MPN method may be applied (Example 4.5).

Example 4.5. The number of iron bacteria in water sample from well was estimated by applying MPN dilution series. Replicates of five test tubes (Winogradsky liquid medium [11],) of each dilution (10^{-1}, 10^{-2} and 10^{-3} mL) of water sample were prepared (Figure 4.10(a)) and incubated at 17 °C for three days. Tubes were scored positive on the basis of bacterial growth and formation of brown precipitate (Figure 4.10(a) and Figure 4.10(b)). According to statistical MPN table, the estimated number of iron bacteria was 1.4×10^3/100 mL of original water sample. From positive test tube culture was transferred onto Winogradsky agar plate to observe the growth of colonies of iron bacteria (Figure 4.10(c)). After incubation, a small amount of culture from one of the colonies was transferred onto slide and cells stained by Meyers method [11]. By examination under the light microscope, cells of iron bacteria were red and deposits of iron were blue. According to formation of chains of cells, isolated bacterium present in the water sample belong to the genus *Leptothrix*.

Various layers in the **atmosphere** can be recognized up to the height of about 1000 km [9]. Although air content and atmospheric pressure vary at different layers, air suitable for the survival of terrestrial plants and terrestrial animals is currently only known to be found in the Earth's troposphere. This layer is the nearest to the Earth. In temperate regions, troposphere extends up to about 11 km while in tropics up to about 16 km. In addition to water droplets, dust particles and other matter, air also contains microorganisms. Atmosphere is not a hospitable environment for microorganisms due to temperature variations, low amount of organic matter and a scarcity of available water. Airborne biological particles are called bioaerosols. They are generated as polydispersed droplets or particles of different sizes (0.5 to 30 μm). Pathway of bioaerosols is as follows: launching into the air from soil or water surfaces, the subsequent transport via dispersion in air and deposition on surface. Transport of bioaerosols can be defined in terms of time and distance and it ranges from minutes to hours or days and from 100 m up to 100 km. The transport and settling of bioaerosols are affected by their physical properties and environmental factors. The size, density and shape of the droplets or particles are the most important physical characteristics, while magnitude of air currents, temperature and relative humidity are environmental factors. Temperature and relative humidity also contribute to the influx of airborne microorganisms. Increased concentrations of some fungal spores (e.g., *Cladosporium* sp. and *Nigrospora*) in outdoor air and increased numbers of bacteria released from plant surfaces have been associated with high temperatures and low relative humidity. In wet weather the rain washes the microorganisms from the air. In order to survive in the air, it is important that these microorganisms adapt to environmental stresses (desiccation, humidity and radiation). While harsh environmental conditions tend to decrease the number of viable airborne microorganisms, there is variability in survival between groups of microorganisms and within genera. In general, fungal spores and enteric viruses are somewhat resistant to environmental stress during transport through air. Bacteria and algae are more susceptible while bacterial endospores (e.g. *Bacillus* sp.) are quite resistant.

For collection of **air samples**, many devices are used on the basis of their sampling methods, but most samplers are based on impaction and impingement [12]. Impingement is the trapping of airborne practices in a liquid matrix by drawing air through the inlet and into a liquid. Impaction is the forced deposition of airborne particles on a solid surface. The impaction method separates particles from the air by utilizing the inertia of the particles to force their deposition onto a solid or semi-solid surface. The collection surface is usually an agar medium such as nutrient agar for bacteria or malt agar for fungi (Figure 4.11).

Figure 4.11: Sampling of bioaerosols by the impaction method: (a) microbial air sampler and after incubation of Petri dishes, (b) bacterial colonies on nutrient agar and (c) fungal colonies on malt agar.

Agricultural practices, wastewater and solid waste treatments (sanitary landfill operations and composting) serve as the origin of bioaerosols in **outdoor environments**. In summer and autumn moulds such as genera of *Cladosporium, Alternaria, Penicillium* and *Aspergillus* may be present in outdoor air in city parks, near busy streets genus *Cladosporium* and near buildings genus *Stachybotrys*. Deterioration of building materials, offensive odours, and adverse human health effects are associated with microbial contamination of **indoor environments**. Bacteria and algae grow in areas with standing water such as humidification systems and sites where water leaking (flooding and condensation) has occurred. Fungi, which have lower water activity requirements than other microorganisms tend to colonize a wide variety of building materials. *Penicillium* sp. and *Aspergillus versicolor* are often isolated on wallpaper and drier margins of wetted walls, while *Cladosporium* sp. proliferates as secondary colonizer. The fungus *Stachybotrys chartarum* is most commonly found in homes or buildings which have sustained flooding or water damage from broken pipes, roof, wall or floor leaks, condensation, etc. Wet conditions are required to initiate and maintain growth. It is most common on the paper covering of gypsum wall board, but can be found on wallpaper, cellulose based ceiling tiles, paper products, carpets with natural fibres, paper covering on insulated pipes, in insulation material, on wood and wood panelling, and on general organic debris. The fungus will usually produce large amounts of conidiophores and conidia giving the substrate a black appearance. This fungus is a serious problem in homes and buildings and one of the causes of the "sick building syndrome". *S. chartarum* produces a mycotoxin that can cause animal and human mycotoxicosis. Some other fungi can attack materials such as leather, cloth and hydrocarbons, but can also cause degradative change in glass and metal because of their ability to produce acid as they grow.

The conditions of **extreme environment** greatly restrict the range of microbial species that can grow in such habitat. The extremes of environmental conditions that microbes must be able to tolerate include high temperatures near to boiling, low and high pH values, high salt concentrations, low water availability, low concentration of nutrients, high concentration of toxic compounds and high irradiation. In hot springs and thermal vents there are obligate thermophilic archaea belonging to the genera *Methanothermus, Sulfolobus, Pirodictium*, and *Pyrococcus*. Acidic environment such as mining waste streams and various mineral oxidizing environments are inhabited by bacteria that belong to genus *Thiobacillus*. Some species oxidize only sulphur compounds, while others, such as *Thiobacillus ferrooxidans*, also oxidize ferrous to ferric iron for ATP generation. Halotolerant organisms require salt concentrations for growth that are higher than those found in sea water. The examples of extreme halophilic bacteria belong to the genera *Halobacterium* and *Halanaerobium*. Very few microorganisms are known to survive in nutrient-free extreme environments (ultra pure water used in semiconductor). It can cause flaws in crystal design of computer chips. *Caulobacter* and *Pseudomonas fluorescens* are known to be present there. Radiation resistance of bacterium *Deinococcus radiodurans* is presented in Example 4.6.

Example 4.6. Bacterial biofilms have been found growing on the surfaces of spent nuclear fuel rods stored in nuclear reactor facilities. An example of bacterium that is unusually resistant to ultraviolet (λ = 100 to 295 nm) and ionizing (up to 5 kGy) radiation is *Deinococcus radiodurans*. This bacterium can survive doses of radiation that would kill other living organisms. Its ability to do these results from an unusually active DNA (deoxynucleic acid) repair system, and over a period of 3–4 hours, overlapping fragments of DNA are spliced together into complete chromosomes, and the cells soon resume normal growth [13]. Attempts to understand the molecular basis of this phoenix-like capability has given rise to numerous hypotheses. Notable among them are proposals that the condensed nature of the *Deinococcus* genome or an unusual capacity to avoid protein oxidation is keys to radiation resistance. Bacterium *D. radiodurans* is of interest to scientists developing new bioremediation strategies because it can survive the radiation found in sites contaminated with high levels of radioactive substances. Cleaning up pollution at such sites manually is expensive, while seeding the site with bacteria, which then carry out the work of decontamination has economic and safety advantages. After cloning to introduce genes into *D. radiodurans* it would allow this bacterium to degrade toxic pollutants such as trichloroethylene or chlorobenzene [14].

4.4 Eutrophication process

Why would the topic of eutrophication be included in environmental microbiology? First, the organisms involved in the eutrophication process and the subsequent degradation of water quality are microorganisms: picocyanobacteria, nitrogen-fixing cyanobacteria, diatoms, dinoflagellates, sulfide-oxidizing bacteria, and the suite of aerobic and anaerobic bacteria involved in biogeochemical processes. Second, the process, the symptoms, and the geographic locations of eutrophication are all increasing worldwide at an accelerated rate. Eutrophication is the increase in the rate of production of carbon or the accumulation of carbon in an aquatic ecosystem. The source of the increased organic carbon may come from within the system (autochthonous) or from outside the system (allochthonous).

Historical course of eutrophication: Changes in nutrients as well as the symptoms of cultural eutrophication follow similar time courses on a global scale among developed countries. This same sequence of events is becoming more evident in developing countries (Figure 4.12). As scientists began documenting the sequence of symptoms progressing toward eutrophication and correlating them with changes in water quality, particularly nutrient loads, the patterns and similar trajectories became evident globally in developed countries.

Given the trajectories of future nutrient loads in both developed and developing countries, principally from terrestrial drainage, atmospheric deposition, and urban discharges, it is likely that coastal eutrophication will continue to expand globally. Populations will continue to expand with higher consumptive requirements for food and fuel, and the current trend is for fertilizer use to escalate as the industrialization of agriculture intensifies in the developed world and spreads even more rapidly in developing countries. Coastal ecosystems in many parts of the world, where conditions are conducive to stratification and retention of water, are at high risk for developing eutrophication [15, 16]. *"Given the growing magnitude of the problem and the significance of the resources at risk, nutrient over-enrichment represents the greatest pollution threat faced by the coastal marine environment"* [17].

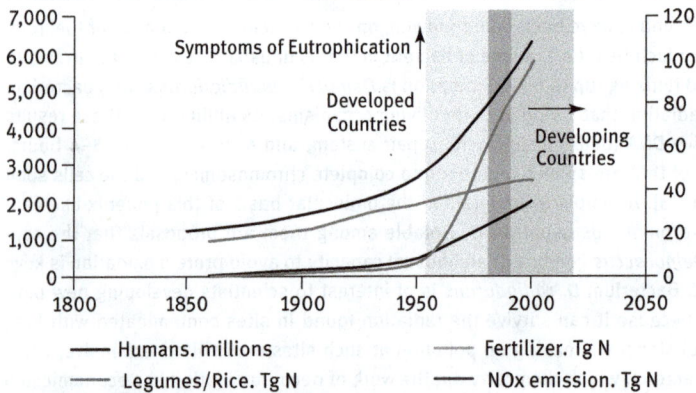

Figure 4.12: Period of the explosive increase in coastal eutrophication in relation to global additions of anthropogenically fixed nitrogen.

Symptoms? The initial response of an estuarine or coastal system to an increase in limiting nutrients is an increase in phytoplankton growth rate and biomass accumulation, growth of filamentous macroalgae, blooms of noxious or toxic algae, reduction in water clarity, shifts in phytoplankton community structure, or combinations of these. A key element of eutrophication is change. Upwelling systems cycle through phases of increased nutrient availability, high primary and secondary productivity, and often, oxygen depletion in the lower water column. The causes may include changes in physical characteristics of the system such as changes in hydrology, changes in biological interactions such as reduced grazing, or an increase in the input of organic and inorganic nutrients. Although the causes may include direct natural or anthropogenic carbon enrichment, eutrophication in the twentieth and twenty-first century is more often caused by excess nutrients that would otherwise limit the growth of phytoplankton [15].

4.5 Microorganisms in biodegradation and bioremediation

The quality of life on Earth is linked inextricably to the overall quality of the environment. In early times, we believed that we had an unlimited abundance of land and resources; today, however, the resources in the world show, in greater or lesser degree, our carelessness and negligence in using them. Due to rapid industrialization and large-scale anthropogenic activities, the pollution level is increasing at a rapid rate, which is a major concern. Though many remediation techniques are available, the use of microorganisms has many advantages like cost effectiveness, few or no by-products, reusability, and more. **Microorganisms** are readily available, rapidly characterized, highly diverse, omnipresent, and can use many noxious elements as their nutrient source. They can be applied in both *in situ* and *ex* *situ* conditions; in addition, many extreme environmental conditions can be cleaned by such entities. Most countries do not restrict industrialization in spite of increased pollution levels; however, these can be minimized using suitable remedial measures, particularly where microorganisms provide a useful tool for a better alternative.

Biodegradation is nature's way of recycling wastes, or breaking down organic matter into smaller compounds by living microbial organisms. In the microbiological sense, "biodegradation" means that the decaying of all organic materials is carried out by a huge assortment of life forms comprising mainly bacteria, and fungi, and possibly other organisms. Complete biodegradation or **mineralisation** involves oxidation of parent compound to form carbon dioxide and water, a process providing both carbon and energy for growth and reproduction of microbial cells. Mineralisation of an organic compound (taking glucose, as an example) under aerobic and anaerobic conditions is shown in Figure 4.13. Irrespective of the structure of the carbon source, each degradation step in the pathway is catalysed by a specific enzyme made by the degrading microbial cell. Lack of appropriate biodegrading enzyme is one common reason for the persistence of organic pollutants in environment. Thus, pollutants that have structures similar to those of natural substrates are normally easily degraded. However, in most cases the term biodegradation is generally used to describe almost any biologically mediated change in a substrate. **Biotransformation** is the metabolic modification of the molecular structure of a compound, resulting in the loss or alteration of some

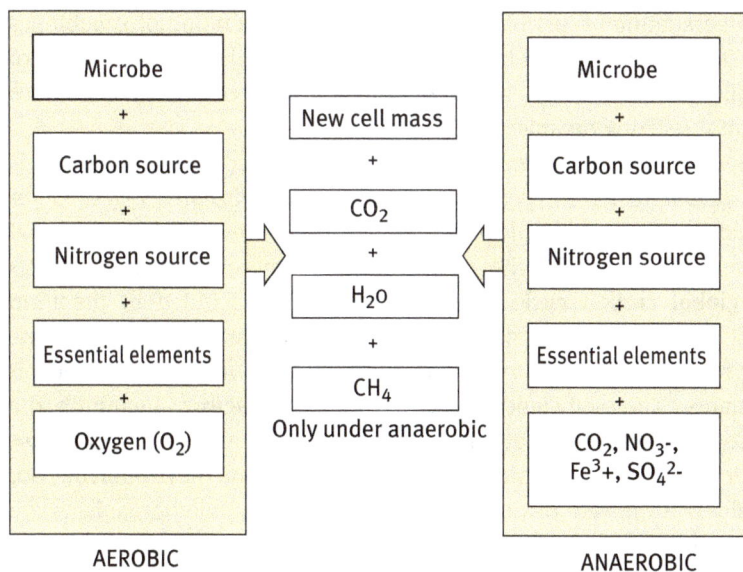

Figure 4.13: Aerobic or anaerobic biodegradation of an organic compound.

characteristic properties of the original compound, with no (or only minor) loss of molecular complexity. Biotransformation may affect the solubility, mobility in the environment, or toxicity of the organic compound. **Bioremediation** is the use of naturally occurring microorganisms to degrade the environmental contaminants into less toxic forms. Bioremediation and biotransformation methods endeavour to harness the astonishing, naturally occurring, microbial catabolic diversity to degrade, transform or accumulate a huge range of compounds including hydrocarbons (e.g. oil), polychlorinated biphenyls (PCBs), polyaromatic hydrocarbons (PAHs), radionuclides and metals [7, 18–21].

In the last few decades, highly toxic organic compounds have been synthesized and released into the environment for direct or indirect application over a long period of time. The problem is worldwide, and the estimated number of contaminated sites is significant. It is now widely recognized that contaminated land is a potential threat to human health, and its continual discovery over recent years has led to international efforts to remedy many of these sites, either as a response to the risk of adverse health or environmental effects caused by contamination or to enable the site to be redeveloped for use.

What are xenobiotics? **Xenobiotic** compounds are chemicals which are foreign to the biosphere. Depending on their fate in **air, water, soil**, or **sediment**, xenobiotic pollutants may become available to microorganisms in different environmental compartments. Actually, the dominant means of transformation and degradation of xenobiotic compounds on Earth resides in microorganisms. In natural habitats, the physicochemical properties of the environment may affect and even control biodegradation performance. Xenobiotics may resist biodegradation, or they undergo incomplete biodegradation or just biotransformation. The definition of xenobiotics as compounds 'foreign to life' exhibiting 'unnatural' structural features does not necessarily imply that xenobiotics are toxic compounds, but many xenobiotics indeed are harmful to living organisms [18].

Why are some compounds degraded by microorganisms? The degradation is done not as a favour to us but because the organisms gain energy needed for growth, reproduction, and other biological functions needed for survival. Microbial activities are very important for the renewal of our environment and maintenance of the global carbon cycle. These activities are included in the term biodegradation. Microorganisms can degrade numerous of organic pollutants owing to their metabolic machinery and to their capacity to adapt to inhospitable environments. Their efficiency depends on many factors, including the chemical nature and the concentration of pollutants, their availability to microorganisms, and the physicochemical characteristics of the environment. So, factors that influence the rate of pollutants degradation by microorganisms are either related to the microorganisms and their nutritional requirements or associated to the environment. A biotic factor is the metabolic ability of microorganisms. The extent to which contaminants are metabolized is largely a function of the specific enzymes involved and their "affinity" for the contaminant and the availability of the contaminant. In addition, sufficient amounts of nutrients and oxygen must be available in a usable form and in proper proportions for unrestricted microbial growth to occur. Other factors that influence the rate of biodegradation are temperature, pH and moisture [7, 18–21].

4.6 Biodegradation of organic pollutants

Many environmental contaminants are subject to chemical or photochemical reactions. However, biological organisms – particularly microorganisms – play a more important role in the removal of many hazardous organics from the environment. Thermodynamically feasible contaminant transformations often do not occur in the absence of a biological catalyst, due to kinetic limitations, but are facilitated by microorganisms via enzymes, which lower the activation energy that must be overcome for a reaction to proceed, and the investment of biochemical energy to convert oxygen and other key co-reactants to more reactive forms [15].

The term biodegradation is often used in relation to ecology, waste management and mostly associated with environmental remediation (bioremediation). A huge number of bacterial and fungal genera possess the capability to degrade organic pollutants. Biodegradation is defined as the biologically **catalysed** reduction in complexity of chemical compounds. The process of biodegradation requires an understanding of the microorganisms that make the process work. The microbial organisms transform the substance through metabolic or enzymatic processes. It is based on two processes: growth and **co-metabolism**. In growth, an organic pollutant is used as sole source of carbon and energy. This process results in a complete degradation of organic pollutants. Co-metabolism is defined as the metabolism of an organic compound in the presence of a growth substrate that is used as the primary carbon and energy source. Compounds that support microbial growth are known as primary substrates. Co-metabolic substrates are called secondary substrates because they do not support growth. Several microorganisms, including fungi and bacteria are involved in biodegradation process. Biodegradation processes vary greatly, but frequently the final product of the degradation is carbon dioxide. Organic material can be degraded aerobically, with oxygen, or anaerobically, without oxygen [19, 21].

Whereas xenobiotics may persist in the environment for months and years, most biogenic compounds are biodegraded rapidly. However, it is not always easy to determine which structural moieties indeed are xenobiotic in the sense of 'foreign to life'. It should be noted that organic chemicals of anthropogenic origin are not necessarily recalcitrant. There are a number of industrial products that are degraded by microorganisms. These compounds obviously are readily recognized by microbial catabolic enzymes. Figure 4.14 summarizes the possible fate of xenobiotic compounds [18].

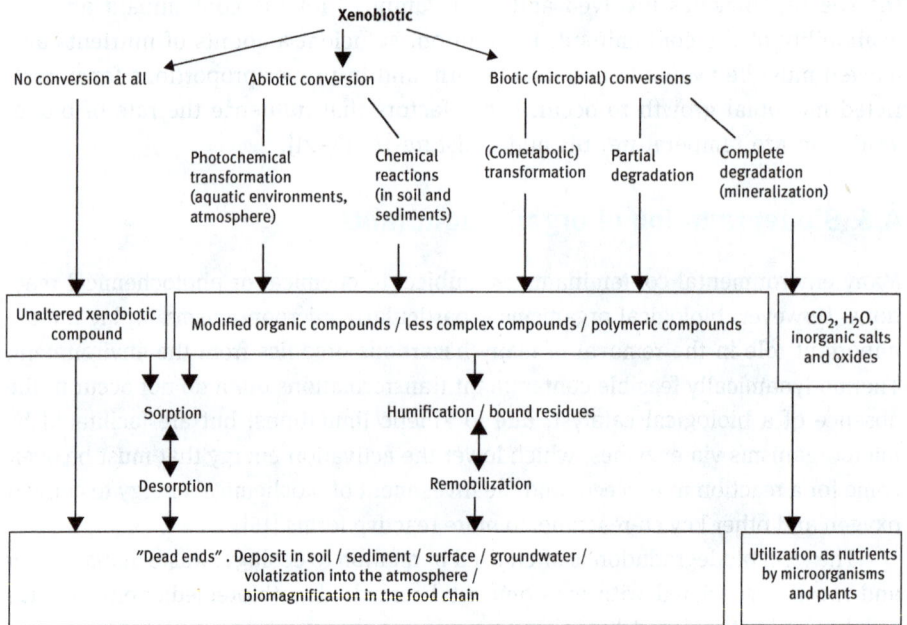

Figure 4.14: Possible environmental fate of xenobiotic compound.

Example 4.7. The products of partial biodegradation, or biotransformation, or co-metabolic conversion of a xenobiotic may be less harmful as the original compound, or they may be as hazardous or even more hazardous than the original compound. For example, tetrachloroethane and trichloroethane can be microbially reduced to vinyl chloride, a known carcinogen, in anoxic habitats. In natural environments, the products of bioconversion processes may be further transformed or degraded by other microorganisms, maybe eventually leading to complete degradation by the microbial consortium. Co-metabolic processes and biodegradation by microbial consortia are thought to be of enormous ecological importance. However, persistent xenobiotics and metabolic dead-end products will accumulate in the environment, become part of the soil humus, or enter the food chain leading to biomagnification. Biomagnification is the increase in concentration of substance that occurs in a food chain. It is a result of, for instance, heavy metal contamination or in other words persistence of contamination which cannot be decomposed by environmental processes [18].

How to describe biodegradation of organic matter? Microbial degradation is generally defined as biological oxidation of organic matter. In natural environments, biodegradation conditions are very complex, and the rate and degree of biodegradation depends on chemical, physical, and biological factors that can differ from one ecosystem to another. Though microbial processes are very complex, some events or groups of events can be presented using a model. Several kinetic models have been developed to describe biological degradation of organic matter. The models are used to gain insight into the applicability and restrictions of treatment processes. The substrate concentration surrounding the microorganisms within the microbial ecosystem is important for the determination of kinetic parameter [15, 21].

Any factor that affects concentration of contaminant, the number of microorganisms present or the expression of specific enzymes by the cells can increase or decrease the rate of contaminant degradation.

Performing the substrate and biomass balance on the batch reactor at constant volume yields the equation for biomass growth rate which is described by the following first-order kinetic eq. (4.4):

$$r_X = \frac{dX}{dt} = \mu X \tag{4.4}$$

where r_X is biomass growth rate, g/L·d; X is the biomass concentration, g/L, t is the time, d, and μ is the specific growth rate of biomass, 1/d. At the same time as the production of the biomass, the substrate is degraded and the equation for the substrate degradation rate is:

$$r_S = -\frac{dS}{dt} = \frac{\mu X}{Y} \tag{4.5}$$

where r_S is substrate consumption rate, g/L·d, S is the limiting substrate concentration, g/L, and Y is the growth yield coefficient, g/g. The substrate degradation rate can also be expressed by the following eq. (4.6):

$$r_S = -\frac{dS}{dt} = qX \tag{4.6}$$

where q is the specific substrate degradation rate, g/g·d, the single parameter which characterises the degradation process. Re-arranging eqs (4.4)–(4.6), the equation for biomass growth rate can be presented by the expressions:

$$r_X = qXY \tag{4.7}$$

$$r_X = r_S Y \tag{4.8}$$

There are several expressions which relate the specific rates (μ and q) to the substrate concentration. How to calculate μ and q is presented in Example 4.8.

Example 4.8. How to calculate specific growth rate and specific substrate consumption rate directly from the experimental data?

Specific growth rate, μ, can be calculated directly from experimental data based on biomass concentration change (X) in time, using eq. (4.9):

$$\mu = \frac{\ln(X_i/X_0)}{(t_i - t_0)} \tag{4.9}$$

Specific substrate consumption q, calculated using eq. (4.10), represents the rate at which the substrate is consumed in relation to biomass growth.

$$q = \frac{(S_0 - S_i)/(t_0 - t_i)}{X_i - X_0} \tag{4.10}$$

In bioprocess modelling, limiting substrate concentration which may occur during the process, may be modelled by various kinetic models. Due to their reliability and simplicity, the Monod model, the modified Monod model with endogenous metabolism, the Haldane model, and the Endo-Haldane model are commonly used [6, 9, 22]. Most frequently applied is the Monod model, which describes the dynamic behaviour of the process, i.e. it shows the relationship between the specific growth rate of the biomass and the limiting substrate:

$$\mu = \frac{\mu_{max} S}{K_S + S} \tag{4.11}$$

In this traditional Monod model, μ_{max} represents the maximum specific growth rate, 1/d, and K_S is the substrate saturation constant, g/L, defined as the concentration of substrate at half the maximum specific growth rate. The biodegradation process results in microbial growth with the removal of substrate. At the end of the process, when most of the organic matter is removed, due to the lack of substrate, a weaker cell population becomes food for the healthier one. Due to biomass decay, the endogenous or decay coefficient, k_d, must be incorporated in the original Monod model. This coefficient corresponds to the endogenous metabolism which involves reactions in cells that consume cell substances.

$$\mu = \frac{\mu_{max} S}{K_S + S} - k_d \tag{4.12}$$

where k_d is the decay coefficient, 1/d. Cell concentration reduction is known as endogenous respiration stage. How to calculate important parameters is presented in Example 4.9.

Example 4.9. How to calculate growth yield coefficient and the biomass decay rate?

Growth yield coefficient Y is one of the most important parameters used in biological kinetics models. It represents biomass concentration produced by unit of removed substrate. Endogenous respiration rate k_d is the biomass decay rate. The dependency of consumed organic substrate to the production of microorganism cells is shown by eq. (13).

$$\mu = Yq - k_d \tag{4.13}$$

Linear regression of μ and q dependency, based on eqs (4.9) and (4.10), produces Y and k_d parameters from the slope and intercept of plot, respectively.

When a substrate inhibits its own biodegradation, the Monod model is inadequate and must be modified by incorporating the inhibition constant K_i:

$$\mu = \frac{\mu_{max}S}{K_S + S + S^2/K_i} \qquad (4.14)$$

This is the Haldane model which takes into account the inhibition constant K_i, g/L·d, which is a measure of sensitivity to inhibition by inhibitory substances. However, some authors have proposed that the decline in cell population, i.e. biomass decay, after complete consumption of the substrate, should be taken into account. Therefore, after attaching the coefficient of microbial decay, k_d, in the expression of the Haldane model, the equation assumes the following form:

$$\mu = \frac{\mu_{max}S}{K_S + S + S^2/K_i} - k_d \qquad (4.15)$$

This is the modified Haldane model, subsequently referred to as the Endo-Haldane model. This model is frequently used because of its ability to assess the effect of inhibition at high concentrations, and cell death and/or maintenance metabolism at low concentrations. The inhibition constant corresponds to the highest substrate concentration at which the specific growth rate equals one-half of the maximum specific growth rate without inhibition.

4.7 Principles of bioremediation

4.7.1 What is bioremediation?

Bioremediation uses microorganisms to reduce pollution through the biological degradation of pollutants into non-toxic or less toxic substances. Bioremediation is an alternative process that involves accelerated natural biodegradation. This can involve either aerobic or anaerobic microorganisms that often use this breakdown as an energy source. Microorganisms might be considered excellent pollutant removal tools in soil, water, and sediments, mostly due to their advantage over other bioremediation procedures. Moreover, bioremediation using biodegradation represents a high impact strategy, but still a low-cost tool for removing pollutants, hence a very viable process to be applied. The principles of bioremediation are based on natural attenuation, bioaugmentation and biostimulation. The simplest method of bioremediation is **natural attenuation**, in which soils are only monitored for variations in pollution concentrations to ensure that the pollutant transformation is active. **Bioaugmentation** is usually applied in cases where natural active microbial communities are present in low quantities or even absent, wherein the addition of contaminant degrading organisms can accelerate the transformation rates. In such cases, the adaptation of exogenous strains that exert highly efficient activities for pollutant transformation to new environments is a key challenge in implementation. The capacity of a microbial population to degrade pollutants can be enhanced also by **biostimulation** of the indigenous microorganisms by addition of nutrients and oxygen [19, 20].

Bioremediation is considered as one of the safest, cleanest, cost effective and environmentally friendliest technology for site decontamination which is contaminated with wide range of pollutants. If the natural attenuation is not quick enough or complete enough, bioremediation will be enhanced either by biostimulation or bioaugmentation.

The popularity of bioremediation is increasing because it often consumes less energy and fewer resources and thus is less expensive and more sustainable than physicochemical treatment approaches, such as landfilling or incineration. Further, many alternative remediation techniques simply transfer organic contaminants to another medium without detoxifying the compounds [15].

History of bioremediation: Bioremediation is not a new concept: microbiologists have studied the process since the 1940s. The first commercial use of naturally occurring microbes to safely and effectively clean up a toxic environmental disaster occurred in the late 1960s following an accidental oil spill in Cat Canyon (located in Santa Barbara, California, USA) after an oil pump shaft broke loose. The spill over went into the drainage system, into a stream and eventually into the nearest drinking water supply. George M. Robinson, assistant county petroleum engineer, treated the oil spill sumps with bacterial cultures that he had isolated in home experiments begun in the 1960s. The new treatment technology of bioremediation grew out of these early studies on petroleum hydrocarbon degradation. The first patent for a biological remediation agent was registered in 1974, a strain of *Pseudomonas putida* that was able to degrade petroleum [19, 20].

The invisible workforce, microorganisms can be isolated from almost any environmental conditions. Microbes can adapt and grow at sub-zero temperatures, as well as extreme heat, desert conditions, in water, with an excess of oxygen and in anaerobic conditions, with the presence of hazardous compounds or on any waste stream. Because of the adaptability of microbes and other biological systems, these can be used to degrade or remediate environmental hazards. The main requirements are an **energy** source and a **carbon** source. Although the microorganisms are present in contaminated soil, they cannot necessarily be there in the numbers required for bioremediation of the site. Their growth and activity must be stimulated. Biostimulation usually involves the addition of nutrients and oxygen to help indigenous microorganisms. These nutrients are the basic building blocks of life and allow microbes to create the necessary enzymes to break down the contaminants. All of them will need nitrogen, phosphorous, and carbon. Carbon is the most basic element of living forms and is needed in greater quantities than other elements. In addition to hydrogen, oxygen, and nitrogen it constitutes about 95 % of the weight of cells. The rate of biodegradation is decreased by roughly one-half for each 10 °C decrease in temperature. Biodegradation can occur under a wide-range of pH; however, a pH of 6.5 to 8.5 is generally optimal for biodegradation in most aquatic and terrestrial systems. Moisture influences the rate of contaminant metabolism because it influences the kind and amount of soluble materials that are available as well as the osmotic pressure and pH of terrestrial and aquatic systems [18–20].

Natural organisms, either indigenous or extraneous (introduced), are the prime agents used for bioremediation. The organisms that are utilized vary, depending on the chemical nature of the polluting agents, and are to be selected carefully as they only survive within a limited range of chemical contaminants. Since numerous types of pollutants are to be encountered in a contaminated site, diverse types of microorganisms are likely to be required for effective mediation (Table 4.1) [19, 23].

Table 4.1: Microorganisms having biodegradation potential for xenobiotics.

Organism	Toxic chemicals
Pseudomonas spp	Benzene, anthracene, hydrocarbons, PCBs
Alcaligenes spp	Halogenated hydrocarbons, linear alkylbenzene sulfonates, polycyclic aromatics, PCBs
Arthrobacter spp	Benzene, hydrocarbons, pentachlorophenol, phenoxyacetate, polycyclic aromatic, aromatics, long chain alkanes, phenol, cresol
Bacillus spp	Halogenated hydrocarbons, phenoxyacetates
Flavobacterium	Naphthalene, biphenyl, aromatics, branched hydrocarbons
Rhodococcus spp	Aromatics
Mycobacterium spp	Hydrocarbons, polycyclic hydrocarbons
Streptomyces spp	Phenoxyacetate, halogenated hydrocarbon, diazinon
Candida tropicalis	PCBs, formaldehyde

4.7.2 Some groups of microbes

1. Aerobic: Examples of aerobic bacteria recognized for their degradative abilities are *Pseudomonas, Alcaligenes, Sphingomonas, Rhodococcus*, and *Mycobacterium*. These microbes have often been reported to degrade pesticides and hydrocarbons, both alkanes and polyaromatic compounds. Many of these bacteria use the contaminant as the sole source of carbon and energy.

2. Anaerobic. Anaerobic bacteria are not as frequently used as aerobic bacteria. There is an increasing interest in anaerobic bacteria used for bioremediation of polychlorinated biphenyls (PCBs) in river sediments, dechlorination of the solvent trichloroethylene (TCE) and chloroform.

3. Ligninolytic fungi. Fungi such as the white rot fungus *Phanerochaete chrysosporium* have the ability to degrade an extremely diverse range of persistent or toxic environmental pollutants. Common substrates used include straw, saw dust, or corn cobs.

4. Methylotrophs. Aerobic bacteria grow utilizing methane for carbon and energy. The initial enzyme in the pathway for aerobic degradation, methane monooxygenase, has a broad substrate range and is active against a wide range of compounds, including the chlorinated aliphatic trichloroethylene and 1,2-dichloroethane.

Microorganisms play a more important role in the removal of many hazardous organics from the environment. Many microorganisms possess the ability to transform hazardous compounds. However, the long-term persistence of many of these contaminants in the environment is a testament to the fact that these naturally occurring processes often do not occur at rates that are fast enough to protect ecosystem and human health. Frequently, the microorganisms are limited by the availability of the pollutant or another key substrate or are not present in sufficient numbers. In many cases, bioremediation can overcome these limitations through careful engineering of the contaminated environment, thereby enhancing the rates of key microbial processes. Thus, successful bioremediation involves the integration of environmental microbiology and engineering techniques with other disciplines, such as geochemistry and hydrology.

Biodegradation is nature's way of recycling wastes, or breaking down organic matter into nutrients that can be used by other organisms. In nature, there is no waste because everything gets recycled. The waste products from one organism become the food for others, providing nutrients and energy while breaking down the waste organic matter. Some organic materials will break down much faster than others, but all will eventually decay. By harnessing these natural forces of biodegradation, people can reduce wastes and clean up some types of environmental contaminants. Through composting, we accelerate natural biodegradation and convert organic wastes to a valuable resource. Wastewater treatment also accelerates natural forces of biodegradation. In this case the purpose is to break down organic matter so that it will not cause pollution problems when the water is released into the environment. Through bioremediation, microorganisms are used to clean up oil spills and other types of organic pollution. Treatments can be either *ex situ* or *in situ*. The technology can involve aerobic and/or anaerobic bioreactors, biofiltration, air sparging, bioventing, composting, landfarming, and biopiles. Intrinsic remediation refers to the combined effects of all natural processes in contaminated environments that reduce the mobility, mass, and risks of pollutants [15, 20].

4.8 Composting

4.8.1 What is composting? Composting is nature's way of recycling

Composting is the controlled aerobic decomposition of organic matter by microorganisms into a stable, humus-like soil amendment. The processes used in composting occur in nature, but systems can be designed and managed to enhance and accelerate the process. The by-product of composting consists of the biomass of dead and living microorganisms, non-degradable raw material, and stable by-products of decomposition. **Compost** is an organic soil conditioner that has been sufficiently stabilized to minimize or eliminate unpleasant odours, substantially reduce or eliminate viable pathogens and weed seeds, facilitate storage and handling without attracting insects and vectors, and provide plants with nutrients that become available throughout the growing season. Composting stabilizes organic matter by converting the readily degradable portion of the organic matter into carbon dioxide and water. Complete stabilization is neither practical nor desirable because complete stabilization would destroy all the slowly degradable organic matter that gives compost its soil-building properties. The degree of stabilization and pathogen destruction desired in compost is dependent on the purpose for composting and intended use of the compost by-product [24, 25].

Microorganisms in composting process decompose the more readily available carbonaceous compounds. Composting involves various types of microorganisms which are all vital to the process. In the process of composting, microorganisms such as bacteria, fungi, and actinomycetes, break down organic matter and produce carbon dioxide, water, heat, and humus, the relatively stable organic end product. Bacteria are most numerous and are the primary biodegraders. The most common microorganisms involved in the composting process are genera *Bacillus, Pseudomonas, Cellulomonas, Trichoderma, Trichosporon, Cladosporium, Thermoactinomyces, Streptomyces.* Under optimal conditions, composting proceeds through three phases: (1) the mesophilic, or moderate-temperature phase, which lasts for a couple of days, (2) the thermophilic, or high-temperature phase, which can last from a few days to several weeks, and finally, (3) a several-month cooling and maturation phase.

Different communities of microorganisms predominate during the various composting phases. Initial decomposition is carried out by mesophilic microorganisms, which rapidly break down the soluble, readily degradable compounds. The heat they produce causes the compost temperature to rapidly rise. As the temperature rises above about 40 °C, the mesophilic microorganisms become less competitive and are replaced by others that are thermophilic, or heat-loving. At temperatures of 55 °C and above, many microorganisms that are human or plant pathogens are destroyed. Because temperatures over about 65 °C kill many forms of microbes and limit the rate of decomposition, compost managers use aeration and mixing to keep the temperature below this point. During the thermophilic phase, high temperatures accelerate the breakdown of proteins, fats, and complex carbohydrates like cellulose and hemicellulose, the major structural molecules in plants. As the supply of these high-energy compounds becomes exhausted, the compost temperature gradually decreases and mesophilic microorganisms once again take over for the final phase of "curing" or maturation of the remaining organic matter.

How does composting happen? For successful composting microbes need nutritious "food", suitable moisture, pH, temperature, and oxygen. The supply of carbon (C) relative to nitrogen (N) is an important quality of compost feedstocks. It is designated as the C:N ratio. The ideal starting range is C:N 25–35:1. Microorganisms such as bacteria, fungi and actinomycetes account for most of the decomposition, as well as the rise in temperature that occurs in the composting process. The ideal moisture content of the compost pile is between 45 and 60 % by weight, and pH 6.5–8.0.

The rate of decomposition or process limitation is a function of microbial activity. As shown in Figure 4.15, temperature rises from ambient to mesophilic and then to thermophilic. As temperature in the compost pile increases, thermophiles (microorganisms that function at temperatures above 40 °C) take over. The temperature in the compost pile typically increases rapidly to 55–65 °C within 24–72 hours of pile formation, which is maintained for several weeks. This is called **the active phase of composting**.

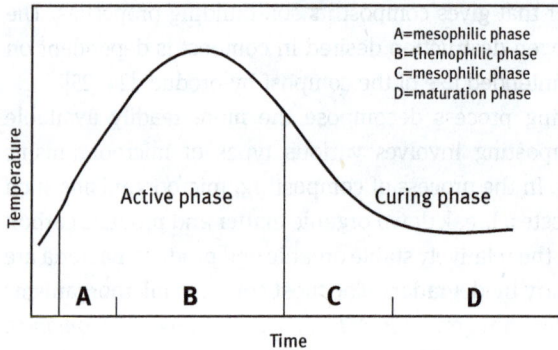

A=mesophilic phase
B=themophilic phase
C=mesophilic phase
D=maturation phase

Active phase Curing phase

A   B   C   D

Time

Figure 4.15: Temperature changes in an average compost pile.

In the active "thermophilic" phase, temperatures are high enough to kill pathogens and weed seeds and to break down phytotoxic compounds (organic compounds toxic to plants). Common pathogens killed in this phase are *E. coli, Staphylococcus aureus, Bacillus subtillus*, and *Clostridium botulinum*. During this phase, oxygen must be replenished through passive or forced aeration, or turning the compost pile. As the active composting phase subsides, temperature gradually declines to under 40 °C. The mesophilic microorganisms recolonize the pile, and the compost enters the **curing phase**. The rate of oxygen consumption declines to the point where compost can be stockpiled without turning. During curing, organic matter continues to decompose and are converted to biologically **stable and mature compost** [25].

Acknowledgements: The authors of this chapter are thankful to Mrs. Marijana Vidakovic on technical support in preparation of photomicrographs for Figure 4 and photographs for Figure 6, Figure 7, Figure 8, Figure 9, Figure 10, Figure 11.

References

[1]   Salyers AA, Whit DD. Microbiology: diversity, disease, and the environment. Bethesda, MA: Fitzgerald Science Press; 2001: 19–32, 439–452.
[2]   Environmental IV. Microbiology for engineers, 2nd ed. Boca Raton, FL: CRC Press, 2010: 1–15, 265–293.

[3]   Nester EW, Anderson DG, Roberts CE, Pearsall NN, Nester MT, Hurley D. Microbiology: a human perspective. New York, NY: McGraw Hill, 2004: 83–103, 785–797.

[4]   Shuler ML, Kargi F. Bioprocess engineering: basic concepts. Upper Saddle River, NJ: Prentice Hall PTR, 2002: 133–155.

[5]   Pepper IL, Gerba CP. Environmental microbiology: a laboratory manual, 2nd ed. Burlington, MA: Elsevier Academic Press, 2004: 17–49, 113–139, 169–174.

[6]   Briški F, Kopčić N, Ćosić I, Kučić D, Vuković M. Biodegradation of tobacco waste by composting: Genetic identification of nicotine-degrading bacteria and kinetic analysis of transformations in leachate. Chem Paper. 2012;66:1103–1110.

[7]   Sharma PD. Environmental microbiology. Harrow, UK: Alpha Science International Ltd., 2005: 66–94.

[8]   Van Elsas JD, Smalla K. Methods for sampling soil microbes. In: (Hurst CJ, editor. in Chief), Manual of environmental microbiology. Washington, DC: ASM Press, 1997: 383–391.

[9]   Briški F Zaštita okoliša (in Croatian), Fakultet kemijskog inženjerstva i tehnologije Sveučilišta u Zagrebu i Element, Zagreb, 2016: 46–50, 87–91, 113–134, 179–185.

[10]   APHA. Standard methods for the examination of water and wastewater, 2nd ed. Washington, DC: American Public Health Association, 1998.

[11]   Rodina AG. Methods in aquatic microbiology. Baltimore, Maryland: University Park Press, 1972: 358–368.

[12]   Stetzenbach LD. Introduction to aerobiology. In: Hurst CJ, Knudsen GR, McInerney MJ, Stetzenbach LD, Walter MV, editors, Manual of environmental microbiology. Washington, DC: ASM Press, 1997: 620–621.

[13]   Slade D, Radman M. Oxidative stress resistance in Deionococcus radiodurans. Microbiol Mol Biol Rev. 2011;75:133–191.

[14]   Lange CC, Wackett LP, Minton KW, Daly MJ. Engineering a recombinant *Deionococcus radiodurans* for organopollutant degradation in radioactive mixed waste environments. Nature Biotechnol. 1998;16:9229–9233.

[15]   Mitchell R, Gu JD. Environmental Microbiology. New Jersey, USA: John Wiley & Sons, Inc., 2010: 115–135, 177–212.

[16]   Rosenberg R. Eutrophication: the future marine coastal nuisance? Mar Pollut Bull. 1985;16:227–231.

[17]   NRC (National Research Council). Clean coastal waters: understanding and reducing the effects of nutrient pollution. Washington, DC: National Academies Press; 2000.

[18]   Doelle HW, Rokem S, Berovic M. Biotechnology- volume X: fundamentals in biotechnology. Oxford, UK: Eolss Publishers Co. Ltd, 2009: 215–246.

[19]   Kumar A, Bisht BS, Joshi V, Dhewa T. Review on bioremediation of polluted environment: a management tool. Int J Environ Sci Technol. 2011;6:1079–1093.

[20]   Adams GO, Fufeyin PT, Okoro SE, Ehinomen I. Bioremediation, biostimulation and bioaugmention: a review. Int J Environ Biorem Biodeg. 2015;3:28–39.

[21]   Chamy R, Rosenkranz F. Biodegradation – life of science. Rijeka: InTech, 2013: 289–320.

[22]   Vuković Domanovac M, Ćosić I, Sojčić M, Briški F. Treatment of tobacco dust leachate by activated sludge- Evaluation of biokinetic parameters. Cabeq. 2013;27:51–56.

[23]   Vidali M. Bioremediation an overview. Pure Appl Chem. 2001;73:1163–1172.

[24]   USDA. National engineering handbook, Part 637, Chapter 2, Composting. Washington, DC: NRCS, 2000.

[25]   Epstein E. Industrial composting: environmental engineering and facilities management. Boca Raton, FL: CRC Press, 2011: 15–24.

A Appendix

Questions and Problems

1. In what ways did the cyanobacteria make possible the later evolutions of living organisms?

2. The prokaryotic members of the microbial world include: 1. algae, 2. fungi, 3, bacteria, 4. archaea, 5. viruses. Which of the following is correct answer?
   A. 1, 2   **B.** 2, 3   **C.** 3, 4   **D.** 1, 5

3. The eukaryotic members of the microbial world include: 1. protozoa, 2. algae, 3. bacteria, 4. viruses. Which of the following is correct answer?
   A. 1, 2   **B.** 2, 3   **C.** 3, 4

4. From the following data calculate the mean generation time if at the beginning of exponential growth $t = 0$ initial cell concentration is $2.5 \cdot 10^3$/mL, and $t = 10$ hours cell concentration is $1 \cdot 10^5$/mL.

5. Describe the role of bacterial species *Nitrosomonas* and *Nitrobacter* in soil.

6. What is the definition of coliform bacteria and name one bacterium which is the member of this group?

7. After bacteriological analysis of water sample from nearby river the number of faecal coliform on mFC plate was $1.7 \cdot 10^2$/100 mL and faecal streptococci on FS plate was $3.0 \cdot 10^1$/100 mL. Calculate the FC:FS ratio and deduce what is the source of contamination.

8. What effect causes the microbial contamination of indoor environment and describe the adverse effect of fungus *Stachybotrys chartarum*.

9. What is eutrophication?

10. What is the difference between biodegradation and biotransformation?

11. What are xenobiotics?

12. How to describe biodegradation of organic matter?

13. How to calculate specific growth rate and specific substrate consumption rate directly from the experimental data?

14. What is bioremediation?

15. What is the difference between bioaugmentation and biostimulation?

16. Describe the process of composting.

Zoran Nakić, Marta Mileusnić, Krešimir Pavlić and Zoran Kovač

5 Environmental geology and hydrology

Abstract: Environmental geology is scientific discipline dealing with the interactions between humans and the geologic environment. Many natural hazards, which have great impact on humans and their environment, are caused by geological settings. On the other hand, human activities have great impact on the physical environment, especially in the last decades due to dramatic human population growth. Natural disasters often hit densely populated areas causing tremendous death toll and material damage. Demand for resources enhanced remarkably, as well as waste production. Exploitation of mineral resources deteriorate huge areas of land, produce enormous mine waste and pollute soil, water and air. Environmental geology is a broad discipline and only selected themes will be presented in the following sub-chapters: (1) floods as natural hazard, (2) water as geological resource and (3) the mining and mineral processing as types of human activities dealing with geological materials that affect the environment and human health.

Keywords: environmental geology, hydrology, floods, groundwater, mine waste, pollution, human health

Environmental geology is scientific discipline dealing with the interactions between humans and the geologic environment. Hence, the impact is mutual. Many natural hazards (e. g. earthquakes, volcanic eruptions, landslides, floods), which have great impact on humans and their environment, are caused by geological settings. On the other hand, human activities have great impact on the physical environment (e. g. contamination of water, soil and air; depletion of mineral resources; imbalance of biogeochemical cycles; climate change; waste production).

Dramatic human population growth significantly increased this mutual impact. Some areas subject to natural disasters, such as fertile soils around rivers prone to flooding or coasts vulnerable to tropical cyclones, are densely populated. Consequently, death toll magnified compared to the similar disasters in the past.

Demand for resources enhanced remarkably, as well as waste production. Many significant resources such as groundwater used for water supply and irrigation are overexploited. In addition, exploitation of mineral resources deteriorate huge areas of land, produce enormous mine waste and pollute soil, water and air.

This article has previously been published in the journal Physical Sciences Reviews. Please cite as: Nakić, Z., Mileusnić, M., Pavlić, K., Kovač, Z. Environmental Geology and Hydrology. *Physical Sciences Reviews* [Online] **2017**, *2* (10). DOI: 10.1515/psr-2016-0119

https://doi.org/10.1515/9783110468038-005

Therefore, the role of environmental geologists is to apply geologic information to predict natural hazards in order to prevent or mitigate disasters, to identify environmental problems caused by humans and remediate the damage and to help land-use planners and policy makers to balance needs for land and resources with their availability, in order to achieve sustainable development.

Environmental geology is a broad discipline and only selected themes will be presented in the following subchapters: (1) floods as natural hazard, (2) water as geological resource and (3) the mining and mineral processing as types of human activities dealing with geological materials that affect the environment and human health.

5.1 Floods as natural hazard

Major natural hazards caused or influenced by geological settings and earth processes are earthquakes, volcanic activity, flooding, mass movements and coastal erosion. It should be emphasised that listed hazards are actually natural phenomena that become hazards when happen in the populated area or when land use changes (e. g. deforestation, urbanisation) influence natural processes. Hence, human population growth and land use changes greatly augment danger of human and material losses caused by natural hazards.

Floods, earthquakes and tropical cyclones were the deadliest natural disasters in the recent history. It is estimated that Cyclone Bhola in Bangladesh killed as many as 500,000 in November 1970 [1]; tsunami caused by earthquake in Southeast Asia in December 2004 killed more than 250,000 people [2], while earthquake that affected Haiti in January 2010 had about 159,000 fatalities reported [3].

Role of environmental geologist is to understand natural processes and help planners and policy makers to develop environmentally sound strategies to minimise their impact on human society. To understand natural processes means to predict natural hazards and to assess their risk. The best strategy is land-use planning based on risk mapping. Good practise is avoidance of locations (e. g. floodplains; active faults) where the hazards are most likely to occur or implementation of rules to minimise damage (e. g. earthquake-prone building law). In the case of populated area, developed warning system, as well as response following disasters, is crucial.

Earthquakes and volcanoes are caused by internal earth processes, i. e. by plate tectonic. Plate boundaries are the places where most of these natural hazards happen. On the other hand, external processes, such as weathering and sedimentation that are caused by sun, water, ice, wind, atmosphere, and organisms, shape the earth surface and cause natural hazards such as floods and mass movements (e. g. landslides, rock falls, subsidence, mudflows). Main activities important for prediction and risk assessment are investigation of past natural processes happened in the area, as well as monitoring of certain parameters (e. g. emanation of gasses in earthquake and volcanic activity prone areas). Hereinafter, flood as natural hazard is presented in more detail.

5.1.1 Hydrological cycle

Hydrological cycle, or water cycle, refers to circulation of water in nature (Figure 5.1). The water moves in different phases through atmosphere, over and below the land to the oceans and back to the atmosphere. As the atmospheric water vapour condense, it precipitates over the land in form of rain, snow, dew, frost etc., and moistens the surface. Some of that water evaporates. Remaining portion of water remains on the surface as surface runoff or enters into the soil as infiltration. Water increases soil moisture in unsaturated zone and finally reaches ground water level and ground-water. As the surface runoff increases, water tends to accumulate in puddles and ponds as depression storage, or in channels and gullies where it forms streamflow. Streamflow also drains from groundwater, and ultimately ends up in large water bodies, such as lakes and oceans. Because of sun radiation, wind and adequate air humidity, surface water evaporates from precipitation, wet vegetation, puddles and lakes, soil, streams and large water bodies back into the atmosphere where it forms clouds, thus closing hydrological cycle.

HYDROLOGICAL CYCLE

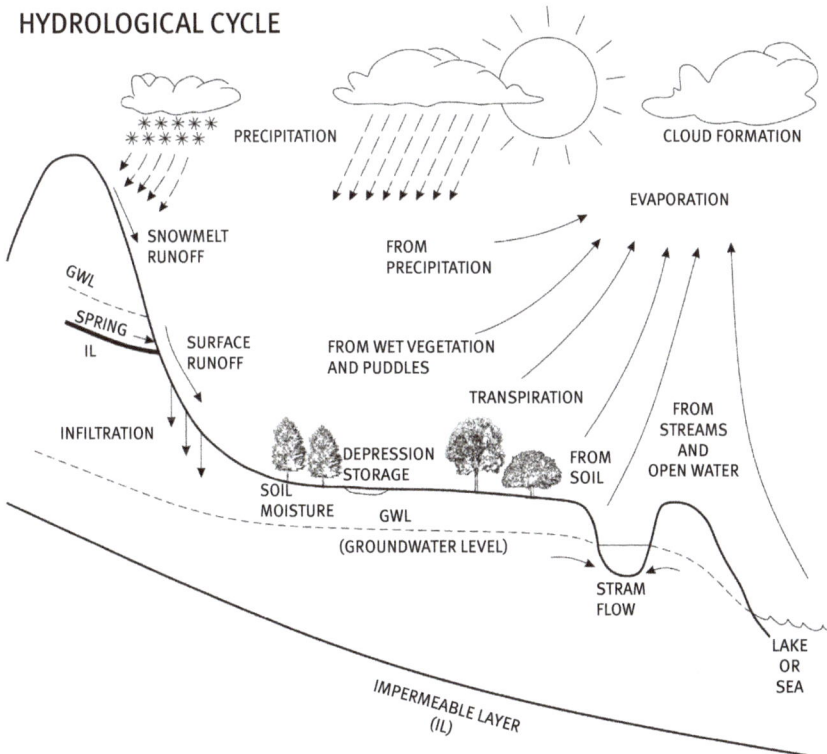

Figure 5.1: Hydrological cycle.

5.1.1.1 Water in the atmosphere

The scientific discipline that studies the water in the atmosphere is called meteorology. It is a branch of earth sciences that study the physical processes and phenomena of the earth's atmosphere. Hydrometeorology is the scientific discipline related to hydrology that studies the physical processes in the atmosphere and on the earth within the hydrological cycle.

The earth atmosphere is approximately 1,000 km high and is divided into four layers, according to temperature change (Figure 5.2).

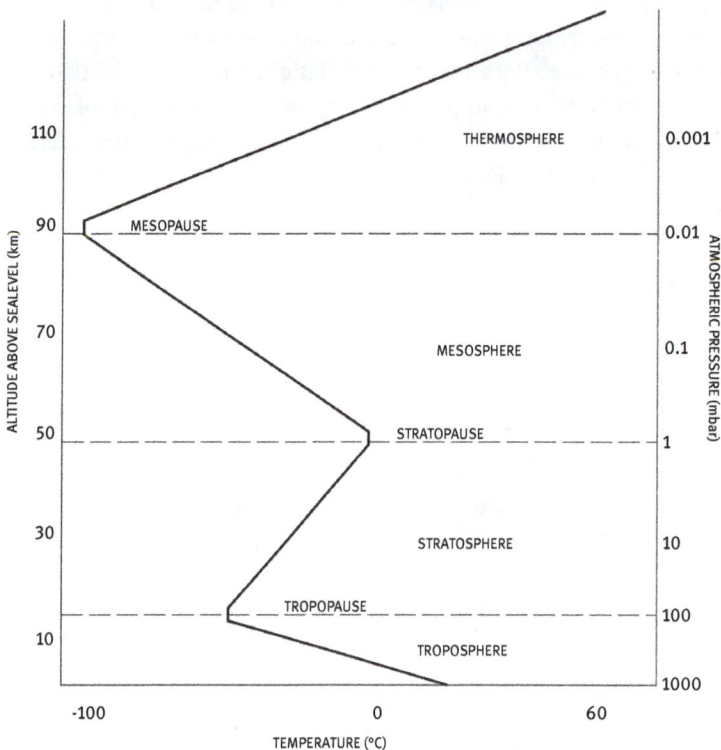

Figure 5.2: Profile of atmosphere according to temperature change.

–   *Troposphere* is lowest layer of atmosphere. In average, it is 12 km high, but thickness ranges from 9 km at poles, to approximately 17 km at equator. It contains about 75 % of all gases and all weather phenomena occurs there. Air circulation is horizontal and vertical. Temperature drops with gradient of about 6.5 °C every kilometre above earth's surface. *Tropopause* is a thin boundary layer with constant temperature at the top of troposphere, extending in the lower part of stratosphere

- *Stratosphere* is layer that ranges from top of troposphere to 50 km above earth's surface. Part of a tropopause is located there where temperature is constant and is about –60 °C. This layer contains ozone layer that absorbs ultraviolet radiation from the sun, causing the temperature to increase with layer height. Air circulation is mainly horizontal, so pollutants such as volcano fumes or freons that reach stratosphere can stay there trapped for months. *Stratopause* is thin layer with constant temperature at the top of stratosphere.
- *Mesosphere* layer ranges from 50 km to 80 km above earth's surface. Temperature drops to about –100 °C and it is the coldest part of the atmosphere. This layer protects earth from meteoroids, where they burn upon entering the atmosphere. At the highest part of mesosphere is layer with constant temperature – *mesopause*.
- *Thermosphere* is a layer above 80 km from earth's surface. The air is very thin there and the temperature reaches about 2,000 °C or more. This layer contains two sublayers:
  - *Ionosphere* is lower part of thermosphere and it extends from 80 km to 550 km from earth's surface. Gas particles absorb sun's ultraviolet and X-ray radiation and gets electrically charged (ionised). Those ions bounce off radio waves back to earth and helps in radio communication. However, solar flares can interfere with transmission of some radio waves.
  - *Exosphere* is the upper part of the thermosphere above 550 km where satellites orbit the earth.
- *Magnetosphere* is outer layer that extends from 1,000 km above earth's surface far to the interplanetary space. It is made of positively charged protons and negatively charged electrons. Those deadly particles are trapped in layers called Van Allen belts, and when large amount are given off by the sun, they collide with each other causing *aurora borealis*.

Since weather phenomena occur in the troposphere, this layer is of most interest from hydrological aspect. Mechanism of precipitation formation consists of several processes. Water-holding capacity decreases as the temperature of air decreases, so the air can be supersaturated by cooling down, for instance by moving warmer air over cooler surface, or by mixing two different air masses. These processes are not very effective and at most can only generate fog or drizzle. More effective mechanism of air cooling is by being lifted to higher altitudes. Air can be lifted to higher altitudes by moving over mountains or by frontal activity when air moves over relatively heavier air. In order water drops to form vapour must condense on solid particles, so another requirement for formation of precipitation is presence of hygroscopic particles, such as dust, smoke or salt. When those two requirements are met, precipitation can occur. Intensity of precipitation depends on stability of air. If the air is lifted gently, so the vertical movement is weak and slow, precipitation that occur is known as stratiform precipitation. On the other hand, if the vertical movement of air is strong and fast, the air is unstable and resulting type of precipitation is called convective precipitation.

Stratiform precipitation has uniform intensity and longer duration, while convective precipitation has shorter duration but stronger intensity. Stratiform precipitation can cause floods if precipitate over large catchment area for long time and convective precipitation can cause flash floods (rapid rises of water in a short amount of time) over smaller and steeper catchment.

5.1.1.2 Water on the earth surface

By the arrival of rainwater to the surface of the earth, the water forms surface runoff in a variety of shapes. Surface runoff is initiated after snowmelt, rainfall or emergence of groundwater at a source. This flow starts as irregular flow, but due to topographical irregularities, water starts to flow in form of rills and rivulets. After merging with other similar surface runoff, water forms rivers, which later end up in larger rivers, lakes and ultimately in seas and oceans.

This part of hydrological cycle investigates hydrology. It is a scientific discipline that investigates water in natural environment. Complete definition of hydrology was made by the end of the twentieth century (e. g. [4]) and is accepted to be science that deals with cycling of water in natural environment investigating: (1) continental water processes and (2) the global water balance [5].

River basin or catchment plays important role for water flow on the earth surface. The terms river basin, catchment, watershed or drainage area are often used as synonyms. A river basin is defined as upstream area that contributes to the open channel flow at a given point along a watercourse. That point along a watercourse is usually taken at points where river flows into larger river, lake or sea, but can also be taken at any other point along watercourse, for instance, where measuring station is positioned. Size of the river basin can be determined by the land surface topography. Hydrogeological properties of the underground are necessary to take into account as well.

Measuring water flow at a point of the watercourse is carried out by measuring water level at that point. Water flow is calculated by combining water-level data with stage–discharge curve. Stage–discharge curve is derived by measuring water speed with known slope of watercourse, roughness and geometry of measuring profile during low, medium and high water levels. Hence, the water flow is derived unit which is more susceptible to errors due to dependence on the stage–discharge curve.

Part of the total amount of precipitation is lost through evapotranspiration and infiltration. Evapotranspiration is combined process of direct evaporation from open water or soil surfaces and transpiration of water from plants. Evaporation is final step of hydrological cycle which returns water from surface back into atmosphere. The amount of evaporation depends on the type of terrain and meteorological factors such as air temperature, wind speed and the amount of moisture in the air. Evaporation is usually higher in urban areas because of ground surface which is less suitable for infiltration of water.

Yearly amount of water that evapotranspirate, D (mm), is calculated using Turc's equation [6]:

$$D = \frac{P}{\sqrt{0.9 + \dfrac{P^2}{L^2}}}$$ (5.1)

where P is yearly amount of precipitation in millimetres (mm), L is function of air temperature given as:

$$L = f(t) = 300 + 25t + 0.05t^3$$ (5.2)

and t is air temperature in degrees Celsius (°C).

Amount of water that can infiltrate into the ground is difference between yearly amount of precipitation and yearly amount of evapotranspiration.

5.1.1.3 Water below the surface

Some of the water that infiltrates into the soil comes again into the atmosphere by transpiration through the plants. The second part, in the case of rapid flow through the shallow soil layers, can come to surface through the springs, and in the case of slow percolation through the unsaturated zone, water reaches into groundwater, which finally seeps into natural river systems and lakes. The dynamics of interaction between groundwater and surface water body depends on the level of groundwater relative to the surface water. Figure 5.3 shows three types of interaction between groundwater and surface water body:
a. effluent flow,
b. influent flow and
c. normal flow.

Effluent rivers get water from groundwater. This kind of flow is directly related to the surface of the ground water (also called water table), and as a consequence, river water level will rise and fall as the water table rise and fall. Effluent rivers become deeper and wider downstream, due to increase in discharge and thus erosion. Good examples of this kind of river are Mississippi River, the Amazon and Columbia River in the Pacific Northwest.

Influent rivers are far less common and are usually in arid climates. They lose water as they flow downstream, because they give water to the groundwater system, which water table is generally lower than river water level. Some rivers lose so much water that they eventually dry up. Good examples of this kind of river are Nile and Colorado River.

a) EFFLUENT FLOW

b) INFLUENT FLOW

c) NORMAL FLOW

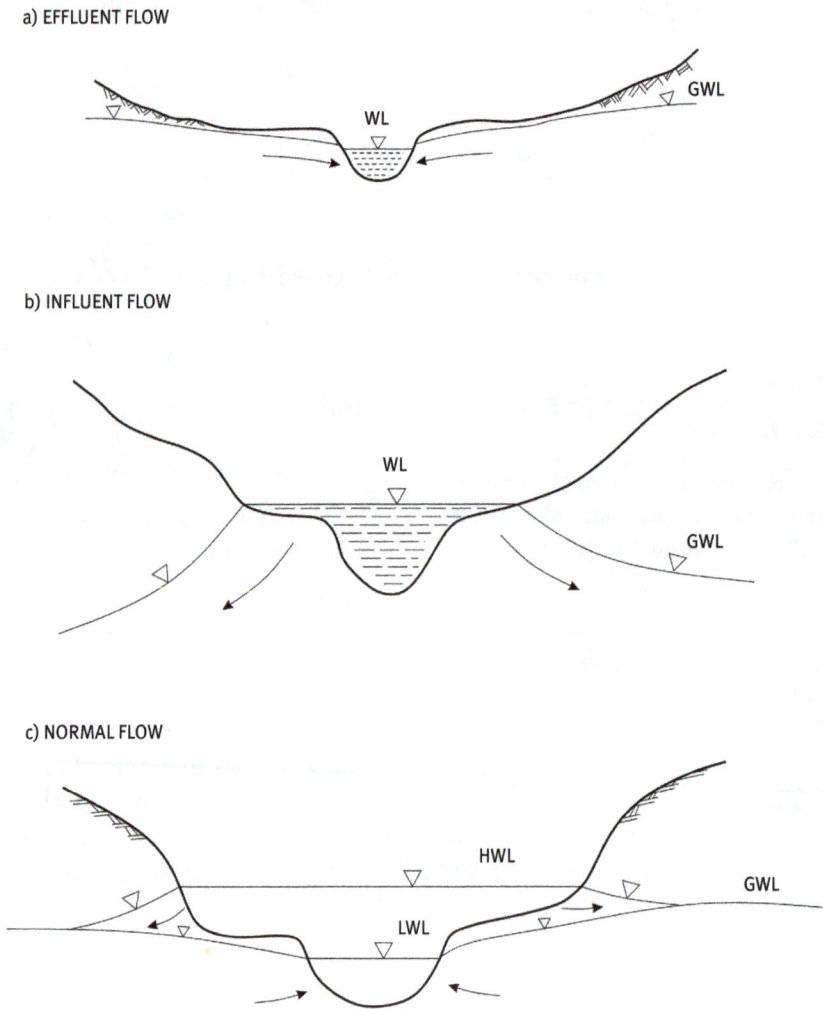

Figure 5.3: Types of interaction between groundwater and surface water body; (a) effluent flow, (b) influent flow and (c) normal flow.

Normal flow is the most common form of interaction between groundwater and river. It can be said that such water flow is a transitional form between the effluent and influent flows, which are extremes.

5.1.2 Extreme events in hydrology

Extreme events in hydrology include maximum and minimum flows, i. e. high and low waters. However, in studying the problem of pollution, high waters are of more

interest. According to UNESCO and WMO's hydrological glossary [7], high water is defined as:
- State of the tide when the water level is highest for any given tidal cycle.
- Highest water level reached in a watercourse or in a lake during a *flood* or reservoir operation.

Flood is defined as:
- Rise, usually brief, in the water level of a stream or water body to a peak from which the water level recedes at a slower rate.
- Relatively high flow as measured by stage height or discharge.

The causes of high water are heavy precipitation, snowmelt or both phenomena together. High waters can also be caused by some extreme events in the river basin, such as landslides in the lake, demolition of dams and embankments. Occurrence and magnitude of high water depend on the distribution of wet and dry seasons throughout the year. In summer, for example, high waters occur mainly due to high intensity rainstorms. Retention of precipitation in the basin in the form of snow, which melts as temperature increase, is typical in winter. Depending on the snow cover thickness and the melting intensity, there is an increase in runoff. The emergence of high waters favours appropriate geological structure of the river basin (permeability), topographical conditions (high coefficient of concentration and a large slope of the river basin) and the overgrown degree of the river basin. Condition of the soil also has a big impact on the magnitude of the high water. Over the frozen soil, and over the water saturated soil, a large part of the rainfall flows into the watercourse. Dry land absorbs precipitation and thus greatly reduces surface runoff. Exception is when a strong summer rainstorm falls on dry land and creates a crust which prevents absorption.

Since the high waters cannot be predicted long term, it is necessary to know the highest water that can occur at any part of the watercourse for any return period. In this case, high waters of different return periods are determined using statistical analysis of measured maximum flows. Different kinds of statistical distributions are used, such as Gauss (or normal), LogNormal, Galton, Pearson III and Gumbel distributions. In extreme cases, when absolute safety is required, probable maximum flood (PMF) is determined. PMF is, according to [7], greatest flood that may be expected, considering, in a deterministic manner, all pertinent factors of location, meteorology, hydrology and terrain.

Example: Big flood in May 2014

The Sava River originates in Slovenia and further flows through Croatia, along the northern border with Bosnia and Herzegovina and in Serbia discharges into the Danube (Figure 5.4).

During 2014, several extreme events occurred in the mid and lower part of the Sava basin. The maximum rainfall that led to the rapid increase in water

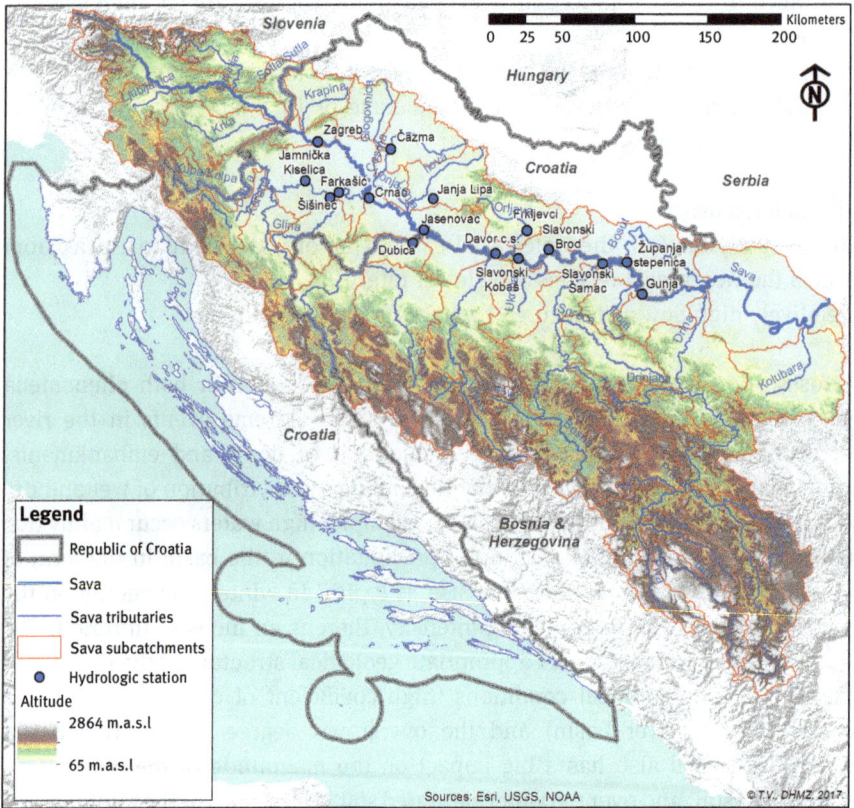

Figure 5.4: Sava River Basin through Slovenia, Croatia, Bosnia and Herzegovina and Serbia.

levels of Kupa River and its tributaries, as well as the upper and middle water-course of the Sava River, was recorded in February. Then the third highest water level since the measurements on the Kupa River was recorded. At the same time, four water waves with flow rates from 1,500 to 2,100 m³/s were recorded at the Sava River [8].

In May, the largest amount of precipitation was recorded concurrently in the lower part of the Sava River Basin in Croatia and in the part of the Sava River Basin in Bosnia and Herzegovina and Serbia. Since the ground was already saturated with water because of prolonged previous rainfall, highest level of the Sava River occurred. Table 5.1 shows water levels recorded in May 2014 at three hydrological stations.

On May 17th, left side embankment rupture occurred at Rajevo Selo and Račinovci and large amount of water poured into the hinterland. Figure 5.5 shows the size of the disaster caused by the flooding.

Table 5.1: Maximum water levels recorded in May 2014 in lower part of Sava River Basin in Croatia [8].

Station	Previous maximum water level		Maximum water level in May 2014	Difference in water level
	(cm)	year	(cm)	(cm)
Slavonski Brod	882	1974	939	+57
Slavonski Šamac	726	1981	891	+165
Županja	1064	1970	1168	+104
Gunja	690	2012	1173	+483

Figure 5.5: The Village Gunja during flooding in May 2014 (courtesy of Agrokom.hr).

5.2 Groundwater as geological resource

All that is required for human life and civilisation and that has value to the individual or society can be defined as resource. The value of the resource can change with time depending on needs, availability and social context. The main geological resources are mineral resources, water and soil. Mineral resources can be divided to energy resources, ore, industrial minerals and stone.

Although geological resources are formed continuously by geological processes, demand for some of them is so high that they cannot be replaced on a human timescale. Hence, they are called non-renewable. Reason for such high demand is exponential population growth and elevated standard of living. The biggest concern is for energy resources (especially petroleum) and critical minerals. Critical minerals are resources that are essential to the economy, and whose supply may be interrupted. Many critical minerals are ores, which metals (e. g. rare earth elements, lithium, indium, tellurium, gallium, platinum group of elements) are important to high-tech sectors, i. e. electronics such as smartphones and tablets, wind turbines and solar panels. Projection of the future demand of various resources can help in prediction of how long supplies will last. However, this is very difficult as many resources are extremely unevenly distributed and its supplies and demands are dependent upon political and economic situation as well.

Exploitation and processing of mineral resources, as well as manufacture, distribution, use, repair and maintenance, and disposal or recycling of its products, hence, all stages of product cycle, have also harmful impact on the environment by polluting soil, water and air.

Water and soil are considered principally renewable resources. However, in some places water is heavily exploited, especially "fossil groundwater", i. e. water that has been stored underground since the time with different climatic conditions (e. g. Nubian Sandstone Aquifer System). Similarly as water, at many places soil is eroded faster than new soil is being formed, making it a non-renewable. Soil erosion is natural process, but human activities (e. g. agriculture, urbanisation, deforestation) enhanced erosion causing loss of top soil which resulted in lower productivity, desertification and dust pollution. Hence, soil erosion is one of the major environmental problems worldwide.

At some places groundwater and soil are polluted, making them unusable. It is difficult to talk about sustainable use of the finite resources. The best what can be done is to conserve and recycle them so they can be used as long as possible.

More about soil as a resource and about soil pollution and remediation is presented in the chapter 9. Hereinafter, water as geological resource is presented in more detail.

5.2.1 Geological occurrence of groundwater

Groundwater is a part of the hydrologic cycle that includes surface and atmospheric waters. Relatively minor amounts of groundwater may enter this cycle from other origins [9], e. g. as fossil interstitial water that has migrated from its original underground location (connate water) or water derived from magma (magmatic water). Groundwater occurs in many types of geologic formations, e. g. gravels, sands, magmatic rocks or karst. Rock that holds enough water and transmits it rapidly enough to be used as a source of a potable water is an aquifer. When the aquifer is overlain only by permeable rocks and/or soil through which may be recharged by infiltration, it is described as an unconfined aquifer [10]. When the aquifer is bounded above and below by low-permeability rocks, which prevent free flow of water to aquifers, it is described as a confined aquifer (Figure 5.6). In the unconfined aquifer the water table presents the surface of atmospheric pressure and appears at the level at which water stands in a well penetrating the aquifer [9]. In the confined aquifer the water level rises above the bottom of the confining bed.

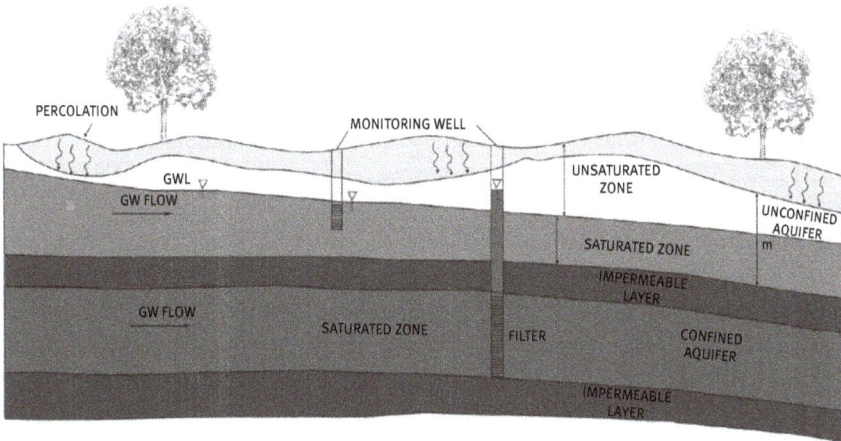

Figure 5.6: Types of groundwater aquifers.

5.2.2 Water flow in unsaturated and saturated zone

When meteoric water percolates downward through soil or rock, it eventually reaches an irregular horizon below which the pore space is saturated with water. This horizon is water table and it separates the zone of saturation (saturated zone) and the unsaturated zone (zone of aeration or vadose zone). The unsaturated zone is typically defined to extend from land surface to the underlying water table or saturated zone [11]. The unsaturated zone consists of voids occupied partially by water and partially by air or some other fluid (e. g. contaminants in pure phase, infiltrated from the surface of terrain). In the zone of saturation all voids are filled with water under hydrostatic pressure. The saturated zone extends from the upper surface of saturation (water table in unconfined aquifer) down to underlying aquitard, low permeability formations which store water but cannot supply production wells.

Water movement in both the zone of saturation and the unsaturated zone can be expressed by Darcy's law:

$$Q = -KA\frac{dh}{dl} \tag{5.3}$$

where K is hydraulic conductivity [L/T], a constant that measure the ease with which water flows through an aquifer or confining bed, A is cross-sectional area [L^2] of the part of aquifer through which water flows at a rate Q and dh/dl is hydraulic gradient.

Hydraulic conductivity (Darcy's proportionality constant) is the function of properties of porous medium and the fluid passing through it [12]. It can be expressed as:

$$K = -\frac{Cd^2\gamma A}{\mu} \tag{5.4}$$

where C is proportionality constant called shape factor, d is diameter of mineral grains [L], γ is specific weight of the fluid [M/L^3] and μ is dynamic viscosity of water [M/LT]. Both d and C are properties of the porous medium, whereas μ and γ are fluid properties [13].

The water flow in the unsaturated zone is influenced by the potential energy of water, in that water will move from the area with high water potential energy to the area of lower potential energy. The total water potential is represented by the sum of its components, the most important being a gravity potential, Z, and moisture potential, ψ [13, 14]. Generally, water kinetic energy can be neglected, since the flow in the unsaturated zone is generally very slow.

Gravity potential Z is an energy per unit volume of water required to move an infinitesimal amount of pure, free water from the arbitrary reference elevation (z_o) to the unsaturated water elevation (z_{unsat}). The moisture potential ψ (sometimes called suction potential, matric potential or capillary pressure) is a negative pressure due to the surface tension on the air–water interface. Generally, in unsaturated medium the water is at lower pressure than the air.

The moisture potential sometimes greatly exceed the gravity potential. Very often, it is the only significant type of energy determining the chief water transport processes in an unsaturated medium. At low moisture potential (much less than zero), only a few pores are filled with water and a large fraction of the total water is in thin films, which adhere onto the solid medium. If this potential is high (close to zero) nearly all pores are filled with water.

The relation between matric pressure and water content, called a water retention curve, depends on the medium (Figure 5.7). Larger pores empty first as the water content decreases. A medium with many large pores will have a retention curve that drops rapidly to low content at high matric pressures [15]. A fine-pore medium will retain more water even at low matric pressure.

In the unsaturated media, hydraulic conductivity, K, depends on the whole set of filled pores, especially the size, shape and connectedness of filled pores [15]. It depends very strongly on the water content. As water content decreases, the large, more conductive, pores empty first. Remaining pores, filled with water, are smaller and therefore less conductive. When the medium is quite dry, only few pores are filled with water, which flows mainly through poorly conducting films adhering to grain surfaces.

The transient nature of the unsaturated flow in a porous medium, which is expressed by changing the water content by flow throughout the medium, leads to continuously changing of hydraulic conductivity of the medium and driving forces (pressures). These processes can be expressed by combining the equation of continuity, which states that the change in the water content in a given volume is due to spatial change in the water flux:

Water retention curve

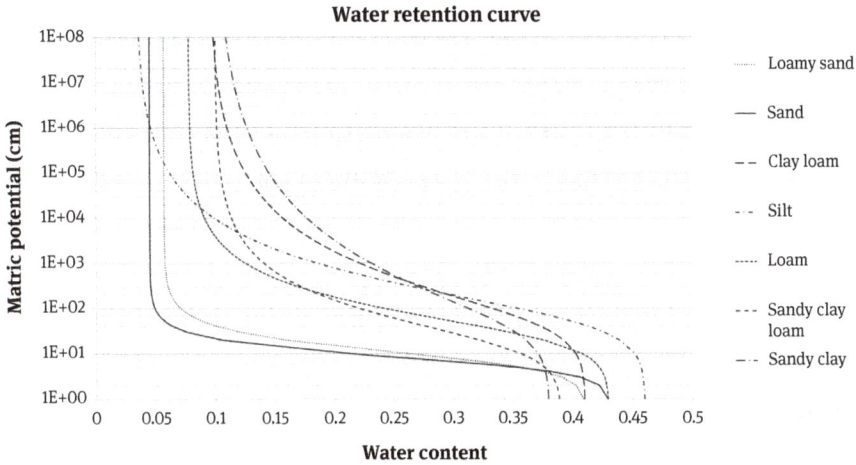

Figure 5.7: Characteristic water retention curves for different materials.

$$\frac{\partial \theta}{\partial t} = \frac{\partial q}{\partial z} \tag{5.5}$$

with Darcy's law to get [16] non-linear partial differential equation. For one-dimensional vertical flow it can be written as:

$$C\frac{\partial \psi}{\partial t} = \frac{1}{\rho g}\frac{\partial}{\partial z}\left[K\frac{\partial \psi}{\partial z}\right] + \frac{\partial K}{\partial z} \tag{5.6}$$

where θ is volumetric water content $[L^3/L^3]$ (volume of water per bulk volume of the medium), q is water flux density $[L/T]$, z is vertical coordinate (positive upwards) $[L]$, C is differential water capacity, a property of the medium defined as $d\theta/d\psi$, ρ is density of water $[M/L^3]$ and K is unsaturated hydraulic conductivity $[L/T]$.

In natural porous media, with significant heterogeneity, it is quite common to observe preferential flowpaths in the unsaturated zone that permit fast movement of fluid through it. Preferential flowpaths transport water and contaminants much faster than would be predicted from medium properties and Richard's equation. Three basic types of preferential flow are [15]: (1) macropores, caused by flow-enhancing features including wormholes, root holes and fractures, (2) funnelled flow, caused by flow-impeding features of the medium, which occurs with contrasting layer or lenses and (3) fingered flow, caused by temporary flow-enhancing conditions of parts of the medium, which typically occurs when downward percolating water does not immediately pass a layer of fine material above coarse material.

Similarly, to the condition in the unsaturated media, the flow of water in the saturated media is controlled by the mechanical energy of water. Because the amounts of energy vary spatially, groundwater will move from the area of the high-energy state to the area of the low energy state, in order to eliminate this energy differentials. The total energy of the unit volume of fluid is the sum of the three components – kinetic, gravitational and fluid pressure energy [13]. For steady flow of water, the sum of the three components is a constant:

$$\frac{v^2}{2} + gz + \frac{P}{\rho} = const.$$ (5.7)

where v is velocity [L/T], g is acceleration of gravity [L/T^2], z is elevation of the centre of gravity of the fluid above the reference elevation [L], P is pressure of water [M/LT^2] and ρ is density of water [M/L^3]. All terms of the equations have the units of [L/T]^2.

If each term in eq. (5.7) is divided by g, then sum of these three factors will be the total mechanical energy per unit weight, known as hydraulic head, h [L], which is usually measured in the field or laboratory. Since kinetic factor is very low, it can be neglected in practical computations and the hydraulic head, h, can be expressed as:

$$h = z + h_p$$ (5.8)

where h is total head, z is elevation head and h_p is pressure head (Figure 5.8).

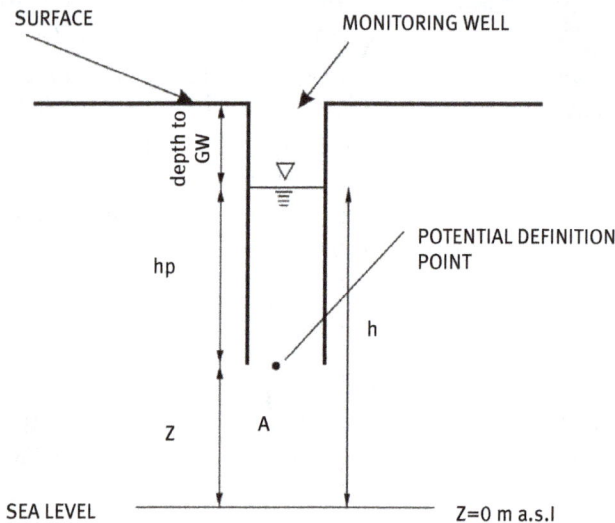

Figure 5.8: Definition of total, elevation and pressure heads.

The flow of water through saturated porous media can be described by partial differential equations in which the spatial coordinates, x, y and z and time, t, are independent variables. In deriving this equations the law of mass conservation or continuity principle are used, which states that there can be no net change in the mass of fluid contained in a small aquifer volume. Any change in mass flowing into the small volume of the aquifer must be balanced by appropriate change in mass flux out of the volume, or a change in the mass stored in the volume, or both [13]. The general equation of flow for confined aquifer (for flow in three dimensions for an isotropic, homogeneous porous medium) can be expressed as:

$$K\left(\frac{\partial^2 h}{\partial x^2} + \frac{\partial^2 h}{\partial y^2} + \frac{\partial^2 h}{\partial z^2}\right) = (\alpha \rho_w g + n\beta \rho_w g)\frac{\partial h}{\partial t} \qquad (5.9)$$

where α is compressibility of the aquifer skeleton [$1/(M/LT^2)$], n is porosity (percentage of the rock that is void material) and β is compressibility of the water [$1/(M/LT^2)$]. The term $\alpha \rho_w g + n\beta \rho_w g$ is called specific storage, S_s, the amount of water per unit volume of saturated formation that is stored or withdrawn from storage due to compressibility of the aquifer (mineral skeleton and pore water) per unit change of hydraulic head.

The general equation of flow for two-dimensional unconfined flow is known as a Boussinesq equation:

$$\frac{\partial}{\partial x}\left(h\frac{\partial}{\partial x}\right) + \frac{\partial}{\partial y}\left(h\frac{\partial}{\partial y}\right) = \frac{S_y}{K}\frac{\partial h}{\partial t} \qquad (5.10)$$

where S_y is specific yield [L^3/L^3], the ratio of the volume of water that can be drained by gravity to the total volume of the rock.

5.2.3 Mass transport in porous medium

Transport of different substances in the porous medium is associated with water flow. Some substances dissolve in water and are transported as solutes through unsaturated and saturated zone. Other substances are only partly soluble in water, so that a dissolved phase as well as non-aqueous phase can be present [17]. Three phase flow may occur in the unsaturated zone with air, water and non-aqueous phase. Two phase flow may occur below the water table with water and dense non-aqueous phase liquid, DNAPL, which have densities that are greater than water. Solute transport can be either conservative (transported solute mass is constant along the flow path) or non-conservative (transported solute mass is usually decreasing along the flow path due to chemical, nuclear and biological processes like sorption, volatilisation, precipitation etc.) [18].

Main processes that have impact on solute transport in the porous medium are advection, mechanical dispersion and diffusion. Advection is the process by which a solute particle is transported with the moving groundwater at the same rate as the average linear velocity of the groundwater. Mechanical dispersion is the process of mixing of the solute-containing water with water that doesn't contain the solute along the flowpath. It results in a dilution of the solute at the advancing edge of flow. Longitudinal dispersion is mixing that occurs along the direction of the flowpath and transverse dispersion occurs in direction normal to flowpath [18]. Diffusion is the process by which solutes randomly move from areas of higher concentration to areas of lower concentration. Advection and mechanical dispersion occur in systems with flow, while diffusion occurs as long as there is a concentration gradient (Figure 5.9).

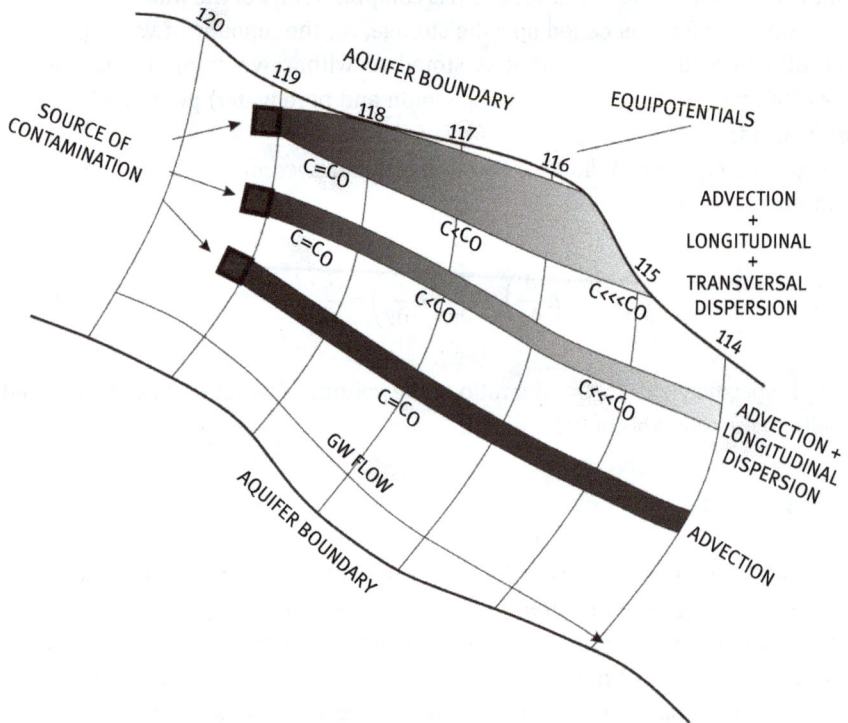

Figure 5.9: Schematic diagram demonstrating advection and dispersion transport mechanisms.

The one dimensional mass flux, F_x [M/L^{2}T^1], due to advection, is equal to the water flow normal to a unit cross sectional area of the porous media times the concentration of dissolved solids [18]:

$$F_x = v_x n_{ef} C \tag{5.11}$$

where v_x is average linear velocity [L/T]:

$$v_x = \frac{K}{n_{ef}} \frac{\partial h}{\partial l} \tag{5.12}$$

K is hydraulic conductivity [L/T], n_{ef} is effective porosity (porosity available for fluid flow), C is solute concentration [M/L³] and $\partial h/\partial l$ is hydraulic gradient [L/L].

The one-dimensional advective transport equation can be expressed as:

$$\frac{\partial C}{\partial t} = -v_x \frac{\partial C}{\partial x} \tag{5.13}$$

Mechanical dispersion (MD) is a function of the average linear velocity. It is equal to the property of the porous medium, called dispersivity, α, times the average linear velocity. Longitudinal mechanical dispersion, which occurs along the principal direction of flow, can be expressed as:

$$MD = \alpha_L v_x \tag{5.14}$$

where α_L is longitudinal dispersivity [L]. Transverse mechanical dispersion, which occurs perpendicular to the principal direction of flow, can be represented by the similar expression.

The diffusion of solute through water in the porous medium under steady-state condition can be described by Fick's law:

$$F = -D_* \frac{\partial C}{\partial x} \tag{5.15}$$

where F is mass flux of solute [M/L²T¹], D_* is effective diffusion coefficient [L²/T], that takes into account that diffusion in porous medium proceed slower than in pure water because solute particles follow longer pathway around mineral grains:

$$D_* = \omega D \tag{5.16}$$

ω is empirical coefficient related to tortuosity and ranges from 0,5 to 0,01 [19] and D is diffusion coefficient in pure water [L²/T].

Mechanical dispersion and diffusion are combined in a single parameter called hydrodynamic dispersion coefficient. It can be expressed by the formulas:

$$D_L = \alpha_L V_x + D^*$$

(5.17)

$$D_T = \alpha_T V_x + D^*$$

(5.18)

Where D_L and D_T are *longitudinal* and *transverse hydrodynamic dispersion coefficients* [L^2/T] respectively.

The most common mathematical description for treating changes in solute fluxes due to velocity and concentration variations is the mass balance equation for a solute, also known as the solute transport equation or advection-dispersion equation (ADE). The ADE for two-dimensional transport of conservative solute, which doesn't interact with porous media or undergo decay, in homogeneous and isotropic medium with a uniform velocity field, can be expressed as:

$$D_L \frac{\partial^2 C}{\partial x^2} + D_T \frac{\partial^2 C}{\partial y^2} - v_x \frac{\partial C}{\partial x} = \frac{\partial C}{\partial t}$$

(5.19)

To solve the ADE, it is necessary to know the value of the hydrodynamic dispersion coefficient. It can be estimated by field or laboratory experiments using a non-reacting compound (tracer) at appropriate scale.

5.2.4 Modelling of water flow and solute transport in porous medium

The equations of the water flow and the advection-dispersion equation can be solved using numerical or analytical models. Analytical models include analytical solution of the partial differential equations and are limited to simple, homogeneous and isotropic aquifers. A number of simplifying assumptions regarding the ground-water system are necessary to obtain an analytical solution. Although these assumptions do not necessarily mean that analytical models cannot be used in "real-life" situations, they do require sound professional judgment and experience in their application to field situations [20]. Numerical models include solution of the partial differential equations by numerical methods of analysis and can be used in modelling water flow and solute transport in very complex, heterogeneous aquifers.

In order to obtain a unique solution of the partial differential equations, it is necessary to specify the initial and the boundary conditions that apply [21]. The initial condition describe the water potential and/or solute concentration distribution at some initial time equal 0. Boundary conditions present conditions specified at the model domain edges which define how the site specific model area interacts with its external environment [18]. Generally, there are two elementary types of boundary conditions [22]. If the head or concentration of the solute is known at the boundary of the model domain, this is Dirichlet or first-type of boundary conditions. If the flux (head or concentration gradient) across the boundary to the model region is specified, this is Neumann or second-type of boundary conditions. If the flux across the boundary changes in response to changes in head within the model area adjacent to the boundary, the flux is specified function of that head and varies during the problem solution as head varies. A variable flux boundary or mixed boundary condition in which a flux across a boundary is related to both the normal derivative and the value is Cauchy or third-type of boundary condition.

A groundwater model can be defined as a simplified version of a real groundwater system that approximately simulates the relevant excitation-response relations of this system. The simplification is introduced as a set of assumptions. Those assumptions express the nature of the system and relate to different factors, i. e. to the geometry of the investigated domain, the way various heterogeneities will be smoothed out, the nature of the porous medium, the properties of the fluids involved, and the type of flow regime under investigation [20]. There are two objectives of hydrogeology where it is needed to rely upon models of a real groundwater system: (1) to understand why a flow system is behaving in a particular observed manner, and (2) to predict how a flow system will behave in the future [13].

The first step in the modelling process is the construction of a conceptual model, which represents the current understanding of the groundwater system based on the knowledge of its natural characteristics. The conceptual model should describe: the geology of the groundwater system, the geometry and the main conditions on the boundaries of the system that express the interaction with its surrounding environment, the initial conditions within model domain, groundwater flow characteristics (one-dimensional, two-dimensional etc.) and flow regime (laminar or non-laminar), sources and sinks of water and contaminants within the model domain and main processes that affect behaviour of contaminants.

The accuracy and complexity of the conceptual models increases with the amount of, and confidence in, the available environmental information, so they become more effective and reliable description of the system [23]. Development of conceptual model usually starts with qualitative description of groundwater system under investigation and gradually ends, through a number of iterations, with quantitative description. Through data collection and identification of knowledge gaps, a conceptual model continuously evolves until it can describe the measured data with adequate certainty and complexity. The availability of field data required for model calibration and parameter estimation dictates the type of conceptual model to be developed and the degree of approximation involved [20].

The next step in the modelling of groundwater system is to define the conceptual model in the form of a mathematical model. All mathematical models starts with groundwater flow model development to compute rate and direction of fluid movement. It can be used for many purposes, e. g. for: interpretation of observed hydraulic heads in aquifers, estimation of water balances (or element of water balances), delineation of wellhead protection zones and catchment areas of wells, preparation of simulation of solute (contaminant) transport etc. Solute transport model consists of solute-transport equations, which are integrated in the flow model to derive movement and retardation values of contaminants [13]. It can be used e. g. for: interpretation of solute concentration data, estimation of mass balance of contaminants, prediction of contaminant plumes, planning of groundwater monitoring strategy, design of contaminated groundwater remediation systems etc.

The mathematical model is defined with the same information as the conceptual one, but expressed as a set of equations with the dependent variables selected for the problem solution. For that purpose, available field or laboratory data are used. Each groundwater model of flow and solute (contaminant) transport consist of the following equations [20]:

- equation(s) that expresses the balance of water mass or solute mass;
- flux equations that relate the flux(es) of water mass or solute mass to the relevant state variables of the problem;
- constitutive equations that define the behaviour of the fluids and solids involved; and
- equation(s) that expresses initial conditions at some initial time and boundary conditions at the model domain edges of the considered groundwater system.

Every groundwater model should be calibrated before it is used as a tool for predicting the behaviour of a certain groundwater system. Most groundwater models are initially calibrated against groundwater heads, measured in piezometers or monitoring wells. During the process of model calibration the values for the aquifer parameters and/or groundwater recharge are tuned until the model closely reproduces the measured head in the groundwater system. During the calibration of the model, it is possible to define how sensitive the model is to changes in aquifer parameters and boundary conditions. This is a part of the sensitivity analysis of the model.

Many mathematical models that simulate flow and contaminant transport in saturated and unsaturated zone have been developed. A large number of computer codes for modelling of water flow and contaminant transport already exists and are available for everyday use. In choosing computer code for modelling a real-world problem it is important to bear in mind that code have to fulfil the purpose of the modelling and must be compatible with existing data. Some codes can solve different problems, while others are developed for particular ones.

Many practitioners and researchers use standardised, widely available groundwater modelling code MODFLOW [24–27]. It is a finite-difference groundwater flow model that simulates groundwater flow conditions. MODFLOW's modular structure provide a framework to simulate coupled groundwater/surface-water systems, solute transport, variable-density flow (including saltwater), aquifer-system compaction and land subsidence, parameter estimation, and groundwater management.

MODPATH [28, 29] is a particle tracking model that works with MODFLOW to calculate groundwater velocities, flowpaths and travel times of water particle through a simulated groundwater system for steady state and transient conditions.

MT3DMS [30] is a modular three-dimensional transport model for the simulation of advection, dispersion, and chemical reactions of dissolved constituents in groundwater systems. It solves solute transport equations on the basis of flow field calculated by MODFLOW.

The widely used codes for modelling flow and contaminant transport in the unsaturated zone are HYDRUS 1D and 2D/3D and MACRO. HYDRUS 1D [31] presents one-dimensional finite-element model for simulation of water movement, heat and multiple solutes in the variably saturated media. HYDRUS 2D/3D [32] presents a software package for simulating water, heat and solute movement in two and three-dimensional variably saturated media. MACRO [33] is one-dimensional dual permeability model for flow of water and reactive solute transport in variably saturated media.

5.2.5 Groundwater quality and pollution

Groundwater quality can be defined as the sum of soil-modified atmospheric inputs, water-rock interaction taking place at the soil-bedrock interface and long-term reactions taking place along flow paths in the saturated zone [34]. Due to growing human impacts on ground water quality, it is important to distinguish between natural and man-made concentrations in groundwater. It can be done using geochemical background criteria as a reference to assess whether groundwater concentrations are natural or influenced by anthropogenic pollution.

Geochemical background was defined by [35] as the normal abundance of an element in barren earth material. European Groundwater Directive (2006/118/EC) explained the term background level as: "the concentration of a substance or the value of an indicator in a body of groundwater corresponding to no, or only very minor, anthropogenic alterations to undisturbed conditions". Natural background levels reflect natural processes unaffected by human activities, but it can be argued that such a background no longer exists due to human influence on the whole planet [36]. In many parts of the world, human activities have been pervasive for such a long time that it may be vainly to attempt to define presettlement background values [37]. To overcome this problem [38], recommended the use of ambient background values under slightly altered conditions, when elevated concentrations are results of long-term human impacts and are no longer natural.

Many approaches have been developed for defining background values. When deciding which one presents the best option, it is important to take into consideration the following issues: the purpose of background values determination, original data quality, usability of methods, and representation of the background values – is it true or ambient background values of concern – and any potential sources of pollution near the sampling points [23].

Probably the best estimate of background values can be done using element concentrations from pristine waters in distant areas [39] or deep aquifers, which are free from anthropogenic influence [40]. However, such data are very scarce and, hence, of limited use. Furthermore, samples taken from groundwater in pristine areas may have substantially different physical, chemical and biological characteristics than the investigated areas [23]. When defining background concentrations using groundwater samples collected in pristine areas, it is very important to ensure that climate, geology and the history of land use are similar between areas compared [37].

Geochemical background is very often defined as a fixed value (mean or median) that represents a hypothetical background concentration without taking into account natural variability [41]. However, it changes both regionally with the basic geology and locally with the type and genesis of the overburden. It is more realistic to view it as a range of values rather than as an absolute value [23, 36].

Model-based objective methods for background values determination take into consideration the natural variability of groundwater chemistry. The theory behind these methods refers to the use of probability graph approach for the partitioning background data, which recognises that a background population can be closely approximated by normal or log-normal density function that results from the summation of natural processes that have produced the background substrate [42]. In essence, using the probability graph approach it is possible to split the overall data distribution into separable components and identify boundary between background (normal) population of concentrations and non-background (anomalous, non-normal) population of concentrations resulting from human impacts. The application of this approach is presented in Figure 5.10, which shows ambient background sodium (3.9–17.2 mg/l Na^+) concentrations in Samobor aquifer calculated by iterative

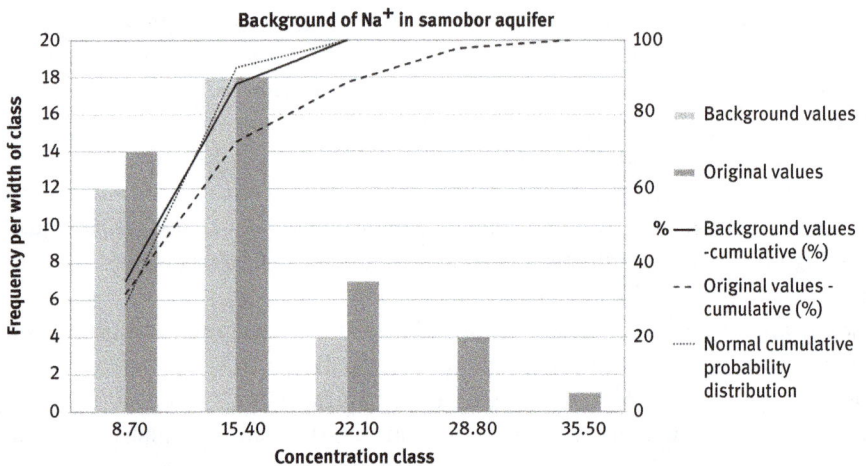

Figure 5.10: Na^+ background concentrations in Samobor aquifer calculated by iterative 2-σ technique.

2-σ technique [43]. This method can be used for the definition of an approximated normal distribution around the mode value of the original data set. It is applicable to unimodal and skewed original data distributions and aims to determine the outliers (anomalous population concentrations) above, as well as below, the lower limit of normal background fluctuation for particular chemical parameter [41].

Increasing pressure from human activities can lead to groundwater pollution by diffuse loading (e. g. agriculture or urban drainage from households and industrial estates) or point sources (e. g. mines, municipal waste water). Groundwater pollution refers to degradation of groundwater quality as measured by biological, chemical or physical criteria. From a public health or ecological point of view, a pollutant is any substance in which an identifiable excess is known to be harmful to desirable living organisms [44]. Pollutants, causing regional or local pollution which, depending on soil properties and hydrogeological conditions, may reach groundwater, can be natural, as well as synthetic substances. Natural substances include (1) potentially toxic metals; (2) radionuclides; (3) road salt; and (4) nutrients (mainly nitrogen); while synthetic substances comprise: (1) pesticides; (2) chlorinated aliphatics and petroleum hydrocarbon contaminants; and (3) organic wastewater contaminants [23]. In addition, pollutants can be certain pathogenic microorganisms, mainly bacteria and viruses. The hazard introduced by a particular groundwater pollutant depends on the volume of pollutant discharged, the concentration or toxicity of the pollutant in the groundwater system and the degree of exposure to humans or ecosystems.

5.3 Impact of mining industry on environment and human health

Alongside problems caused by natural hazards and excessive demand for geological resources, we discussed about previously, environmental geology deal also with other relationships of geology to specific function of society (e. g. land use planning and environmental impact assessment, geologic aspects of waste management, pollution, environmental health, climate change). All these issues are in some respect interconnected. Good site selection, as well as assortment of measures for minimizing environmental impact of any anthropogenic activity and waste disposal, decreases pollution. Climate change, as a global environmental problem, is caused by high level of air pollutants. Pollution has also negative impact on health. The adverse effects of pollutants on living organisms are topic of toxicology; however, the study of the effects of geologic materials and processes on human, animal and plant health, with both good and hazardous results is subject of medical geology.

Given the limited space of subchapter and a large number of topics, we will limit the subject to waste, pollution and health related to mining activities. As a direct consequence of the tremendous demand for the mineral resources, mining activities produce significant share of the total amount of waste produced annually worldwide. Often this waste contain notable amount of hazardous substances that can be released and have impact on the environment and human health. Hence, monitoring, treatment and secure disposal of such waste is necessity.

5.3.1 MineWaste

Waste is a general term for material that currently has little or no economic value. It can be in solid, liquid or gaseous state. The principal sources of solid waste are mining and agricultural activities with incomparably lower share of industrial and municipal waste. Still, world production of municipal solid waste is around 1.3 billion tonnes per year and it is expected to increase to 2.2 billion tonnes by 2025 [45]. As agricultural waste is not highly toxic, except when polluted with agrochemicals, and as it is not usually collected for disposal, we will concentrate on the waste from mining activities that mainly consists of geological materials. Approximately 15–20 billion tonnes of mine waste are produced annually [46], which is 10–15 times more than municipal waste produced. Considering the amount of mine waste produced, long-distance transportation and sophisticated treatment are uneconomical. Such waste is usually handled and disposed on site.

Mine wastes are solid, liquid and gaseous by-products of the process in the mining industry, which consists of three main activities: mining, mineral processing and metallurgical extraction [46]. Mining is extraction of material from mineral deposit, mineral processing is physical separation and concentration of the ore minerals (e. g. crushing, grinding, gravity, magnetic or electrostatic separation, flotation), while metallurgical extraction is seclusion of metal or compound from ore mineral (e. g. heap leaching; vat leaching). Hence, mine waste is classified as [46]: mining waste (e. g. waste rocks, overburden, mining water); processing waste (e. g. tailings, sludge, mill water); and metallurgical waste (e. g. slags, roasted ores, process water, atmospheric emissions). Mine waste often contains high concentrations of elements and compounds that can have a detrimental impact on the ecosystem and people [47]. Each mine produces its own unique waste, due to differences in the composition of the ore and the great variety of mining and mineral processing methods applied.

Solid mining waste includes overburden and waste rock excavated in open pits or underground mines dumped in large piles near the mine (Figure 5.11). Physical and chemical properties of the solid mining waste depend on the mineral and chemical composition, particle size and moisture content of the excavated material, as well as on the type of mining equipment. It is heterogeneous geological material, which may contain particles of different sizes (from clay particles size to the large blocks of rock), different rock types and/or soil.

Mineral processing wastes are crushed, ground, washed or treated excavated material left after segregation of concentrated ore minerals. Physical and chemical properties of this waste depend on the mineral and chemical composition and

Figure 5.11: Unrehabilitated rock waste heap at Berg Aukas mine in Namibia (photo: Mileusnić).

particle size of the processed materials, as well as the type of processing technology and chemical treatment. The particle sizes of this waste are nanometer to centimeter in size. Most of the waste is disposed of near the mine in the form of tailings (Figure 5.12). In the case of the ores, due to a rather low concentration factors for profitable mining, great amount of original mined material may eventually become tailings. Consequently, tailings have the greatest volume comparing to other mine wastes.

Metallurgical waste is unwanted residue of leached (hydrometallurgical extraction) or smelted ore concentrate (electrometallurgical and pyrometallurgical processes) (Figure 5.13).

Mine waters are groundwater or meteoric water, whose composition is altered by mineral-water reactions in mines or in mine waste.

Gaseous wastes comprise particulate matter and sulphur oxides (SO_x). They are mainly produced during smelting (high-temperature chemical processing) and vary in their composition depending on the mineral deposit.

Mine waste may become commodity with increasing need for specific mineral resource or because of improved technology. There are many examples of exploitation of historical mining waste.

Figure 5.12: Unsecured and unrehabilitated tailings dam with visible water erosion (gullies) at Kombat mine in Namibia [48].

5.3.2 Impact on the environment

Contamination/pollution is the introduction of substances or energy (contaminants/pollutants) into different environmental compartments (e. g. soil, water, air). The difference between these two terms is that contamination represents elevated concentrations of substances without harmful effects, while in the case of pollution, presence of pollutants cause toxicity, disrupt ecological processes, inflict damage to infrastructure or pose a hazard to human health [49]. Elevated concentration of some elements in soil, water and air can be result of natural processes. Such cases cannot be called contamination/pollution. Sources of pollution can be divided into point sources (e. g. smelter) and nonpoint sources (e. g. acid drainage from an abandoned strip mine). Prevention of pollution hazard is rehabilitation of mine wastes and mine sites. The weathering of mine waste can be a significant environmental media (soil, water, air) pollution hazard, depending on the type of the rocks and ore mined. Principal pollutants are potentially toxic metals and metalloids, radionuclides, sulphuric acid and process chemicals. Removal of pollutants from environmental media is called remediation.

Tailings are especially prone to both, physical and chemical weathering due to their small particle size (Figure 5.12 and Figure 5.14). Furthermore, the deposited

Figure 5.13: Unrehabilitated slag dump at Berg Aukas mine in Namibia with local people collecting scrap metal (photo: Mileusnić).

tailings undergo chemical change, most often a function of exposure to atmospheric oxidation, and tends to make previously, perhaps safely held contaminants mobile and available [50]. Therefore, accepted practice is storage of tailings behind the dams and in isolated impoundments under water where contact with atmosphere is substantially reduced. However, dams frequently miscarry, releasing enormous quantities of tailings into river catchments. Hence, proper isolation of tailings to prevent them from entering soil, surficial- and ground-waters, rivers, as well as air by means of the wind must be priority for a responsible mining organisation [50]. Capping with impermeable layer of solid, inert material and/or phytostabilisation, i. e. covering tailings with soil and planting vegetation, is an effective strategy to isolate tailings.

Acid mine drainage (AMD) is term used for process where mine water become acidic (low in pH) due to sulphide minerals, particularly pyrite, accelerated oxidation. It occurs very often in the sulphides and coal mines and their mine waste (especially tailings). Simplified chemical equation for AMD is:

$$4FeS_2 + 15O_2 + 14H_2O4Fe(OH)_3 + 8SO^{2-}_4 + 16H^+ \qquad (5.20)$$

Figure 5.14: The tailings dam wind erosion (Kombat, Namibia) [48].

However, for the formation of AMD, beside presence of both oxygen and water, for accelerated oxidation, significant role plays: (1) ferric iron (Fe^{3+}), product of pyrite oxidation by oxygen, which oxidised sulphides in an oxygen-independent reaction when pH is above 4; and (2) acidophilic chemotrophic bacteria when pH is below 4 [51, 52]. Harmful impact on the environment is not only because of acidity, but also because of high concentration of potentially toxic metals (e. g. Pb, Cd) and metalloids (e. g. As) that are easily leached from ore and mine waste at the low pH. Acid mine drainage is easily recognisable due its bright yellow to red colour. Preventing the formation of AMD (e. g. flooding/sealing of underground mine; underwater storage of mine tailings; application of anionic surfactants) would be the preferable option. However, this is not feasible in many locations. Therefore, in such cases, it is necessary to collect, treat, and discharge mine water. There are various options available for remediating AMD, i. e. to neutralise it and remove metals from solution. They may be divided into those that use either chemical (e. g. aeration and lime addition or anoxic limestone drains) or biological mechanisms (e. g. aerobic wetlands) and both include those that are classed as "active" (i. e., require continuous inputs of resources to sustain the process) or "passive" (i. e., require relatively little resource input once in operation) [51]. Which option to use depends on a number of economic (e. g. transportation of neutralising agent; land area needed) and environmental factors (e. g.

topography). Products of AMD remediation can be a resource. For instance, iron oxide sludge recovered from a drainage channel at an abandoned coal mine in Pennsylvania has been used to manufacture burnt sienna pigment [53].

Uncontrolled release of mine waste through artisanal mining, i. e. application of primitive mining and processing techniques that account for approximately 15–20 % of the world's non-fuel mineral production and employs 11.5 to 13 million people worldwide [54] should not be left out. Many developing nations have sizable gold ore deposits, making small artisanal gold mining a major source of employment in the world. Poverty drives vulnerable, rural populations into gold mining because of social and economic instabilities [55]. The main method of extraction used in artisanal gold mining, which in 2012 accounted for approximately 12 % of all the gold produced in the world, is gold amalgamation, a process that accounts for the release of between 1,000 and 1,600 tonnes of metallic mercury every year worldwide [56]. This is longstanding problem, as dispersed mercury causes adverse health effects and environmental and social ramifications. Despite elevated price of mercury in a restricted market due to international policies, gold amalgamation is still in wide use around the world. Metallic mercury dispersed in the environment can be converted to severely toxic methylmercury, organic compound of mercury produced by microorganisms in water. Methylmercury accumulates up the food chain that is called biomagnification, i. e. increasing concentration of a substance in the tissues of organisms at successively higher levels in a food chain.

5.3.3 Impact on human health

As mentioned in introduction, interdisciplinary scientific field studying the relationship between natural geological factors and their effects on human and animal health is called medical geology or geomedicine. Some of the well-known examples of medical geology topics are: exposure to arsenic rich groundwater in Southeast Asia; deficiency of selenium in parts of the China; exposure to radionuclide radon in areas with uranium rich rocks; deficiency of iodine; and exposure to volcanic emissions [57]. In its broader sense, impact of mine waste on human health, although result of anthropogenic activities, could be consider topic of medical geology as the mine waste is mostly composed of geologic materials (rests of rocks, minerals, ores).

People living near mining or mineral processing sites are exposed to mining waste in the following ways [58]: (1) incidentally ingestion of solid mine waste or soil contaminated with mine waste (hand to mouth; especially children and farmers) (Figure 5.15); (2) inhalation of dust from the solid mine waste (especially tailings); (3) inhalation of gases or airborne particles resulting from smelting or roasting; and (4) by eating food (meat, vegetables, fruits) grown in the contaminated area.

Factors important for geological material impact on human health are: the intensity and duration of exposure (dose); exposure routes (e. g. inhalation; ingestion); physical and chemical properties of material; presence of microbes or other pathogens

Figure 5.15: Ingestion of soil by hand to mouth transfer in the area of Berg Aukas mine complex (photo: Mileusnić).

in the material; biosolubility, bioresistivity, bioaccessibility and bioreactivity of materials in body fluids; immunological reaction of the organism; physiological processes that control adsorption, distribution, metabolism and excretion of toxic substances; other factors such as age, gender, genetics, personal habits (e.g., smoking), personal socio-economic-, health- and nutritional status and other factors that can enhance or neutralise the toxic effects of exposure to mining waste [59].

Potential health problems are diverse, including cancer, respiratory diseases, neurological diseases, systemic excretion and secondary diseases such as heart failure, increased susceptibility to pathogen infections and many others [60].

Many medical methods are used to assess the potential effects of exposure to mining waste on health [61] such as biological monitoring (e.g. testing of blood, urine or tissues of individuals), epidemiological studies (disease rates assessments), and pathological studies (testing of tissue samples collected by biopsy or autopsies). Material toxicity testing can be performed *in vitro* (cell cultures) and *in vivo* (animal testing). *In vitro* bioaccessibility tests indicate the release of toxic chemical elements from materials in simulated (synthesised) body fluids (e.g. gastric, intestinal, pulmonary fluid) [58]. It should be emphasised that bioaccessible concentration of toxic elements is lower than concentration absorbed by human body (i. e. bioavailable concentration).

Bioavailable concentration reaches the systemic circulation where it may cause adverse effects on human health [62].

Example: Impact of mining, mineral processing and smelting on the environment and human health in Otavi Mountainland (Namibia)

Indigenous people in the Otavi Mountainland mined copper deposits for generations before the arrival of European explorers. Mining, processing and smelting of sulphide ore found in carbonate rocks played a significant role in economic development of the wider region since the beginning of the twentieth century. World famous Pb-Cu mine Tsumeb having ore bodies with more than 250 different minerals, V-Zn mine Berg Aukas with renowned descloizite minerals, Cu mine Kombat, and many smaller mines such as Abenab and Khusib Springs are situated here. Due to weak environmental awareness and economic pressures, adverse impact on the environment and human health could not be avoided in this region.

Long history of mining [63] resulted in huge amount of mine waste [64, 65]. Mining in Tsumeb lasted for 90 years. During that time, approximately 25 millions tons of ore were excavated, leaving large amount of waste covering 64 hectares of land. Some old slags and tailings in Tsumeb were reprocessed and re-extracted. The Tsumeb smelter is yet in operation although the mining activities ceased. It is, one of only a few in the world, able to treat complex copper concentrates that contain arsenic. Consequently, the ore from Bulgaria, Chile, Peru and Namibia is smelted here. In Berg Aukas, where mining activities lasted for 36 in total, there are: two waste heaps 96,880 m^3 in volume covering area of 1.3 hectares; two tailings dam 343,500 m^3 in volume covering area of 6.9 hectares m^2; and one slag heap 1,756,055 m^3 in volume covering area of 3.5 hectares. These waste accumulations are nor secured from erosion and sliding, neither fenced, enabling children to dwell in that area. Slags were used for roads as well. Mining and ore processing activities in the Kombat area, which lasted for 46 years, left 300 million tons of processing waste in the form of tailings covering an area of 15 hectares. Metallurgical waste is not present there, as the ore was not smelted on site.

Pollution of soil by potentially toxic metals (e. g. Pb, Zn, Cu, Cd) and metalloids (e. g. As, Sb) in this area is result of abandoned, improperly disposed solid mine waste (Figure 5.11–Figure 5.14), especially tailings, as well as smelting operations. The presence of highly soluble phases in slags might be responsible for the significant release of toxic elements in the Tsumeb area, through their rapid dissolution during thunderstorm events occurring between October and March [66]. Tailings, with very fine particles are prone to wind and water erosion. Consequently, adjacent arable soil have concentrations of some potentially toxic elements exceeding guideline values for agricultural land use [48, 67]. Considering the neutral and slightly alkaline (and lowly organic) character of soils around the tailings, Vaněk and co-authors [68] consider Cu and As as contaminants with the highest potential for rhizospheric mobilisation and subsequent vertical mobility in local soils and/or transfer to plants. Modelling of the dispersion of dust and SO_2 emissions from the Tsumeb smelter was used to delineate the contaminated area [69].

Pollution of the area has the impact of the human health. Anomalous lead and arsenic concentration were found in urine and blood of inhabitants in Tsumeb area [70]. Polluted soil influenced the concentrations of toxic elements in plants (e. g. in Berg Aukas [67]) used by locals as food. Residents are in danger not only because of food consumption, but also because of inhalation of small particles of mine waste and gasses from smelter, as well as ingestion of soil and dust through hand to mouth transfer. Studies of main soil pollutant bioaccessibility pointed out the problem of children playing around houses, especially in the area of tailings [71, 72]. After the closure of the Berg Aukas mine, mine buildings were used for National Youth Service Organisation where students practised agriculture, but thanks to research of Mapani and co-authors [67], the organisation has moved.

5.4 Highlights

Environmental geology is scientific discipline dealing with the interactions between humans and the geological environment.

Hydrology is the scientific discipline dealing with water cycle (distribution and movement), as well as environmental sustainability of water resources in terms of quantity and quality.

Enormous human population growth with increasing standard of living has huge impact on the environment.

Natural phenomena, such as floods, become hazards when happen in the populated area or when land use changes. Hence, human population growth and land use changes greatly augment danger of human and material losses caused by natural hazards.

Groundwater presents the main source of potable water in the world. Due to huge anthropogenic influence groundwater quality and quantity have been endangered.

Demand for mineral resources enhanced remarkably with increasing population growth rate and standard of living. Exploitation of mineral resources deteriorate huge areas of land, produce enormous mine waste and pollute soil, water and air having tremendous impact on environment and human health.

The role of environmental geologists is (1) to apply geologic information to asses risk from natural hazards in order to prevent or mitigate disasters, (2) to identify environmental problems caused by humans and remediate the damage, and (3) to help land-use planners and policy makers to balance needs for land and resources with their availability, in order to achieve sustainable development.

References

[1]   Hurricanes: science and society, 1970 The Great Bhola Cyclone. Available at: http://www. hurricanescience.org/history/storms/1970s/greatbhola/. Accessed: 25 Feb 2017.

[2]   World Health Organization, Regional Office for South-East Asia. Tsunami 2004: a comprehensive analysis Vol. I. New Delhi: WHO, 2013:337

[3]   Mortality, crime and access to basic needs before and after the Haiti earthquake: a random survey of Port-au-Prince households. Taylor Francis Online. Accessed: 21 Feb 2017.

[4]   Eagleson PS. Opportunities in the hydrologic sciences, committee on opportunities in the hydrologic sciences, national research council. Washington, DC: National Academy Press, 1991.

[5]   Brutsaert W. Hydrology, an introduction. Cambridge, UK: Cambridge University Press, 2005.

[6]   Turc L. Le bilan d`eau des sols: relations entre les précipitations, l`évaporation et l`écoulement. Paris: Institut national de la recharge agronomique, 1953:1–252.

[7]   International glossary of hydrology. World Meteorological Organization and United National Educational, Scientific and Cultural Organization, No. 385, Genève, Switzerland, 2012.

[8]   Kuspilić N, Oskoruš D, Vujnović D. Catastrophic Sava river floods in 2014 and flood hazards. Građevinar 2014;66(7):653–661.

[9]   Todd DK. Groundwater hydrology, 2nd ed. New York: Wiley & Sons, 1980:535.

[10] Montgomery CW. Environmental geology, 7th ed. New York: McGraw Hill, 2004:540.

[11] Tindall JA, Kunkel JR. Unsaturated zone hydrology for scientists and engineers, 1st ed. Prentice Hall, NJ: Pearson Education, 1998:624.

[12] Hubbert MK. Darcy's law and the field equations of flow of underground fluids. Trans Am Inst Min Metall Eng 1956;207:222–239.

[13] Fetter CW. Applied hydrogeology, 4th ed. Prentice Hall, NJ: Pearson Education, 2001:598.

[14] Childs EC. Soil moisture theory. In: Chow VT, editor. Advances in hydroscience, vol. 4. New York: Academic Press, 1967:73–117.

[15] Nimmo JR. Unsaturated zone flow processes. In: Anderson MG, Bear J, editors. Encyclopedia of hydrological sciences: part 13-groundwater, vol. 4. Chichester, UK: Wiley, 2005:2299–2322.

[16] Richards LA. Capillary conduction of liquids through porous materials. Physics 1931;1:318–333.

[17] Schwille F. Dense chlorinated solvents in porous and fractured media. Model experiments. Chelsea, MI: Lewis Publishers Inc, 1988:146.

[18] Filipović V, Ondrašek G, Filipović L. Modelling water dynamics, transport processes and biogeochemical reactions in soil Vadose zone. In: Javaid MS, editor. Groundwater – contaminant and resource management. Rijeka, Croatia: InTech, 2016:200.

[19] Freeze RA, Cherry JA. Groundwater, 1st ed. Prentice Hall, NJ, 1979:604.

[20] Bear J, Beljin MS, Ross RR. Fundamentals of ground-water modeling, EPA Ground-Water Issue 1992, EPA/540/S-92/005, 11.

[21] Fetter CW. Contaminant hydrogeology, 2nd ed. Prentice Hall, NJ, 1999:500.

[22] Wang HF, Anderson MP. Introduction to groundwater modeling – finite difference and finite element methods. Amsterdam: Academic Press Inc., 1982:237.

[23] Balderacchi M, Benoit P, Cambier P, Eklo OM, Gargini A, Gemitzi A, et al. Groundwater pollution and quality monitoring approaches at European-level. Crit Rev Environ Sci Technol 2013; 43(4):323–408.

[24] McDonald MG, Harbaugh AW. A modular three dimensional finite difference ground-water flow model. US Geological Survey Techniques of Water Resources Investigation Report, 1988, 06-A1.

[25] Harbaugh AW, McDonald MG. User's documentation for MODFLOW-96, an update to the US geological survey modular finite-difference ground-water flow model, U.S. Geol. Survey Open-file Report 1996, 96–485.

[26] Harbaugh AW, Banta ER, Hill MC, McDonald MG. MODFLOW-2000, The U.S. geological survey modular ground-water model – user guide to modularization concepts and the ground-water flow process, U.S. Geol. Survey Open-file Report 2000, 0–92.

[27] Harbaugh AW. The U.S. geological survey modular ground-water model – the ground-water flow process, U.S. Geological Survey Techniques and Methods 2005, 6-A16.

[28] Pollock DW Documentation of computer programs to compute and display pathlines using results from the U.S. geological survey modular three-dimensional finite-difference ground-water flow model, U.S. Geol. Survey Open-file Report 1989, 89–381.

[29] Pollock DW. User's guide for MODPATH/MODPATH-PLOT version 3: a particle tracking post processing package for MODFLOW, the U.S. geological survey finite-difference ground-water flow model, U.S. Geol. Survey Open-file Report 1994, 94–464.

[30] Zheng C, Wang PP. MT3DMS: a modular three-dimensional multispecies model for simulation of advection, dispersion and chemical reactions of contaminants in groundwater systems; Documentation and user's guide, Contract Report SERDP-99-1, U.S. Army Engineer Research and Development Center 1999, Vicksburg.

[31] Šimunek J, Šejna M, van Genuchten M. The HYDRUS-1D software package for simulating the one-dimensional movement of water, heat, and multiple solutes in variably-saturated, Version 2.0, IGWMC – TPS-70 ed. Golden: International Ground Water Modeling Center, Colorado School of Mines, 1998:202.

[32] Šimunek J, van Genuchten M, Šejna M. Development and applications of the HYDRUS and STANMOD software packages and related codes. Vadose Zone J Special Issue"Vadose Zone Modeling" 2008;7(2):587–600.

[33] Jarvis NJ. The MACRO model (Version 3.1). Technical description and sample simulations. reports and dissertations, 19th ed. Uppsala, Sweden: Swedish University of Agricultural Science, 1994:51.

[34] Nakić Z, Posavec K, Parlov J. Model-based objective methods for the estimation of groundwater geochemical background. AQUA Mundi 2010;1:65–72.

[35] Hawkes HE, Webb JS. Geochemistry in mineral exploration. New York: Harper & Row, 1962:70–238.

[36] Nakić Z, Posavec K, Bačani A. A visual basic spreadsheet macro for geochemical background analysis. Ground Water 2007;45:642–647.

[37] Kelly WR, Panno SV. Some considerations in applying background concentrations to ground water studies. Ground Water 2008;46:790–792.

[38] Reimann C, Garrett RG. Geochemical background – concept and reality. Sci Total Environ 2005;350:12–27.

[39] Kilchmann S, Waber HN, Parriaux A, Bensimon M. Natural tracers in recent groundwaters from different Alpine aquifers. Hydrogeol J 2004;12:643–661.

[40] Hernández-García ME, Custodio E. Natural baseline quality of madrid tertiary detrital aquifer groundwater (Spain): a basis for aquifer management. Environ Geol 2004;46:173–188.

[41] Matschullat J, Ottenstein R, Reimann C. Geochemical background – can we calculate it? Environ Geol 2000;39:990–1000.

[42] Reimann C, Filzmoser P. Normal and lognormal data distribution in geochemistry: death of a myth. Consequences for the statistical treatment of geochemical and environmental data. Environ Geol 2000;39:1001–1014.

[43] Kovač Z, Nakić Z, Posavec K, Parlov J, Bačani A. Ambient background concentrations of chemical parameters in groundwater of Samobor aquifer. Waters in sensitive and protected areas, 2013 (ISBN: 978-953-96071-3-3), 163–166.

[44] Keller EA. Environmental geology, 8th ed. Prentice Hall, NJ, 1999:562.

[45]   Hoornweg D, Bhada-Tata P. WHAT A WASTE – a global review of solid waste management. Urban
       Development Series, Vol. 15, Urban Development & Local Government Unit, 2012, World Bank, 98.
[46]   Lottermoser BG. Mine wastes: characterization, treatment, environmental impacts. New York:
       Springer, 2007.
[47]   Hudson-Edwards KA, Jamieson HE, Lottermoser BG. Mine wastes: past, present, future.
       Elements 2011;7:375–380.
[48]   Mileusnić M, Mapani BS, Kamona AF, Ružičić S, Mapaure I, Chimwamurombe PM. Assessment
       of agricultural soil contamination by potentially toxic metals dispersed from improperly dis-
       posed tailings, Kombat mine, Namibia. Journal of Geochemical Exploration 2014;
       144(Part C):409–420.
[49]   Thornton I, Ramsey M, Atkinson N. Metals in the global environment: facts and misconceptions.
       Int Counc Metals Environ. International Council on Metals and the Environment: Canada
       1995:103.
[50]   Kossoff D, Dubbin WE, Alfredsson M, Edwards SJ, Macklin MG, Hudson-Edwards KA. Mine
       tailings dams: characteristics, failure, environmental impacts, and remediation. Appl
       Geochemistry 2014;51:229–245.
[51]   Johnson DB, Hallberg KB. Acid mine drainage remediation options: a review. Sci Total Environ
       2005;338:3–14.
[52]   Rimstidt JD, Vaughan DJ. Acid mine drainage. Elements 2014;10/2:153–154.
[53]   Hedin RS. Recovery of marketable iron oxide from mine drainage in the USA. Land Contam
       Reclam 2003;11:93–97.
[54]   Kafwembe BS, Veasey TJ. The problem of artisanal mining and mineral processing. Mining
       Environ Manag 2001:17–21.
[55]   Zolnikov TR. Limitations in small artisanal gold mining addressed by educational components
       paired with alternative mining methods. Sci Total Environ 2012;419/1:1–6.
[56]   Veiga MM, Angeloci-Santos G, Meech JA. Review of barriers to reduce mercury use in artisanal
       gold. Extr Ind Soc 2014;1/2:351–361.
[57]   Selinus O, Alloway B, Centeno J, Finkelman RB, Fuge R, Lindh U, et al. Essentials of medical
       geology – impacts of the natural environment on public health. Amsterdam Elsevier, 2005.
[58]   Plumlee GS, Morman SA. Mine wastes and human health. Elements 2011;7:399–404.
[59]   Plumlee GS, Ziegler TL. The medical geochemistry of dusts, soils, and other earth materials.
       Treatise Geochem 2003;9:263–310.
[60]   Plumlee GS, Morman SA, Ziegler TL. The toxicological geochemistry of earth materials: an
       overview of processes and the interdisciplinary methods used to understand them. Rev Mineral
       Geochem 2006;64:5–57.
[61]   Sullivan JB, Krieger GR, editors. Clinical environmental health and toxic exposures, 2nd ed.
       Philadelphia: Lippincott Williams & Wilkins, 2001.
[62]   Cave MR, Wragg J, Denys S, Jondreville C, Feidt C. Oral bioavailability. Chapter 7 In: Swartjes FA.
       Dealing with contaminated sites. Netherlands: Springer Science+Business Media,
       2011:287–324.
[63]   Schneider GI. The history of mining in Namibia. Namibia Brief 1998:19–31.
[64]   Kříbek B, Kamona F. Assessment of the mining and processing of ores on the environment in
       mining districts of Namibia. Final report of the Project of the Development Assistance
       Programme of the Czech Republic to the Republic of Namibia for the year 2004 –Tsumeb Area.
       Czech Geol. Surv. 2005, Record Office – File Report No. 2/2005, Prague, Czech Republic. 94
[65]   Kříbek B, Kamona F. Assessment of the mining and processing of ores on the environment in
       mining districts of Namibia. Final report of the Project of the Development Cooperation
       Programme of the Czech Republic to the Republic of Namibia for the years 2004 –2006. Czech
       Geol. Surv. 2006, Record Office – File Report No. 1/2006, Prague, Czech Republic.

[66] Ettler V, Johan Z, Kříbek B, Šebek O, Mihaljevič M. Mineralogy and environmental stability of slags from the Tsumeb smelter, Namibia. Appl Geochemistry 2009;24:1–15.

[67] Mapani B, Ellmies R, Kamona F, Kříbek B, Majer V, Knésl I, et al. Potential human health risks associated with historic ore processing at Berg Aukas, Grootfontein area, Namibia. J Afr Earth Sci 2010;58:634–647.

[68] Vaněk A, Ettler V, Skipalová K, Novotný J, Penížek V, Mihaljevič M, et al. Environmental stability of the processing waste from sulfide mining districts of Namibia – A model rhizosphere solution approach. J Geochem Explor 2014;144:421–426.

[69] Kříbek B, Majer V, Knésl I, Keder J, Mapani B, Kamona F, et al. Contamination of soil and grass in the Tsumeb smelter area, Namibia: modeling of contaminants dispersion and ground geochemical verification. Appl Geochemistry 2016;64:75–91.

[70] Hahn L, Ellmies R, Mapani B, Schneider G, Amaambo W, Mwananawa N, et al. Blood and Urine tests for heavy metals in Tsumeb.- Ministry of mines and energy. Environ Monit 2008. Series No. 12.

[71] Mileusnić M, Tomašek I, Mapani B. Human bioaccessibility of Pb, Zn and Cd in residential soils from Berg Aukas, Namibia. In: Kribek B, Davies T, editors. Proceedings of the Closing Workshop of the IGCP/SIDA Projects 594 and 606. Prag: Czech Geological Survey, 2014:95–98.

[72] Mileusnić M, Tomašek I, Mapani B. Health risk assessment from the exposure of children to soil contaminated by copper mining and smelting, Nomtsoub township, Tsumeb, Namibia. In: Kribek B, Davies T, editors. Proceedings of the Closing Workshop of the IGCP/SIDA Projects 594 and 606. Prag: Czech Geological Survey, 2014:91–94.

Questions

1. If yearly amount of precipitation over a watershed is $P = 0.800$ m, and yearly mean air temperature is $t = 13\,°C$, calculate amount of water in millimetres available for infiltration using Turc's equation.

2. Top of the monitoring well is located at an altitude of 250 m a.s.l. Depth to groundwater is 10 m. Bottom of filter is located at depth of 40 m. Calculate total head, pressure height and pressure at depth of 50 m.

3. Use Darcy`s law to calculate total flow if hydraulic conductivity is 4.5E-4 m/s, cross-sectional area 2 km^2 while potential drop is 3 m on 10 km.

4. Based on what you have learned about acid mine drainage, explain why mine water in the area Otavi Mountainland where sulphide mineralisation was exploited (See example), is of acceptable quality and can be used for drinking and irrigation purposes. Which toxic chemical element/s could be of concern in this water? Explain.

5. Explain difference between terms contamination and pollution.

6. Describe main pathways through which pollutants can enter human body.

7. Explain difference between terms bioaccessibility and bioavailability.

Aleksandra Sander, Jasna Prlić Kardum, Gordana Matijašić
and Krunoslav Žižek

6 Transport phenomena in environmental engineering

Abstract: A term *transport phenomena* arises as a second paradigm at the end of 1950s with high awareness that there was a strong need to improve the scoping of chemical engineering science. At that point, engineers became highly aware that it is extremely important to take step forward from pure empirical description and the concept of unit operations only to understand the specific process using phenomenological equations that rely on three elementary physical processes: momentum, energy and mass transport. This conceptual evolution of chemical engineering was first presented with a well-known book of R. Byron Bird, Warren E. Stewart and Edwin N. Lightfoot, *Transport Phenomena*, published in 1960 [1]. What transport phenomena are included in environmental engineering? It is hard to divide those phenomena through different engineering disciplines. The core is the same but the focus changes. Intention of the authors here is to present the transport phenomena that are omnipresent in treatment of various process streams. The focus in this chapter is made on the transport phenomena that permanently occur in mechanical macroprocesses of sedimentation and filtration for separation in solid–liquid particulate systems and on the phenomena of the flow through a fixed and a fluidized bed of particles that are immanent in separation processes in packed columns and in environmental catalysis. The fundamental phenomena for each thermal and equilibrium separation process technology are presented as well. Understanding and mathematical description of underlying transport phenomena result in scoping the separation processes in a way that ChEs should act worldwide.

Keywords: momentum transfer, heat transfer, mass transfer, sedimentation, filtration, fluidization, thermal separation process, equilibrium separation process

This article has previously been published in the journal Physical Sciences Reviews. Please cite as:
Sander, A., Kardum, J. P., Matijašić, G., Žižek, K. Transport Phenomena in Environmental Engineering. *Physical Sciences Reviews* [Online] **2018**, *3* (1). DOI: 10.1515/psr-2016-0121

https://doi.org/10.1515/9783110468038-006

6.1 Conservation laws

Conservation laws can be expressed in differential or integral form, representing microscopic or macroscopic balances. Macroscopic balances are usually applied to determine the relationship of input and output variables in the process. On the other hand, microscopic balances enable calculation of velocity, temperature and concentration profiles within system.

A process control volume is now observed to establish macroscopic balances of mass, energy and momentum (Figure 6.1).

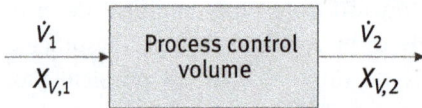

Figure 6.1: Schematic of process control volume with inputs and outputs.

General macroscopic balance can be written as:

$$\text{ACCUMULATION} = \text{INPUT} - \text{OUTPUT} \pm \text{GENERATION} \tag{6.1}$$

or as simple equation:

$$V \cdot \frac{dX_V}{dt} = \dot{V}_1 \cdot X_{V,1} - \dot{V}_2 \cdot X_{V,2} \pm V_r \tag{6.2}$$

where X_V is mass, energy or momentum per unit volume, V is volume, $\dot{V}$ is volume flow rate and indexes 1 and 2 denote inlets and outlets of the system.

6.1.1 Macroscopic mass balance

Let us assume that fluid enters to the system at plane 1 with cross-section S_1 and leaves at plane 2 with cross-section S_2. The corresponding average velocities are v_1 and v_2.

If the total mass is applied to the general balance equation eq. (6.2), assuming that generation is equal to zero and $X_V = m/V$ macroscopic mass balance is obtained:

$$\frac{dm}{dt} = \dot{m}_1 - \dot{m}_2 \tag{6.3}$$

Equation 6.3 is called the law of conservation of mass.

For steady state where $dm/dt = 0$, conservation law is written as:

$$\dot{m}_1 = \dot{m}_2 \tag{6.4}$$

where $\dot{m}$ is mass flow rate of the fluid at plane 1 or 2. At steady state, where no accumulation occurs, the mass flow rate of fluid into the process unit (control volume) is counterbalanced by the outlet mass flow rate.

Mass flow rate can be written as volumetric flow rate multiplied by density:

$$\dot{m} = \dot{V} \cdot \rho = v \cdot A \cdot \rho \tag{6.5}$$

Law of conservation of mass is now expressed in form:

$$v_1 \cdot A_1 \cdot \rho_1 = v_2 \cdot A_2 \cdot \rho_2 \tag{6.6}$$

where v is the average velocity of the fluid.

Equation (6.6) is valid for steady-state, single-input single-output system, density does not change across the surface plane and the velocity is perpendicular to the surface. For isothermal flow of incompressible fluids, density is constant along the planes 1 and 2, and the continuity equation is observed through the balance of volume flow rates.

6.1.2 Macroscopic energy balance

Fluid of given mass, m, carries certain amount of energy. The total energy of the system is the sum of internal energy (U), kinetic energy ($1/2m \cdot v^2$) and potential energy due to gravity ($m \cdot g \cdot h$):

$$E = U + \frac{1}{2}m \cdot v^2 + m \cdot g \cdot z \tag{6.7}$$

Internal energy comprises pressure energy, perceptible and latent heat energy. According to general balance equation eq. (6.2), volumetric concentration of energy (E_V) needs to be defined and it equals X_V. Equation (7), expressed per unit volume, gives volumetric concentration of energy:

$$E_V = p + \rho \cdot c_p \cdot \Delta T + \sum \rho \cdot \Delta H + \frac{1}{2}\rho \cdot v^2 + \rho \cdot g \cdot z \tag{6.8}$$

When there is no change in temperature of the observed process, and there is no phase transition, energy equation is reduced to mechanical energy:

$$E_V = p + \frac{1}{2}\rho \cdot v^2 + \rho \cdot g \cdot z \tag{6.9}$$

Equation (6.9) is now combined with general balance equation (eq. (6.2)). If the total kinetic and potential energy in the system is not changing with time, steady-state macroscopic energy balance is obtained:

$$\dot{V}_1 \cdot E_{V,1} - \dot{V}_2 \cdot E_{V,2} + W - Q = 0 \tag{6.10}$$

where E_V is defined in eq. (6.9), W is added mechanical energy and Q is heat flow. $(W - Q)$ represents the generation term.

For the systems without added mechanical energy with no heat exchange, and with respect to the continuity equation (eq. (6.6)) where $\dot{V}_1 = \dot{V}_2$, energy balance for the flow of ideal non-viscous fluid can be written as well-known Bernoulli's equation:

$$p_1 + \frac{1}{2}\rho \cdot v_1^2 + \rho \cdot g \cdot z_1 = p_2 + \frac{1}{2}\rho \cdot v_2^2 + \rho \cdot g \cdot z_2 \tag{6.11}$$

However, real processes always include flow of the viscous fluid where viscous energy dissipation term, due to the fluid viscosity, should be included in eq. (6.11).

$$p_1 + \frac{1}{2}\rho \cdot v_1^2 + \rho \cdot g \cdot z_1 = p_2 + \frac{1}{2}\rho \cdot v_2^2 + \rho \cdot g \cdot z_2 + E_{vis} \tag{6.12}$$

Energy dissipation term is expressed as:

$$E_{vis} = \rho \cdot g \cdot h_w \tag{6.13}$$

By dividing eq. (6.12) with $(\rho \cdot g)$, all the energy terms in equation become expressed as elevations:

$$z_1 + \frac{p_1}{\rho \cdot g} + \frac{v_1^2}{2 \cdot g} = z_2 + \frac{p_2}{\rho \cdot g} + \frac{v_2^2}{2 \cdot g} + h_w \tag{6.14}$$

where z is geodetic height at the fluid entry (1) and exit (2).

6.1.3 Macroscopic momentum balance

The last balance to be examined is the momentum balance. As in the case of mass balance, a momentum per unit volume expressed as $X_V = (m \cdot v)/V = v \cdot \rho$ is now considered.

$$V \cdot \frac{d(v \cdot \rho)}{dt} = \dot{V}_1 \cdot (v \cdot \rho)_1 - \dot{V}_2 \cdot (v \cdot \rho)_2 + V_r \tag{6.15}$$

where the product $\dot{V} \cdot (v \cdot \rho)$ is the force and generation rate term is sum of all forces.

Therefore, the eq. (6.15) is now written as the sum of all forces and represents the momentum-generation term according to Newton's second law of motion:

$$\sum_{i=1}^{n} F_i = \dot{V}_2 \cdot (v \cdot \rho)_2 - \dot{V}_1 \cdot (v \cdot \rho)_1 + V \cdot \frac{d(v \cdot \rho)}{dt} \tag{6.16}$$

Since there is no accumulation term, the steady-state momentum balance is written as:

$$\sum_{i=1}^{n} F_i = \dot{V}_2 \cdot (v \cdot \rho)_2 - \dot{V}_1 \cdot (v \cdot \rho)_1 \tag{6.17}$$

where $\sum_{i=1}^{n} F_i$ is the sum of external forces acting on the system.

In the case where the sum of all external forces is equal to zero, the steady-state momentum balance is written as:

$$\dot{V}_1 \cdot (v_1 \cdot \rho_1) = \dot{V}_2 \cdot (v_2 \cdot \rho_2) \tag{6.18}$$

6.2 Motion of particles in fluid

Environmental engineering is permanently dealing with solid–liquid separations in general and especially gravitational sedimentation where solid particles settle in liquids. Such separation is commonly used in wastewater and water treatments.

If one considers motion of the particles in fluids, different scenarios are pointed out. Non-flocculent particles in a dilute suspension tend to settle freely, without any obstruction from other particles. On the other hand, flocculent particles in concentrated suspensions will be subjected to the high extent of interparticle forces and consequently hindered settling. Modelling of such systems is far more empirical than theoretical.

When particle is moving in a fluid, there are three acting forces: external force, F_e, (gravity or centrifugal force), buoyant force, F_b, and drag force, F_D. Resulting acceleration force on the particle is then:

$$m \cdot \frac{dv}{dt} = F_e - F_b - F_D \tag{6.19}$$

where m is mass of the particle and v is velocity of the particle relative to the fluid.

Very slow motion of the particle in fluid is called creeping flow and hydrodynamically is defined as laminar flow. This is so-called Stokes' law region where total drag force resisting motion is originally obtained by Stokes:

$$F_D = 3 \cdot \pi \cdot \mu \cdot d_p \cdot v \tag{6.20}$$

where μ is the fluid viscosity and d_p is the particle size.

Net drag force is a result of surface friction (2/3) due to the fluid viscosity and form drag (1/3) due to the shape of a particle.

Stokes' law is valid for low Reynolds numbers ($Re < 0.3$) [2] where single-particle Reynolds number can be defined as:

$$Re = \frac{v \cdot d_p \cdot \rho}{\mu} \tag{6.21}$$

Increasing the particle velocity, form drag starts to dominate and the drag force is well described using Newton's law:

$$F_D = C_D \cdot A \cdot \rho \cdot \frac{v^2}{2} \tag{6.22}$$

where A is a cross-section of particles in a plane perpendicular to the flow direction, ρ is density of the fluid and C_D is a drag factor.

For spherical particles, cross-section is defined as:

$$A = \frac{d_p^2 \cdot \pi}{4} \tag{6.23}$$

Combining eqs. (6.20), (6.22) and (6.23), drag factor for Stokes' region is obtained:

$$C_D = \frac{24 \cdot \mu}{v \cdot d_p \cdot \rho} = \frac{24}{Re} \tag{6.24}$$

Equation (24) represents a straight line in the drag curve that represents the relationship between drag factor and Reynolds number when Re < 0.3 (Figure 6.2).

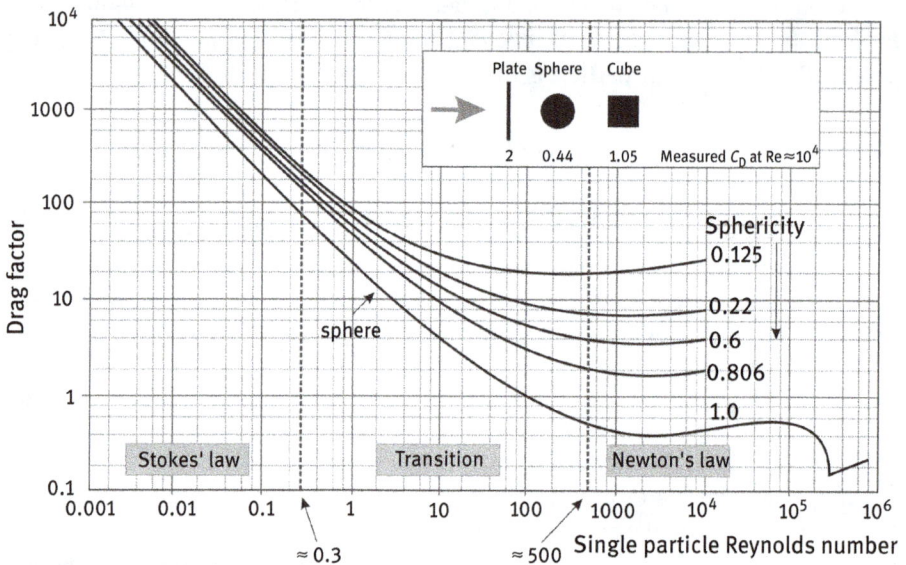

Figure 6.2: Drag factor versus Reynolds number.

In transition region ($0.3 < Re < 500$), drag factor is affected by both, Reynolds number and shape of the particle (sphericity) and can be efficiently evaluated using empirical equations. When Reynolds number exceeds 500, inertia starts to prevail and drag factor is mainly constant over the range of Reynolds number ($500 < Re < 2 \times 10^5$) but significantly vary with particle sphericity. Finally, for Reynolds number above 2×10^5, the flow in the boundary layer changes from streamline to turbulent and the separation takes place nearer to the rear of the particle. Drag factor suddenly drops due to the boundary layer separation.

Particle terminal velocity is the maximal velocity that particle reaches in the fluid. At that point, acceleration reduces to zero and force balance is achieved. From eq. (6.19), taking into account spherical particles that settle down under gravity only, force balance can be written as follows:

$$\frac{d_p^3 \cdot \pi}{6} \cdot \rho_p \cdot g - \frac{d_p^3 \cdot \pi}{6} \cdot \rho \cdot g - C_D \cdot \frac{d_p^2 \cdot \pi}{4} \cdot \rho \cdot \frac{v^2}{2} = 0 \tag{6.25}$$

where ρ_p is the particle density.

Equation (6.25) gives the following expression for the single-particle terminal velocity:

$$v = \sqrt{\frac{4}{3} \cdot \frac{\left(\rho_p - \rho\right) \cdot g \cdot d_p}{C_D \cdot \rho}} \tag{6.26}$$

Regarding eqs. (6.24) and (6.26), terminal velocity of the particle in Stokes' region is given by:

$$v = \frac{d_p^2 \cdot \left(\rho_p - \rho\right) \cdot g}{18 \cdot \mu} \tag{6.27}$$

Given equations are applicable only for the settling of a single particle in an infinite volume of fluid. Underlying phenomenon of free settling can be easily transferred to the sedimentation of particles in dilute suspensions that will take place in a primary clarifier in wastewater treatment [3].

In hindered settling, particle trajectories overlap and practical usage of Stokes' or Newton's law is out of the question. Velocity of uniformly sized particles in the case of hindered settling is much lower than the velocity of a single particle of the same size. Settling velocity of spherical particles in hindered settling (v_h) is defined with a well-known Richardson–Zaki equation [4, 5, 6]:

$$v_h = v_t \cdot \varepsilon^n = v_t \cdot (1 - c)^n \tag{6.28}$$

where v_t is the terminal velocity of a single particle, ε is voidage or porosity and c is volumetric concentration. Richardson–Zaki index, n, is an empirical parameter that depends on single-particle Reynolds number and wall effect expressed as particle to column diameter ratio. However, simple correlation given by eq. (6.27) is not able to take into account all aspects of real sedimentation process of polydisperse particles of different shapes and with strong particle–particle and particle–fluid interactions. Therefore, design of sedimentation basins is mostly based on experimental procedures that follow calculation of the sedimentation rate from recorded sedimentation curve using Kynch method. Hindered settling can be found in a final clarifier for the activated sludge process [3].

Example 6.1. Spherical particle with a density of 1,370 kg m^{-3} is settling under gravity only in a fluid of density 750 kg m^{-3} and viscosity 1 mPa s. Estimate:
a.   terminal velocity of the particle having diameter 0.1 mm,
b.   hindered velocity in a suspension built up of the same particles with 400 kg of solid per m^3 of suspension. Richardson–Zaki index is 4.65.

Solution
a.   **Terminal velocity of the particle** Assuming Stokes' law, terminal velocity can be calculated from eq. (6.27)

$$v = \frac{d_p^2 \cdot (\rho_p - \rho) \cdot g}{18 \cdot \mu} = \frac{(0.1 \times 10^{-3})^2 \cdot (1370 - 750) \cdot 9.81}{18 \cdot 1 \times 10^{-3}} = 3.4 \times 10^{-3} \text{ m s}^{-1}$$

Since the assumption of Stokes' law was made, the validity should be properly checked by calculating the Reynolds number from eq. (6.21).

$$Re = \frac{v \cdot d_p \cdot \rho}{\mu} = \frac{3.4 \times 10^{-3} \cdot 0.1 \times 10^{-3} \cdot 750}{1 \times 10^{-3}} = 0.26$$

Calculated value of Reynolds number is 0.26 which falls in the Stokes' law region (Re < 0.3).

b.   **Hindered velocity in suspension** The first step in calculating settling velocity of particles in suspension, from eq. (6.28), is to find the volumetric concentration of suspension, c. Volumetric concentration of suspension is volumetric fraction of solid and can be calculated from the following equation:

$$c = \frac{V_{solid}}{V_{suspension}} = \frac{\frac{m_{solid}}{\rho_{solid}}}{V_{suspension}} = \frac{\frac{400}{1370}}{1} = 0.29$$

With Richardson–Zaki index 4.65 and calculated terminal velocity of one particle, hindered settling velocity can be calculated from eq. (6.28) as follows:

$$v_h = v_t \cdot (1 - c)^n = 3.4 \times 10^{-3} \cdot (1 - 0.29)^{4.65} = 6.9 \times 10^{-4} \text{ m s}^{-1}$$

Obtained results confirmed that settling velocity in a suspension is almost five times smaller than the settling velocity of single particle. The higher the concentration of suspension is, the greater the decrease of settling velocity.

Example 6.2. A particle having sphericity 0.22 and density 2,500 kg m^{-3} falls under gravity in a fluid of density 1,000 kg m^{-3} and viscosity 10^{-5} Pa s. Calculate the minimal size and terminal velocity of the particle that would achieve turbulent settling.

Solution
For turbulent settling, Reynolds number should be greater than 500 (Re > 500). From Figure 6.1, for particle having sphericity 0.22 and for Re > 500, drag factor becomes constant and has value 9.

If one takes that minimal value of Reynolds number in a turbulent regime is 500, from eq. (6.21) follows:

$$Re = \frac{v \cdot d_p \cdot \rho}{\mu} = 500 \text{ Rearranging the equation:}$$

$$\frac{v \cdot d_p \cdot 1000}{10^{-5}} = 500 \text{ or } d_p = 5 \times 10^{-6} \frac{1}{v}$$

Terminal velocity is expressed using Newton's law for $500 < Re < 2 \times 10^5$. From eq. (6.26) using obtained relationship for particle size and velocity, terminal velocity of the particle can be calculated:

$$v = \sqrt{\frac{4}{3} \cdot \frac{(\rho_p - \rho) \cdot g \cdot d_p}{C_D \cdot \rho}} = \sqrt{\frac{4}{3} \cdot \frac{(\rho_p - \rho) \cdot g \cdot 5 \times 10^{-6}}{C_D \cdot \rho \cdot v}}$$

$$v = \sqrt[3]{\frac{4}{3} \cdot \frac{(\rho_p - \rho) \cdot g \cdot 5 \times 10^{-6}}{C_D \cdot \rho}} = \sqrt[3]{\frac{4}{3} \cdot \frac{(2500 - 1000) \cdot 9 \cdot 81 \cdot 5 \times 10^{-6}}{9 \cdot 1000}}$$

$$v = 2.22 \times 10^{-2} \text{ m s}^{-1}$$

Now, particle size can be calculated:

$$d_p = 5 \times 10^{-6} \frac{1}{v} = 5 \times 10^{-6} \frac{1}{2.22 \times 10^{-2}}$$

$$d_p = 2.25 \times 10^{-4} \text{ m} = 0.225 \text{ mm}$$

Minimal particle size that would settle in a turbulent regime, with $Re = 500$, is 0.225 mm. Larger particles, having the same density, will certainly fall into the region where Newton's law is valid ($500 < Re < 2 \times 10^5$).

6.3 Flow through porous media

The flow of fluid through porous media fundamentally occurs in many technologies in almost every domain of human productiveness. This flow phenomenon is omnipresent in petroleum engineering, in food and pharmaceutical sectors, as well as in environmental engineering. The focus in here is on the flow phenomenon that governs all filtration processes, and that is the core for each separation process carried out in a packed column and for many environmental catalysis problems that acquire fluidization as an operation for small- and large-scale waste treatment. Undoubtedly, the outcome of aforementioned processes in a real environment, effective the process will be or not, is highly dependent on the engineer's understanding of underlying flow phenomenon. Scoping these processes has to start with its inherent phenomenon, with the flow of fluid through porous media.

Two types of porous media can occur and both of them are of enormous relevance in environmental technology. These are the fixed and the fluidized bed. In the first, constituents are fixed and are not moved by fluid (as in filtration and in packed columns), while in the second the flow of fluid will lift up the solid bed of particles and phenomenologically convert it to the fluidized state. There are so many advantages for such occurrence of solid state and this is why fluidization nowadays is enormously implemented in environmental technologies.

Filtration is the process of separating dispersed particles from a suspension using porous media that will retain solids only and pass a liquid phase purely, a filtrate. Phenomenologically, filtration as a size exclusion separation process, highly employed in water treatments, is basically the flow of liquid through a fixed porous structure of a filter cake and a filter medium as well. Similarly, fluidization as the process of converting solids in a fluidized state is the flow of gas through a mobile bed of particles. If one considers additionally the fact that many separation processes are nowadays efficiently provided and carried out in packed columns, even in the simultaneous presence with a chemical reaction, everyone can see that there is a strong connection between this flow phenomenon with many technological systems in environmental engineering.

In following, a mathematical description of underlying phenomenon will be provided in a general form for a fixed-bed type using the perspective of the flow that is driven by some driving force, but also with omnipresent resistances in such hydrodynamic systems. In addition, this descriptive scope of the flow phenomenon will be concisely applied to macrofiltration process. Also, it will be highlighted that this flow equation is used for predicting the minimum fluidization velocity.

First, let us profoundly view the flow of fluid through a fixed-bed type of porous media. This bed may be considered as a group of particles of various sizes and shapes. These bed particles will create many interparticle voids (pores) of irregular shapes with various lengths and diameters. In such a bed, there are many non-cylindrical interparticle voids that can be considered as highly tortuous channels. Therefore, this porous medium can be considered as a truly complex hydrodynamic system, especially for its geometrical description. For its simplicity, it is assumed that a fixed bed is built up of spherical particles only.

Like any other flow phenomena, the flow through bed of particles is also driven with some driving force and opposite with resistances. This fluid flow will occur only if some pressure difference will be acquired in such a porous structure. During fluid flow through bed of particles, an irreversible energy loss arises too due to the presence of certain resistances in such hydrodynamic system. These are viscous friction (due to the viscosity of a fluid) and the resistance of particles packed in a bed and expressed by package density and geometry of porous medium. Irreversible energy loss upon this flow phenomenon is frequently expressed as a dissipated energy per unit of mass. In the following derivation, a common measure of pressure drop will be used for indicating the dissipated energy in a bed.

Now, let us consider what will be the dissipated energy in terms of pressure drop as a function of the superficial velocity, v_A. This velocity is closely related to the cross-section of a bed at its boundary. In fact, it is the velocity by which fluid approaches to its bound surface. The actual velocity of fluid in the pores of a bed, v_p, significantly differs from the superficial velocity. It is of course higher than v_A according to the equation of continuity, eq. (6.29):

$$v_A \cdot A_{bed} = v_p \cdot A_p \tag{6.29}$$

The idea is to mathematically describe the energy that will be dissipated around one spherical particle as a function of the superficial velocity, v_A. Using the perspective that each flow of fluid through a fixed bed of particles can be fundamentally considered as an accompanied flow phenomenon around obstacles, the force that the flow exerts on one spherical particle can be additionally determined:

$$F_D = C_D \cdot \frac{d_p^2 \pi}{4} \cdot \frac{v^2 \rho}{2} \tag{6.30}$$

where C_D is a drag factor, obviously different from one that is found for one sphere in an infinite volume of fluid (see Figure 6.2.).

Furthermore, the total energy dissipated in a porous medium will be obtained if the energy dissipated around one particle only is multiplied with the total number of particles in the bed. Total number of packed particles can be easily calculated as the ratio of the volume of all particles in a bed and the volume of a single spherical particle:

$$n_p = \frac{A_{bed} \cdot L \cdot (1 - \varepsilon_{bed})}{\frac{d_p^3 \pi}{6}} \tag{6.31}$$

where ε_{bed} is the bed porosity, in fact a volume content of pores (interparticle space) in a porous bed and L is the height of the bed. Consequently, the total force exerted by the flow on the whole bed:

$$n_p \cdot F_D = \frac{A_{bed} \cdot L \cdot (1 - \varepsilon_{bed})}{\frac{d_p^3 \pi}{6}} \cdot C_D \cdot \frac{d_p^2 \pi}{4} \frac{v_p^2 \rho}{2} \tag{6.32}$$

is balanced with the pressure force according to the law of conservation of momentum in an observed hydrodynamic system:

$$\Delta p \cdot A_p = \Delta p \cdot A_{bed} \cdot \varepsilon_{bed} \tag{6.33}$$

Finally, this force balance easily results in the estimation of dissipated energy in terms of pressure drop for the flow of fluid through a fixed bed of spherical particles:

$$\Delta p = \frac{3}{2} \cdot C_D \cdot \frac{1 - \varepsilon_{bed}}{\varepsilon_{bed}^3} \frac{L}{d_p} \frac{v_A^2 \rho}{2} \tag{6.34}$$

This expression provides an estimation of the energy dissipation for any hydrodynamic regime in the system (where the drag factor C_D is a function of Reynolds number) and therewith is valid for all kinds of pores in the bed, both wide and narrow. Regarding this, it should be noted that geometry of such porous structures strongly affects the hydrodynamic regime in the system.

As the pores in the medium are small, sometimes even capillary, there will be a high resistance to the flow of fluid in such hydrodynamic system. Consequently, the flow rate of fluid will be low irrespective of applied driving force and laminar flow conditions will occur. Contrary, in the presence of wide pores, laminar and turbulent flow conditions may occur depending on the pressure difference applied.

The Reynolds number for this characteristic flow is defined as follows:

$$Re = \frac{v_p \cdot d_{eq} \cdot \rho}{\mu} \tag{6.35}$$

Velocity of fluid in the pores of a bed, v_p, can be expressed using the equation of continuity, eq. (6.29) as:

$$v_p = \frac{v_A}{\varepsilon_{bed}} \tag{6.36}$$

In eq. (6.35) stands an equivalent diameter of pores, d_{eq}, as a relevant one because the pores are of irregular, non-circular cross-sections. In this, hydraulic radius can be used. It is defined as the ratio of cross-sectional area of the pore to its circumference:

$$d_{eq} = 4\frac{A_{pore}}{S_{pore}} = 4\frac{\varepsilon_{bed}}{S_V} \tag{6.37}$$

Regarding aforementioned and the fact that specific surface area of pores is approximately equivalent to the sum of external surfaces of all particles building a porous medium per unit of bed volume:

$$S_V = \frac{n_p \cdot S_{particle}}{V_{bed}} = \frac{6 \cdot (1 - \varepsilon_{bed})}{d_p} \tag{6.38}$$

a relevant Reynolds number for the flow through bed of particles can be derived easily:

$$Re = \frac{2}{3}\frac{v_A \cdot d_p \cdot \rho}{(1 - \varepsilon_{bed})\mu} \tag{6.39}$$

For sufficiently high velocities in the bed (for high Re number values), turbulent flow will occur and the drag factor is constant. Ergun found that for Re>700, the drag factor is equal to 2.3 [3]. On the other hand, it was discovered that for Re<1, the flow is purely laminar and the drag factor can be estimated with [4]:

$$C_D = \frac{152}{Re} \tag{6.40}$$

Adding both contributions (for laminar and turbulent flow) in the dissipated energy description eq. (6.34), Ergun equation is obtained for the entire Re number range:

$$\frac{\Delta p}{L} = 170 \frac{(1-\varepsilon_{bed})^2}{\varepsilon_{bed}^3} \frac{v_A \mu}{d_p^2} + 1.75 \frac{(1-\varepsilon_{bed})}{\varepsilon_{bed}^3} \frac{v_A^2 \rho}{d_p} \qquad (6.41)$$

Ergun eq. (6.41) is a fundamental description of the flow resistances for both hydrodynamic regimes in terms of irreversible energy loss and therewith in a pressure drop that will invariably arise. It should also be emphasized that this quantifying of pressure drop is valid for a single-phase fluid flowing only.

This consideration can be extrapolated for a fluidized-bed type of porous media and in such form is efficiently applied in procedures of reaching the right process properties for operating fluidization in both regimes, particulate and aggregative fluidization. The velocity at which the fixed bed converts into the fluidized bed can be easily described using the Ergun equation with the help of the balance of forces that are present in a fluid bed process control volume, the pressure force and the apparent weight of the particles.

The first part in Ergun equation is for the laminar flow only and is called Carman–Kozeny equation. This flow equation can efficiently describe an idealized cake filtration process that is running at laminar flow conditions only. In fact, those hydrodynamic conditions are almost invariably obtained in filtration process due to the presence of small pores, sometimes even capillary, and consequently due to the low flow rate of a filtrate irrespective of the pressure difference applied. An idealized description of filtration process strongly connotes that the Carman–Kozeny equation works well for:

a. incompressible cakes,
b. constant driving force, i. e. filtration pressure,
c. a bed built-up of spherical particles only,
d. no resistances to the flow from filter medium,
e. narrow porosity ranges.

A major critic to the Carman–Kozeny equation is that velocity incorrectly tends to infinity when the bed porosity approaches 1 [7].

Carman–Kozeny equation represents a purely theoretical approach to the flow through porous media as it strongly relies on a theory of momentum transfer during the flow phenomenon. Completely different approach for mathematical description of the flow through porous media is given by Darcy's law. It was experimentally derived in 1856 on the basis of monitoring a pressure drop over sand filters in a water filtration process. Also, it is the fundamental relation between the pressure drop and the flow rate of a liquid passing through a packed bed of solids.

A modern filtration theory prefers the Ruth form of Darcy's law [7,8]:

$$\frac{dV}{dt} = \frac{A \cdot \Delta p}{\mu \cdot R_{bed}} \tag{6.42}$$

where R_{bed} is known as the bed resistance and A is filtration area. In cake filtration process, the bed resistance consists of the filter medium resistance, R_M, and the resistance of the deposited solids from a suspension on a filter medium, i. e. a filter cake resistance, R_C. Developing this mathematical description, the basic equation for cake filtration analysis can be obtained:

$$\frac{dV}{dt} = \frac{A \cdot \Delta p}{\mu \cdot \left(\frac{\alpha \cdot G \cdot V}{A} + R_M\right)} \tag{6.43}$$

where G is solid mass per unit volume of filtrate and α is specific filter cake resistance.

This equation enables us to quantify resistances, R_M and R_C, carrying out filtration test experiments on laboratory scale filters.

Example 6.3. Estimate the pressure drop for a packed column under the following conditions:
Gas velocity, $v_G = 1\times10^{-2}\,\mathrm{m\ s^{-1}}$
Gas density, $\rho_G = 2\,\mathrm{kg\ m^{-3}}$
Kinematic viscosity of a gas, $v_G = 1\times10^{-4}\,\mathrm{m^2\ s^{-1}}$
Packing; Berl saddles, 24 mm ($S_V = 406\,\mathrm{m^2\ m^{-3}}$, $\varepsilon = 0.32$).

Solution
The flow of gas through a porous medium (a packed column)
I Equivalent diameter of the packing from eq. (6.38)

$$d_p = \frac{6 \cdot (1-\varepsilon_{bed})}{S_V} = \frac{6 \cdot (1-0.32)}{406} = 1.005 \times 10^{-2}\,\mathrm{m}$$

II Reynolds number for the flow using eq. (6.39)

$$Re = \frac{2 \cdot v_G \cdot d_p}{3 \cdot (1-\varepsilon_{bed}) \cdot v_G} = \frac{2 \cdot 1\times 10^{-2} \cdot 1.005 \times 10^{-2}}{3 \cdot (1-0.32) \cdot 10^{-4}} = 0.985$$

For Re numbers below 1, a laminar flow is observed.
  III Drag factor for a system
  For Re<1 eq. (6.40) can be arguably applied

$$C_D = \frac{152}{Re} = \frac{152}{0.985} = 154.31$$

IV Pressure drop for a column according to eq. (6.34)

$$\Delta p = \frac{3}{4} C_D \frac{1 - \varepsilon_{bed}}{\varepsilon_{bed}^3} \frac{L}{d_p} v_G^2 \cdot \rho_G$$

$$= \frac{3}{4} \cdot 154.31 \cdot \frac{1 - 0.32}{0.32^3} \cdot \frac{1}{1.005 \times 10^{-2}} \cdot \left(1 \times 10^{-2}\right)^2 \cdot 2 = 47.79 \, \text{Pa}$$

per meter of packed height

Example 6.4. Suspension built-up of spherical particles, all in the size range of 125–180 μm, and a water (of density 1×10^3 kg m^{-3} and viscosity of 1×10^{-6} m^2 s^{-1}) has to be treated by vacuum filtration. Filter area on a rotary drum is 3 m^2. Vacuum filtration is strictly carried out under constant pressure of 0.2 bars. Before filtration, a specific surface area of solids was estimated at 12,344 m^2 m^{-3} on the basis of measured size distributions. If there is an assumption that deposited filter cake of an averaged height of 13 cm is almost incompressible and that filter medium will not provide any resistance to the flow of filtrate, estimate the volume of a suspension that can be treated per one hour under this drum conditions.

Solution
The flow of liquid through a porous medium (a filter cake)
Carman–Kozeny equation works well for:
a.   incompressible cakes,
b.   constant filtration pressure,
c.   a bed built-up of spherical particles only,
d.   no resistances to the flow from filter medium.

All aforementioned assumptions are fulfilled. Regarding this and the fact that filtration process is practically always carried out under laminar flow conditions, the Carman–Kozeny equation can be arguably used for estimation of a suspension volume.
I Mean diameter of solid particles in a suspension
All particles are in the size range of 125–180 μm. So, for sufficiently narrow size range, it can be assumed that particles are of mean diameter of 152.5×10^{-6} m.
II Filter cake porosity
Filter cake will contain spherical particles only because there are no others in treated suspension. So, porosity of a filter cake can be estimated efficiently using eq. (6.38).

$$S_V = \frac{6 \cdot (1 - \varepsilon_{bed})}{d_p}$$

Easily, bed porosity can be calculated using the:

$$(1 - \varepsilon_{bed}) = \frac{S_V \cdot d_p}{6} = \frac{12344 \cdot 152.5 \times 10^{-6}}{6} = 0.3137$$

Estimated porosity of a filter cake is 0.6863.
III Estimate the velocity of a filtrate using Carman–Kozeny eq. (6.41)

$$\frac{\Delta p}{L} = 170 \frac{(1 - \varepsilon_{bed})^2}{\varepsilon_{bed}^3} \frac{v_F \cdot \mu}{d_p^2}$$

$$\frac{0.2 \times 10^5}{13 \times 10^{-2}} = 170 \frac{(1 - 0.6863)^2}{0.6863^3} \frac{v_F \cdot 1 \times 10^{-6} \cdot 1 \times 10^3}{\left(152.5 \times 10^{-6}\right)^2}$$

$$v_F = 0.0691 \text{m s}^{-1}$$

IV Volume flow rate of a filtrate

$$\dot{V} = v_F \cdot A = 0.0691 \cdot 6.3 = 0.4355 \text{m}^3 \text{ s}^{-1}$$

There is a possibility to treat 0.4355 m^3 of a suspension per unit of time under these conditions.
  V Estimated volume of a suspension

$$V = \dot{V} \cdot t = 0.4355 \cdot 3600 = 1,567.95 \text{m}^3$$

Under given conditions, it is possible to treat 1,567.95 m^3 of a suspension per one hour in a rotary drum filter.

6.4 Heat transfer

Almost all processes important in chemical, petrochemical, pharmaceutical and food industries as well as environmental engineering include heat transfer, when fluids must be either heated or cooled in a wide range of plant. The control of the heat flow at desired rate forms one of the most important areas of chemical engineering.

The heat transfer occurs from a medium at higher temperature to the lower temperature, and will stop when both mediums reach the equal temperature.

Three different modes of heat transfer exist:
a. Conduction – occurs at molecular level when a temperature gradient exists in a medium
b. Convection – in fluid in motion (forced or natural)
c. Radiation – when heat energy is transferred by electromagnetic waves between surfaces at different temperatures.

6.4.1 Heat conductivity

Heat is transferred from the region of higher temperature to the region of lower temperature until the equilibrium is reached. The molecules at a higher temperature are moving faster, and transfer its energy to the molecules that are at lower temperatures and move slowly. Resistances to heat transfer from one place to another depend on the materials through which heat is transferred. Different materials store and conduct heat differently. Some thermo-physical properties of materials can be defined: specific heat c_p as a measure of a material's ability to store thermal energy, likewise, the thermal conductivity, k, is a measure of a material's ability to conduct heat. Another material property is the thermal diffusivity, α, which represents how fast heat diffuses through a material and is defined as:

$$\alpha = \frac{k}{\rho \cdot c_p} \tag{6.44}$$

A material of a high thermal conductivity or a low heat capacity will obviously have a large thermal diffusivity. The larger the thermal diffusivity, the faster is the propagation of heat into the medium. A small value of thermal diffusivity means that heat is mostly absorbed by the material and a small amount of heat is conducted further [9].

6.4.2 Heat transfer by conduction

Heat transfer by conduction is a molecular mechanism between particles which are in direct contact. Conduction occurs in solid bodies and in stationary fluids when two points have different temperatures. In gases and liquids, conduction results from the collisions and diffusion of the molecules during their random motion. In solids, it is due to the combination of vibrations of the molecules in a lattice and the energy transport by free electrons. The rate of heat conduction through a medium depends on physical properties, geometric characteristics and local differences in temperature.

Fourier established that the rate of heat conduction through a plane layer is proportional to the temperature difference across the layer and the heat transfer area, but is inversely proportional to the thickness of the layer.

I. Fourier's law of heat conduction is valid for a steady state:

$$\dot{Q} = -k \cdot A \cdot \frac{\partial T}{\partial x} \tag{6.45}$$

where $\partial T/\partial x$ is a temperature gradient in the direction of heat flow. The minus sign is inserted so that the second principle of thermodynamics will be satisfied; heat is conducted in the direction of decreasing temperature.

For steady-state conditions, the temperature does not change with time and problem can be easily solved by integrating eq. (6.45). If the temperature of solid is changing with time, or if there are heat sources or sinks, calculation becomes more complex [10].

For the element of thickness dx (Figure 6.3), the following energy balance may be written:

Change in internal energy = energy conducted in − energy conducted out
± generated heat within element

$$\rho \cdot c_p \cdot A \cdot \frac{\partial T}{\partial t} \cdot dx = -k \cdot A \cdot \frac{\partial T}{\partial x} - A \cdot \left[-k \cdot \frac{\partial T}{\partial x} + \frac{\partial}{\partial x} \left(-k \cdot \frac{\partial T}{\partial x} \right) dx \right] \pm \dot{Q} \tag{6.46}$$

$$\rho \cdot c_p \cdot \partial T \partial t \cdot dx = \partial \partial x (k \cdot \partial T \partial x) \pm \dot{Q}$$

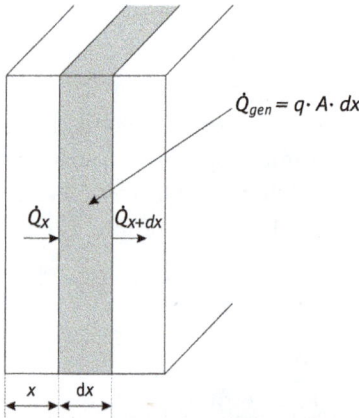

Figure 6.3: Elemental volume for one-dimensional heat conduction analysis.

For heat flow in all direction, all three coordinate directions have to be considered:

$$\rho \cdot c_p \cdot \frac{\partial T}{\partial t} = \frac{\partial}{\partial x}\left(k \cdot \frac{\partial T}{\partial x}\right) + \frac{\partial}{\partial y}\left(k \cdot \frac{\partial T}{\partial y}\right) + \frac{\partial}{\partial z}\left(k \cdot \frac{\partial T}{\partial z}\right) \pm \frac{\dot{Q}}{V} \tag{6.47}$$

For constant thermal conductivity and without generation inside system:

$$\frac{\partial T}{\partial t} = \alpha \cdot \left[\frac{\partial^2 T}{\partial x^2} + \frac{\partial^2 T}{\partial y^2} + \frac{\partial^2 T}{\partial z^2}\right] \tag{6.48}$$

Equation (6.48) is called II. Fourier's law.

6.4.3 Heat transfer by convection

Convection is the mode of energy transfer between a solid surface and the adjacent fluid that is in motion, and it involves the combined effects of conduction and fluid motion. If the fluid velocity is increased, the rate of heat transfer by convection will be increased as well.

Convection is classified as natural and forced convection, depending on the way how the fluid motion is initiated. In natural convection, any fluid motion is caused by density differences associated with temperature changes generated by heating or cooling. Fluid flow is induced by buoyancy forces. The heat transfer itself generates the flow which conveys energy away from the point at which the transfer occurs. In forced convection, the fluid is forced to flow over a surface or in a pipe by external means such as a pump or a fan.

The heat transfer across the fluid/solid interface is based on the Newton's law of cooling:

$$\dot{Q} = U \cdot A \cdot \Delta T \tag{6.49}$$

where U is the overall heat transfer coefficient, A is the surface area through which convection heat transfer takes place and ΔT is a driving force for heat transfer, usually difference between temperatures of solid body and temperature in the fluid bulk or between two fluids.

6.4.3.1 Overall heat transfer coefficient

When heat flows from one fluid to another through the wall with convection in both fluids, (Figure 6.4), overall heat transfer coefficient can be calculated as reciprocal value of the sum of all thermal resistances:

$$U = \frac{1}{\left(\frac{1}{h_{c1}} + \frac{l}{k} + \frac{1}{h_{c2}}\right)} = \frac{1}{\sum\limits_{i=1}^{n} R} \tag{6.50}$$

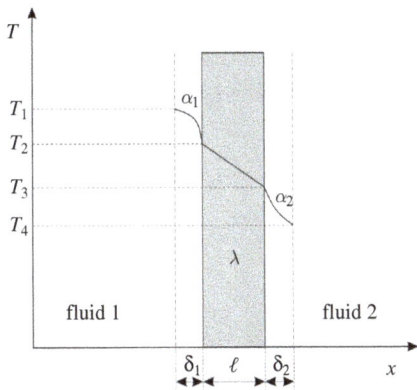

Figure 6.4: Heat flow from one fluid to another through the wall with convection in both fluids.

Note that thermal resistances in fluids are calculated from convective heat transfer coefficient, k.

The convective heat transfer coefficient is not a property of the fluid. It is an experimentally determined parameter whose value depends on all the variables influencing convection such as the surface geometry, the nature of fluid in motion, the properties of the fluid and the bulk fluid velocity.

$$h_c = f\left(v, \delta, l, k, \mu, c_p, \rho\right) \tag{6.51}$$

Due to the influence on several parameters, convective heat transfer coefficient can be calculated using correlation equations which are obtained by experiments, numerical methods (dimensional analysis) or from analytical solutions.

For forced convection, the heat transfer correlation can be expressed as:

$$Nu = k \cdot Re^m \cdot Pr^n \tag{6.52}$$

or for natural convection in which fluid motion is caused by buoyancy effects, the Reynolds number is replaced with a Grashof number:

$$Nu = k \cdot Gr^m \cdot Pr^n \tag{6.53}$$

6.4.4 Thermal radiation

Heat transfer by thermal radiation is a separate mechanism from convection and conduction. Thermal radiation is an electromagnetic phenomenon which occurs as a result of a body's absolute temperature. All objects give out and take in thermal radiation, which is also called infrared radiation. It is propagated between two surfaces at different temperature by electromagnetic waves through space which allow travelling of waves. An electromagnetic wave is transferred through the vacuum at the speed of a light. Thermal radiation refers to the range of wave length between 0.1 and 100 μm.

When the body is exposed to radiation, the overall radiation energy will be divided into three parts. Part of that energy is reflected, part is absorbed and part is transmitted:

$$\dot{Q} = \dot{Q} \cdot \rho + \dot{Q} \cdot \alpha + \dot{Q} \cdot \tau \tag{6.54}$$

$$\rho + \alpha + \tau = 1 \tag{6.55}$$

where

$\dot{Q} \cdot \alpha$ – is absorbed part of energy

$\dot{Q} \cdot \rho$ – is reflected part of energy

$\dot{Q} \cdot \tau$ – is transmitted part of energy

Intensity of radiation and relation between absorbed, reflected and transmitted energy depend on the nature of the surface, the temperature of the surface and wavelengths/frequency of radiation.

Blackbody which is an ideal thermal radiator will emit energy at a rate proportional to the forth power of the absolute temperature of the body and to its surface area (A) according to Stephan–Boltzmann's law of thermal radiation:

$$\dot{Q} = \sigma \cdot A \cdot T^4 \tag{6.56}$$

Equation (6.56) can be applied only to black bodies. The radiative heat energy emitted by a real surface is lower than emitted by a black body and can be calculated by introducing radiative property of the surface, called emissivity, ε:

$$\dot{Q}_{grey\ body} = \sigma \cdot \varepsilon \cdot A \cdot (T_1 - T_2)^4 \tag{6.57}$$

In general, net radiative energy interchange between two real non-black bodies at different temperatures depends on many factors: their surface properties, geometry and orientation with each other [11].

Radiation exchange between two parallel planes

Two or more bodies exchange heat by radiation depending on their temperature (Figure 6.5). Although all bodies radiate, heat flow always goes from warmer to a colder body. Heat transfer depends on the position of bodies. Radiation heat exchange between two parallel planes can be expressed by equation:

$$\dot{Q}_{12} = \dot{Q}_1 - \dot{Q}_2 = \varepsilon_1 \cdot \sigma \cdot A_1 \cdot T_1{}^4 - \varepsilon_2 \cdot \sigma \cdot A_2 \cdot T_1{}^4 \tag{6.58}$$

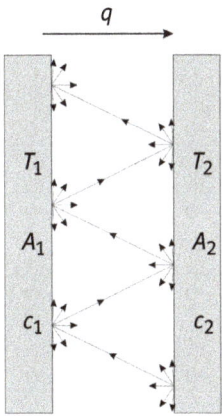

Figure 6.5: Heat transfer by radiation between two parallel planes.

For the same surface area eq. (6.59) is obtained:

$$\dot{Q}_{12} = \varepsilon_{12} \cdot \sigma \cdot A \cdot (T_1 - T_1)^4 \tag{6.59}$$

where ε_{12} is reduced emissivity and can be calculated from equation:

$$\varepsilon_{12} = \frac{1}{1\varepsilon_1 + 1\varepsilon_2 - 1} \tag{6.60}$$

Radiation in an enclosure

If heat transfers surface at temperature T_1 that is completely enclosed by much larger surface at T_2, then the net radiant exchange can be calculated using the following equation:

$$\dot{Q}_{12} = \varepsilon_1 \cdot \sigma \cdot A_1 \cdot (T_1 - T_2)^4 \qquad (6.61)$$

It can be noticed that surface of enclosed body is relevant for exchanged heat.

Example 6.5. Calculate the heat loss due to the forced convection and radiation from a horizontal hot cylinder (diameter: 0.015 m; height: 0.07 m) to the surrounding air. The velocity of air is 5 m s^{-1}. The cylinder is made of stainless steel. Temperatures of the cylinder and the surrounding air are 65 and 25 °C. (Properties of air at the average temperature: $v = 1.82512 \times 10^{-5}$ m^2 s^{-1}; $k = 0.02760$ W m^{-1} K^{-1}; Pr = 0.704.)

Solution

Total heat loss: $Q_{tot} = Q_c + Q_r$

Heat loss due to the forced convection: $Q_c = h_c \cdot A \cdot (T_s - T_a)$

Heat loss due to the radiation: $Q_r = h_r \cdot A \cdot (T_s - T_a) = \varepsilon \cdot \sigma \cdot A \cdot (T_s^4 - T_a^4)$

For the calculation of convective heat transfer coefficient, the following correlation equation can be used:

$$\text{Nu} = 0.3 + \frac{0.62 \cdot \text{Re}^{0.5} \cdot \text{Pr}^{0.33}}{\left(1 + \left(\frac{0.4}{\text{Pr}}\right)^{0.66}\right)^{0.25}} \cdot \left(1 + \left(\frac{\text{Re}}{28,2000}\right)^{0.5}\right) \text{ for Re > 400,000}$$

$$\text{Nu} = 0.3 + \frac{0.62 \cdot \text{Re}^{0.5} \cdot \text{Pr}^{0.33}}{\left(1 + \left(\frac{0.4}{\text{Pr}}\right)^{0.66}\right)^{0.25}} \text{ for Re < 10,000}$$

Properties of air are taken for the average temperature:

$$T_{film} = \frac{T_s + T_a}{2} = \frac{65 + 25}{2} = 45 \text{ °C}$$

Reynolds number: $\text{Re} = \frac{v \cdot D}{v} = \frac{5 \cdot 0.015}{1.82512 \cdot 10^{-5}} = 4,109.32$

Nusselt number: $\text{Nu} = 0.3 + \frac{0.62 \cdot 4109.32^{0.5} \cdot 0.704^{0.33}}{\left(1 + \left(\frac{0.4}{0.704}\right)^{0.66}\right)^{0.25}} = 31.35$

Convective heat transfer coefficient is now: $h_c = \frac{\text{Nu} \cdot k}{D} = \frac{31.35 \cdot 0.02760}{0.015} = 57.68$ W m^{-2} K^{-1}

Area of the cylinder: $A = \frac{D \cdot \pi}{L} = \frac{0.015 \cdot \pi}{0.07} = 0.673$ m^2

Heat loss due to the forced convection: $Q_c = 57.68 \cdot 0.673 \cdot (65 - 25) = 1,552.75$ W

Emissivity of the stainless steel is 0.59.

Heat loss due to the radiation: $Q_r = 0.59 \cdot 5.67 \cdot 10^{-8} \cdot 0.673 \cdot (338^4 - 298^4) = 116.29$ W

Total heat loss: $Q_{tot} = 1552.75 + 116.29 = 1,669.04$ W

6.5 Mass transfer

When a system contains two or more components whose concentrations vary from point to point, there is a natural tendency for mass to be transferred, minimizing the concentration differences within a system. The transport of one constituent from a region of higher concentration to that of a lower concentration is called mass transfer.

Mass transfer within a fluid mixture or across a phase boundary is a process that plays a major role in many industrial processes. Mass transfer occurs during distillation, absorption, drying, extraction, etc.

The primary driving force for mass transfer is the concentration gradient, although mass transfer can take place because of other driving forces: pressure diffusion due to the existence of pressure gradient; forced diffusion is a result of an external force field such as an electric or magnetic field and thermal diffusion occurs due to the existence of temperature gradient.

6.5.1 Molecular diffusion

When mass is being transferred from one distinct phase to another or through a single phase, the basic mechanism is the same whether the phase is a gas, liquid or solid. The phenomenon of diffusion is a result of the motion of molecules in a fluid.

Fick's law deals with mass transfer within a medium due to the difference in concentration between various parts of it. This is very similar to Fourier's law of heat conduction. Fick's law defines molecular diffusion process of component A in an isothermal, isobaric system. For diffusion in only one direction, the Fick's rate equations are:

$$\text{Mole basis}: J_A = -D_{AB}\frac{dc_A}{dy} \; [\text{mol s}^{-1}\text{m}^{-2}] \tag{6.62}$$

$$\text{Mass bases}: J_A = -D_{AB}\frac{d\rho_A}{dy} \; [\text{kg s}^{-1}\text{m}^{-2}] \tag{6.63}$$

The concentration of species can be defined in several ways such as density, mass fraction, molar concentration and mole fraction, and thus Fick's law can be expressed mathematically in many ways [9].

6.5.2 Coefficient of mass diffusivity

The constant of proportionality in Fick's law is defined as another transport property (similar to the thermal conductivity in Fourier's law) called the coefficient of mass diffusivity, D_{AB}. The coefficient of mass diffusivity is analogue to other, already mentioned transport properties such as kinematic viscosity and thermal diffusivity and depends on the temperature, pressure and nature of the components of the system. The diffusion coefficients are highest in gases and lowest in solids. Due to the complex nature of mass diffusion, the diffusion coefficients are usually determined experimentally [9].

6.5.3 Steady-state molecular diffusion in binary systems

6.5.3.1 General case for diffusion and convection

Mass diffusion can occur in a stationary medium, when the only motion is of molecules due to the differences in concentration, without motion of the mixture as a whole. On the other hand, many practical problems involve diffusion in a moving fluid where the bulk motion is caused by an external force. Mass diffusion in such cases is far more complicated due to the fact that chemical species are transported both by diffusion and by the bulk motion of the medium (i. e. convection) [9].

The velocities and mass flow rates of species in a moving medium consist of two components: one due to molecular diffusion and one due to convection. Flux of mixture that consists of components A and B can be expressed with velocity, v of mixture:

$$J_A + J_B = v \cdot c \tag{6.64}$$

For component A:

$$(J_A + J_B) \cdot \frac{c_A}{c} = v \cdot c_A \tag{6.65}$$

Diffusivity of component A (or B) could be written in terms of relative velocity of component A:

$$J_A = c_A \cdot (v_A - v) \tag{6.66}$$

Diffusivities of components A and B in the mixture are:

$$J_A = c_A \cdot (v_A - v) = -D_{AB} \cdot \frac{dc_A}{dy} \tag{6.67}$$

$$J_B = c_B \cdot (v_B - v) = -D_{BA} \cdot \frac{dc_B}{dy} \tag{6.68}$$

By combining eqs. (6.65). and (**67**), expression for steady-state molecular diffusion in binary systems is obtained:

$$J_A = -D_{AB} \cdot \frac{dc_A}{dy} + (J_A + J_B) \cdot \frac{c_A}{c} \tag{6.69}$$

The first term on the right-hand side of this equation is diffusional molar flux of component A and the second term is flux due to the bulk motion. The term can be solved with the knowledge of the relationship between molar fluxes of components A and B.

There are two boundary cases:

1. Equimolar counter diffusion ($J_A + J_B = 0$),
2. Diffusion of component A through stagnant B ($J_B = 0$).

6.5.4 Equimolar counter diffusion

Equimolar counter diffusion takes place in the thermal separation processes where the mass transfer occurs through the boundary layer in both directions. A good example of such diffusion is the two-component distillation with similar latent heat of evaporation for both components when concentrations at any point in the gas mixture remain constant with time. Concentration gradient is linear. Diffusivity of A in B, called the coefficient of mass diffusivity or diffusion coefficient, D_{AB}, is equal to diffusivity of B in A but in opposite direction. The process is described by Fick's low for liquid:

$$J_A = -D_{AB} \cdot \frac{c_{A2} - c_{A1}}{y_2 - y_1} = -D_{AB} \cdot \frac{\Delta c}{\Delta y} \tag{6.70}$$

For gas, eq. (6.70) should be modified using the ideal gas law:

$$J_A = \frac{D_{AB}}{RT\Delta y} \cdot (p_{A1} - p_{A2}) \tag{6.71}$$

6.5.5 Diffusion of component A through stagnant component B

Many mass transfer operations involve the diffusion of one gas component through another non-diffusing component. Absorption and humidification are typical operations defined by these equations. To derive the case for diffusing of component A in a stagnant component B, $N_B = 0$ is substituted into the general eq. (6.71).

$$J_A = -D_{AB} \cdot \frac{dc_A}{dy} + (J_A + 0) \cdot \frac{c_A}{c} \tag{6.72}$$

Keeping the total pressure, p constant, substituting: $c = p/(R \cdot T)$, instead of concentrations, partial pressures may be used for calculation of the rate of mass transfer in gas:

$$J_A = -D_{AB} \cdot \frac{dp_A}{R \cdot T \cdot dy} + J_A \cdot \frac{p_A}{p} \tag{6.73}$$

Rearranging and integrating Stephan's diffusivity law is obtained:

$$J_A = -\frac{D_{AB} \cdot p}{RT_{GF}\delta_D \cdot \bar{p}_B} \cdot (p_{A1} - p_{A2}) \tag{6.74}$$

6.5.6 Transient diffusion

Fick's first law applies to steady-state systems, where concentration keeps constant with time, but in many cases of diffusion, the concentration will significantly change with time. In the case of mass transfer under non-steady conditions, at a given point, the concentration is changed over time due to the fluctuation in flow density.

If the situation illustrated in Figure 6.6 is considered, at two positions along the y axis at y_1 and y_2, a distance Δy apart, the concentration of the solute is c_1 and c_2, respectively, Fick's first law can be used to determine the amount of solute per unit time which enters this element at y_1 and which leaves it at y_2. Because matter must be conserved, the difference between these must be equal to the rate at which solute accumulates within the element:

$$\frac{\partial c}{\partial t} = \frac{\partial J}{\partial y}$$

(6.75)

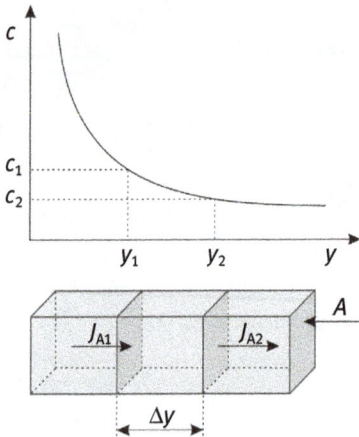

Figure 6.6: A schematic for concentration profile through an element.

Fick's law can be substituted into eq. 6.75, which gives:

$$\frac{\partial c}{\partial t} = -\frac{\partial}{\partial y}\left(-D\frac{\partial c}{\partial y}\right) = \frac{\partial}{\partial y}\left(D\frac{\partial c}{\partial y}\right)$$

(6.76)

If the diffusion coefficient does not depend on the concentration of diffusing species Fick's second law can be simplified to:

$$\frac{\partial c}{\partial t} = D\frac{\partial^2 c}{\partial y^2}$$

(6.77)

This equation is called Fick's second law, and it is an alternative form of first Fick's law which contains an explicit dependence on time [12].

6.5.7 Convective Mass Transfer

Convective mass transfer takes place between a moving fluid and a solid body or between two relatively immiscible moving fluids. The rate of convective mass transfer is directly proportional to the driving force which is the difference between the concentration at the phase boundary, C_{AS} (a solid surface or a fluid interface), and the concentration at some arbitrarily defined point in the fluid medium, C_A, and convective mass transfer coefficient, K:

$$J_A = K \cdot (c_{AS} - c_A) \tag{6.78}$$

Similar to convective heat transfer, mass transfer coefficient is a function of physical properties of fluid, hydrodynamic conditions and geometry of the system. The influence of the hydrodynamic conditions on the developed concentration boundary layer is the same as on the hydrodynamic or heat boundary layers, whenever is mass transferred between a surface and fluid in motion. It is obvious that analogies between mass, momentum and energy transfers exist. In convective mass transfer, the non-dimensional numbers corresponding to Prandtl and Nusselt numbers of convective heat transfer are Schmidt and Sherwood numbers.

6.5.8 Interphase mass transfer

In the separation processes like absorption, stripping, distillation or extraction, mass transfer occurs across the interface between two phases: liquid–liquid or gas–liquid. Among several theoretical approaches developed for description of the mass transfer between two phases (film theory – Nernst; penetration theory – Higbie; surface-renewal theory – Danckwerts; film-penetration theory – Toor and Marchello). Two-film theory (Whitman) has been widely accepted for the modelling of steady-state separation processes [13]. Matter is transferred by turbulent convection in the bulk of the phase and by diffusion through the concentration boundary layers formed at both sides of the phase boundary. Since the mass transfer is very fast in the bulk of both phases, the concentrations are uniform. The basic assumption is that at the interface, phase equilibrium is achieved. The main resistances occurring in the concentration boundary layers are shown in Figure 6.7 [14].

The rate of the steady-state mass transfer of component A from gas to liquid is given by the following equation:

$$J_A = \frac{D_{AB}}{\delta_G} \cdot (c_{Ab} - c_{AI})_G = \frac{D_{AB}}{\delta_L} \cdot (c_{AI} - c_{Ab})_L \tag{6.79}$$

Since the thickness of the boundary layer is unknown, instead of diffusion coefficient the rate of mass transfer can be written in terms of mass transfer coefficient.

Boundary layers

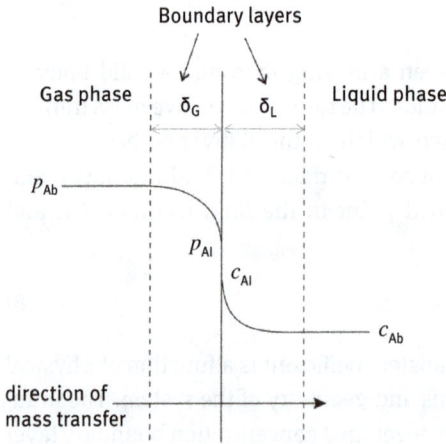

Figure 6.7: Concentration gradients.

The mass flux through the gas phase:

$$J_A = k_p \cdot (p_{Ab} - p_{AI}) \tag{6.80}$$

Similarly, the mass flux through the liquid phase is:

$$J_A = k_c \cdot (c_{AI} - c_{Ab}) \tag{6.81}$$

Due to the assumption of phase equilibrium:

$$p_{AI} = H_A \cdot c_{AI} \tag{6.82}$$

If it is assumed that the concentration of the liquid phase, $c_A^* = p_{Ab}/H_A$, is in equilibrium with the concentration in the bulk gas, and the concentration of the gas phase, $p_A^* = H_A \cdot c_{Ab}$, is in equilibrium with the concentration in the liquid phase, mass flux equations can be written in terms of the overall driving forces and overall mass transfer coefficients as follows:

$$J_A = K_G \cdot (p_{Ab} - p_A^*) = \frac{(p_{Ab} - p_A^*)}{(1/k_p) + (H_A/k_c)} \tag{6.83}$$

$$J_A = K_L \cdot (c_A^* - c_{Ab}) = \frac{(c_A^* - c_{Ab})}{(1/H_A \cdot k_p) + (1/k_c)} \tag{6.84}$$

If $k_p \gg k_c$, the rate of mass transfer is controlled by the liquid phase resistance and $K_L = k_c$. When $k_c \gg k_p$, the rate of mass transfer is controlled by the gas phase resistance and $K_G = k_p$.

In the real thermal separation processes, the interface between two phases constantly changes due to the phase flow. Nevertheless, two-film theory is widely used for evaluation and prediction of the interphase mass transfer.

The general forms of the empirical correlations for the calculation of mass transfer coefficient for the liquid and gas phases are given with following equations [15]:

$$Sh_L = C_1 \cdot Re_L^m \cdot Sc_L^n \cdot Ga_L^q \tag{6.85}$$

$$Sh_G = C_2 \cdot Re_G^r \cdot Sc_G^s \tag{6.86}$$

where m, n, q, r and s are empirical exponents.

Sherwood number, Sh, is the ratio of the convective mass transfer to the rate of diffusive mass transport:

$$Sh = \frac{k \cdot L}{D} \tag{6.87}$$

Reynolds number, Re, is the ratio of the internal and viscous forces:

$$Re = \frac{v \cdot L}{v} \tag{6.88}$$

Schmidt number, Sc, is the ratio of the momentum and mass diffusivity:

$$Sc = \frac{v}{D} \tag{6.89}$$

Galileo number, Ga, is a dimensionless group representing a ratio of forces present in the flow of viscous fluids, or Ga = Re · gravitational force/viscous force:

$$Ga = \frac{g \cdot L^3}{v^2} \tag{6.90}$$

Example 6.6. Mixture of ammonia and air enters a column absorber (at 1 atm and 25 °C) filled with ceramic Raschig rings ($\varepsilon = 0.68$; $a = 190$ m^2 m^{-3}; $d_p = 25$ mm) at a total flow rate of 18 kmol h^{-1} (feed density: 1.182 kg m^{-3}). Water enters countercurrently at the flow rate of 1,360 dm^3 h^{-1}. Feed contains 5 mol% of ammonia, and 90 % of the ammonia is removed from the gas phase. Column inside diameter is 0.42 m. Evaluate the mass transfer coefficient of the ammonia on the gas and liquid side. Average collision diameter, σ, is equal to 3.3055 A; collision integral, Ω, is equal to 1.2258, molar masses of air and ammonia are 28.97 g mol^{-1} and 17.07 g mol^{-1}; molar volume of ammonia is 25.8 cm^3 mol^{-1}; association factor of water is 2.6; dynamic viscosity of water is 0.89 cP. Calculate the rate of mass transfer on the gas side.

Solution

For estimation of gas phase mass transfer coefficient and Raschig rings:

$$Sh_G = 0.407 \cdot Re_G^{0.655} \cdot Sc_G^{1/3},$$

With characteristic length: $L = 4 \cdot \frac{\varepsilon}{a} = 4 \cdot \frac{0.7}{330} = 0.0085$ m$L = 4 \cdot \frac{\varepsilon}{a} = 4 \cdot \frac{0.68}{190} = 0.014$ m

Diffusion coefficient D_{AB} on the gas side can be evaluated by the Wilke and Lee equation [16]:

$$D_{AB} = \frac{\left[3.03 - \left(\frac{0.98}{M_{AB}^2}\right)\right] \times 10^{-3} \cdot T^{3/2}}{p \cdot M_{AB}^{1/2} \cdot \sigma_{AB}^2 \cdot \Omega_D} = \frac{\left[3.03 - \left(\frac{0.98}{21.48^2}\right)\right] \times 10^{-3} \cdot 298^{3/2}}{1.01325 \cdot 21.48^{1/2} \cdot 3.3055^2 \cdot 1.2258} = 0.248 \text{ cm}^2 \text{ s}^{-1}$$

In this equation, M_{AB} is calculated from: $M_{AB} = 2 \cdot \left(\frac{1}{M_A} + \frac{1}{M_B}\right)^{-1} = 2 \cdot \left(\frac{1}{28.97} + \frac{1}{17.07}\right)^{-1} = 21.48$

Molar mass of the gas mixture: $M = 0.95 \cdot 28.97 + 0.05 \cdot 17.07 = 28.38$ kg kmol^{-1}

Volume flow rate of feed: $\dot{V} = \frac{G \cdot M}{\rho_G} = \frac{(18/3600) \cdot 28.38}{1.182} = 0.12$ m^3 s^{-1}

Column cross-sectional area: $A = \frac{D^2 \cdot \pi}{4} = \frac{0.42^2 \cdot \pi}{4} = 0.139$ m^2

$$Re_G = \frac{v \cdot L}{v} = \frac{(0.12/0.139) \cdot 0.014}{1.575 \times 10^{-5}} = 767.39$$

$$Sc_G = \frac{v_G}{D_G} = \frac{1.575 \times 10^{-5}}{0.248 \times 10^{-4}} = 0.635$$

$$Sh_G = 0.407 \cdot Re_G^{0.655} \cdot Sc_G^{1/3} = 0.407 \cdot 767.39^{0.655} \cdot 0.635^{1/3} = 27.14$$

$$k_G = \frac{S}{h_G} \cdot D_G L = \frac{27.62 \cdot 0.248 \cdot 10^{-4}}{0.014} = 0.0478 \text{ m s}^{-1}$$

Mass transfer coefficient is estimated for liquid phase and Raschig rings:

$$Sh_L = 0.32 \cdot Re_L^{0.59} \cdot Sc_L^{0.5} \cdot Ga_L^{0.17},$$

With characteristic length equal to nominal diameter of packing: $L = d_p = 25 \times 10^{-3}$ m

Diffusion coefficient on the liquid side can be estimated using Wilke and Chang equation:

$$D_{AB} = 7.4 \times 10^{-8} \cdot \frac{T}{\mu_1} \cdot \frac{(C \cdot M_1)^{0.5}}{V_2^{0.6}} = 7.4 \times 10^{-8} \cdot \frac{298}{0.89} \cdot \frac{(2.6 \cdot 18)^{0.5}}{25.8^{0.6}} = 2.411 \times 10^{-5} \text{ cm}^2 \text{ s}^{-1}$$

$$Re_L = \frac{v \cdot L}{v_L} = \frac{((1,360 \cdot 10^{-3}/3,600)/0.139) \cdot 0.025}{0.89 \times 10^{-6}} = 76.34$$

$$Ga_L = \frac{g \cdot L^3}{v_L^2} = \frac{9.81 \cdot 0.025^3}{(0.89 \cdot 10^{-6})^2} = 1.98 \times 10^8$$

$$Sc_L = \frac{v_L}{D_L} = \frac{0.89 \cdot 10^{-6}}{2.411 \cdot 10^{-9}} = 369.14$$

$$Sh_L = 0.32 \cdot Re_L^{0.59} \cdot Sc_L^{0.5} \cdot Ga_L^{0.1} = 0.32 \cdot 76.34^{0.59} \cdot 369.14^{0.5} \cdot (1.98 \cdot 10^8)^{0.1} = 536.09$$

$$k_L = \frac{Sh_L \cdot D_L}{L} = \frac{536.09 \cdot 2.411 \times 10^{-9}}{0.025} = 0.0000517 \text{ m s}^{-1}$$

Partial pressure of ammonia in air: $p_{AB} = 0.05 \cdot 1 = 0.05$ atm

The amount of absorbed ammonia: $n_A = 0.9 \cdot 0.05 \cdot (18000/3600) = 0.225$ mol s^{-1},

and molar concentration: $c_A = \frac{n_A}{V} = \frac{0.225}{1,360/3,600} = 0.596$ mol dm^{-3}

Concentration of gas phase in equilibrium with liquid: $p_A^* = H_A \cdot c_A = 0.0161 \cdot 0.596 = 0.0096$ atm

Resistances to mass transfer: $\frac{1}{k_p} = \frac{1}{k_G/(R \cdot T)} = \frac{1}{0.0478/(8.314 \cdot 298)} = 51,832.1$ s m^2 Pa mol^{-1}

$$\frac{H_A}{k_c} = \frac{0.0161 \cdot 101,325}{0.0000517} = 31553,820.1 \text{ s m}^2 \text{ Pa mol}^{-1}$$

The rate of mass transfer:

$$J_A = K_G \cdot (p_{Ab} - p_A^*) = \frac{(p_{Ab} - p_A)}{(1/k_p) + (H_A/k_c)} = \frac{(0.05 - 0.0096) \cdot 101,325}{51,832.1 + 31553,820.1}$$
$$= 1.295 \cdot 10^{-4} \text{ mol m}^{-2} \text{ s}^{-1}$$

Symbols

A	cross-sectional area (pipe, particle, filter, etc.), m^2
A	surface area of heat transfer, m^2
A_{bed}	bed cross-sectional area, m^2
A_p	pore area, m^2
C	association factor, –
c	volumetric concentration, m^3 m^{-3}
c	molar concentration, mol dm^{-3}
c_{Ab}	concentration of compound A in the bulk of the phase, mol dm^{-3}
c_{Ai}	concentration of compound A at the phase interface, mol dm^{-3}
c_A^*	equilibrium concentration, mol dm^{-3}
c_p	specific heat for constant pressure, J kg^{-1} K^{-1}
C_D	drag factor, –
D	diameter, m
D, D_L, D_G, D_{AB}	diffusion coefficients (liquid, gas, binary), m^2 s^{-1}
d_{eq}	equivalent diameter of pores, m
d_p	nominal diameter of packing, m
d_p	particle diameter, m
E_V	energy per unit volume, J m^{-3}
E_{vis}	energy dissipation due to viscosity, J m^{-3}
F	force, N
F_b	buoyant force, N
F_D	drag force, N
F_e	external driving force, N
G	solid mass per unit volume of filtrate, kg m^{-3}
G	molar flow rate for gas, mol s^{-1}
Ga	Galileo number
Gr	Grashof number
g	gravitational acceleration, m^2 s^{-1}
H	enthalpy, J kg^{-1}
H_A	Henry constant, Pa mol^{-1} dm^3
h_c	convective heat transfer coefficient, W m^{-2} K^{-1}
h_r	radiation heat transfer coefficient, W m^{-2} K^{-1}
J	diffusion rate, mol m^{-2} s^{-1}
K_G, K_L	overall mass transfer coefficient, W m^{-2} K^{-1}
k	thermal conductivity, W m^{-1} K^{-1}

k_L, k_G	mass transfer coefficient (liquid, gas), m s^{-1}
L	bed height, m
L	characteristic length, m
l	linear dimension, m
M	molar mass, kg mol^{-1}
m	mass, kg
$\dot{m}$	mass flow rate, kg s^{-1}
Nu	Nusselt number
n	Richardson–Zaki index, –
n_A	mole flow rate, mol s^{-1}
n_p	total number of packed particles, –
p	pressure, Pa
p_{Ab}	partial pressure of compound A in the bulk of the phase, Pa
p_{Ai}	partial pressure of compound A at the phase interface, Pa
Q	heat per unit volume, J m^{-3}
$\dot{Q}$	heat flow rate, W
Pr	Prandtl number
R	gas constant, 8.314 J mol^{-1} K^{-1}
R_{bed}	bed resistance, m^{-1}
R_C	filter cake resistance, m^{-1}
R_M	filter medium resistance, m^{-1}
Re	Reynolds number
S	cross-sectional area of the plane, m^2
$S_{particle}$	external surface of spherical particle, m^2
S_{pore}	pore circumference, m
S_V	specific surface area of pores, m^2 m
Sc	Schmidt number
Sh	Sherwood number
T	temperature, K
t	time, s
x, y, z	rectangular coordinates, m
U	internal energy, J
U	overall heat transfer coefficient, W m^{-2} K^{-1}
V	volume, m^3
V_{bed}	bed volume, m^3
$\dot{V}$	volume flow rate, m^3 s^{-1}
v	velocity, m s^{-1}
v_A	superficial velocity of the fluid, m s^{-1}
v_h	hindered velocity of particles in suspension, m s^{-1}
v_p	velocity of fluid in the pores of a bed, m s^{-1}
v_t	terminal velocity of the particle, m s^{-1}
W	mechanical energy, J m^{-3}
X_V	mass, energy or momentum per unit volume, kg m^{-3}, mol m^{-3}, J m^{-3}, kg m^{-2} s^{-1}
z	geodetic elevation, m

Greek symbols

α	absorptivity, –
α	specific filter cake resistance, m kg^{-1}
α	thermal diffusivity, m^2 s^{-1}

δ_G, δ_L	thickness of the diffusion layer (gas, liquid), m
ε_{bed}	bed porosity and packing porosity, –
ε	emissivity, –
μ	fluid dynamic viscosity, Pa s
ρ	fluid density, kg m^{-3}
ρ_p	particle density, kg m^{-3}
ρ	the reflectivity,
Δp	pressure difference, pressure drop and filtration pressure, Pa
v	kinematic viscosity, m^2 s^{-1}
σ	collision diameter, m
σ	Stephan-Boltzmann constant, W m^{-2} K^{-4}
Ω	collision integral, –
τ	transmissivity, –

References

[1] Byron Bird R, Stewart WE, Lightfoot EN. Transport phenomena. UK: John Wiley & Sons Ltd; 1960.
[2] Rhodes M. Introduction to particle technology. UK: John Wiley & Sons Ltd; 2007. p. 29–46.
[3] Reynolds TD, Richards PA. Unit operations and processes in environmental engineering. USA: PWS Publishing Company; 1996. p. 219–82.
[4] Beek WJ, Muttzall KM. Transport phenomena. UK: John Wiley & Sons Ltd; 1975. p. 98–108.
[5] Badlock TE, Tomkins MR, Nielsen P, Hughes MG. Settling velocity of sediments at high concentrations. Coastal Eng. 2004;51:91–100.
[6] Di Felice R, Kehlenbeck R. Sedimentation velocity of solids in finite size vessels. Chem Eng Technol. 2000;23:1123–26.
[7] Svarovsky L. Solid–liquid separation. 3rd ed. UK: Butterworth & Co. Ltd; 1990. p. 693–95.
[8] Svarovsky L. Solid–liquid separation. UK: Butterworth & Co. Ltd; 1977. p. 173–75.
[9] Çengel YA. Heat transfer: A practical approach. USA: The McGraw–Hill Science; 1998.
[10] Holman JP. Heat transfer. Singapur: McGrow Hill Book Company; 1989. p. 1–7.
[11] Dixit DK. Heat and mass transfer. India: McGraw Hill Education Pvt. Ltd; 2015.
[12] Wilkinson DS. Mass transport in solids and fluids. UK: Cambridge University Press; 2000. p. 47–49.
[13] Sattler K, Feindt HJ. Thermal separation processes, principles and design. Weinheim, USA: VCH; 1995. p. 60–113.
[14] Cussler EL. Diffusion, mass transfer in fluid systems. UK: Cambridge University Press; 2009.
[15] Seader JD, Henley EJ. Separation process principles. USA: John Wiley and Sons; 2006.
[16] Wilke CR, Lee CY. Estimation of diffusion coefficients for gases and vapors. Ind Eng Chem. 1955;47:1253–57.

Examples for practice

Example 1:
A particle of equivalent volume diameter 1 mm and sphericity 0.6 falls freely under gravity in a fluid of density 1,000 kg m^{-3} and viscosity 1 mPa s. The measured terminal velocity is 0.02 m s^{-1}.

Calculate: (a) Reynolds number and drag factor; (b) particle density.

Solution

 (a) Re = 500, C_D = 4; (b) ρ_p = 3, 446 kg m^{-3}

Example 2:

Wastewater of concentration 350 kg m^{-3} is settling in sedimentation basin under gravity. Particles are of uniform size 40 μm, having density 1,500 kg m^{-3} and are assumed to be spherical. Fluid has density 1,000 kg m^{-3} and viscosity 0.001 Pa s. Calculate hindered velocity and porosity of the settling suspension if Richardson–Zaki index is 4.

Solution

$v_h = 5 \times 10^{-5}$ m s^{-1}; $\varepsilon = 0.77$

Example 3:

Calculate the pressure drop for an ion exchange bed that is efficiently used to deionize a liquid of density and viscosity 1,215 kg m^{-3} and 1.3 mPa s, respectively. The design flow rate is 4 m^3 h^{-1}. Particles are packed with a porosity of 0.24 and with a specific surface area of 300 m^2 m^{-3}. The bed is cylindrical and in diameter of 10 cm.

Solution

$\Delta p = 1.23 \times 10^5$ Pa per meter of bed height

Example 4:

Continuous rotary disc has been used for vacuum filtration of a suspension that contains particles of maximum sphericity and of an averaged size of 213 μm. Suspension owns viscosity of 2.3×10^{-6} m^2 s^{-1} and density of 1.323 g cm^{-3}. Filtration is carefully carried out under constant filtration pressure of 1 bar on the filtration area of 2 m^2. If there will be an incompressible filter cake of a height of 8 cm and of surface area 1 $\times 10^4$ m^2 m^{-3}, predict the time needed for treatment of 1×10^4 l of a suspension on a filter medium with no resistance.

Solution

t = 21.42 s

Example 5:

Calculate the total heat loss with data from Example 6.5. If the temperature of the cylinder is 529 °C and compare the obtained results ($v = 4.434 \cdot 10^{-5}$ m^2 s^{-1}; $k = 0.0436$ W m^{-1} K^{-1}; Pr = 0.680).

Solution

Q_{tot} = 28,806.43 W

Example 6:

Calculate the heat loss for a semi-detached house. Total area and the corresponding U-values are as follows: external walls: 84 m², 0.25 W m⁻² K⁻¹; roof: 63 m², 0.13 W m⁻² K⁻¹; floor: 63 m², 0.20 W m⁻² K⁻¹; windows: 31.5 m², 2.20 W m⁻² K⁻¹; door: 1.85 m², 3.00 W m⁻² K⁻¹. Temperature inside the house is 24 °C, while the outside temperature is 5 °C. Evaluate the major source of heat loss.

Solution

Q = 2,332.8 W; The major sources of heat loss are windows.

Example 7:

Calculate the rate of mass transfer using data from Example 6.6. The column is filled with Berl saddles ($\varepsilon = 0.68$; $a = 260$ m² m⁻³; $d_p = 25$ mm). Equations for evaluation of mass transfer coefficients are:

$Sh_G = 5.23 \cdot Re_G^{0.7} \cdot Sc_G^{1/3} \cdot (a \cdot d_p)^{-2}$, with characteristic length $L = 1/a$.

$Sh_L = 0.25 \cdot Re_L^{0.59} \cdot Sc_L^{0.5} \cdot Ga_L^{0.1}$, with characteristic length $L = d_p$.

Solution

$J_A = 5.31 \times 10^{-3}$ mol m⁻² s⁻¹

Example 8:

Mixture of ethanol and water is separated in the continuous distillation column at the atmospheric pressure. At some point in the column, the concentration of ethanol in the vapour is 0.68 (mole fraction). The equilibrium concentration of ethanol in the liquid phase is 0.58 (mole fraction), and the temperature is 79.5 °C. At this temperature, partial pressure of pure ethanol is 800 mmHg and diffusion coefficient is 0.182 cm² s⁻¹. It can be assumed that the film thickness is 0.1 mm. Calculate the rate of mass transfer.

Solution

$J_A = 0.437$ mol m⁻² s⁻¹

This article is an excerpt from the textbook *Environmental Engineering: Basic Principles* (978-3-11-046801-4).

Karolina Maduna and Vesna Tomašić

7 Air pollution engineering

Abstract: Air pollution is an environmental and a social problem which leads to a multitude of adverse effects on human health and standard of human life, state of the ecosystems and global change of climate. Air pollutants are emitted from natural, but mostly from anthropogenic sources and may be transported over long distances. Some air pollutants are extremely stable in the atmosphere and may accumulate in the environment and in the food chain, affecting human beings, animals and natural biodiversity. Obviously, air pollution is a complex problem that poses multiple challenges in terms of management and abatements of the pollutants emission. Effective approach to the problems of air pollution requires a good understanding of the sources that cause it, knowledge of air quality status and future trends as well as its impact on humans and ecosystems. This chapter deals with the complexities of the air pollution and presents an overview of different technical processes and equipment for air pollution control, as well as basic principles of their work. The problems of air protection as well as protection of other ecosystems can be solved only by the coordinated endeavors of various scientific and engineering disciplines, such as chemistry, physics, biology, medicine, chemical engineering and social sciences. The most important engineering contribution is mostly focused on development, design and operation of equipment for the abatement of harmful emissions into environment.

Keywords: Air pollution, Pollutants, Air pollution sources, Particulate matter, Gaseous air contaminants, Air pollution control techniques

> Who does not learn from the past will not succeed in the future.
> Paul Crutzen, Nobel Prize Winner in Chemistry (1995)

7.1 Introduction

7.1.1 The complexities of the air pollution protection

Economic, industrial and population growth in the world have had a significant impact on urban and rural air quality. Today, air pollution presents a substantial environmental risk, especially in some areas that are frequently exposed to

This article has previously been published in the journal Physical Sciences Reviews. Please cite as: Maduna, K., Tomašić, V. Air Pollution Engineering. *Physical Sciences Reviews* [Online] **2017**, *2* (12). DOI: 10.1515/psr-2016-0122

https://doi.org/10.1515/9783110468038-007

elevated concentrations of toxic air pollutants. Emissions from power generation, industrial sources, traffic and transport vehicles, as well as from other industrial and anthropogenic activities are significant contributors to the air quality. In addition to human activities, air pollution can also be caused by natural disasters (e. g. volcanic eruptions, forests fires), by sea-salt emissions, natural processes and biological processes. According to some estimates, there are about 1,500 active volcanoes around the world, 50 of which erupt every year and 2012 was declared as *The Year in Volcanic Activity*. However, it is generally accepted that the major contributors to the increasing emissions into atmosphere are mostly the man-made or anthropogenic activities [1].

The US National Oceanic and Atmospheric Administration (NOAA) recorded 2015 as the hottest year since records began in 1880 and the 16 warmest years recorded are in the 1998–2015 period [2]. In 2015, the five largest emitting countries (including the European Union), which together account for two thirds of total global emissions, were: China (with a 29 % share in the global total), the United States (14 %), the European Union (EU-28) (10 %), India (7 %), the Russian Federation (5 %) and Japan (3.5 %) [2]. Climate model projections indicated that during the twenty-first century the global surface temperature is likely to rise a further 0.3 to 1.7 °C for their lowest emissions scenario and 2.6 to 4.8 °C for the highest emissions scenario [3]. Increased temperature, especially in the Arctic, melts ice caps and glaciers and affects the oceans, resulting in habitat changes for many species of plants and animals. Extreme weather, such as increased rainfall in some areas, droughts and heat waves in others, affects the growth of food crops and other plants. Climate change – a result of global warming, which have been observed and scientifically proved in the last few decades, depend significantly on the specific problems related to the air pollution. There is no doubt that human activities are the critical cause of the climate change that Earth is experiencing since the Industrial Revolution in the mid-eighteenth century. It is well known that Industrial Revolution marked the beginning of a dramatic increase in the use of fossil fuels. Fuel combustion is also directly related to the number of people and their standard of living and energy consumption. According to the UN calculations, today there are more than 7 billion living humans on the Earth [4]. Between 1900 and 2000 the increase in world population was three times greater than the entire previous history of humanity with an increase from 1.5 to 6.1 billion in just 100 years. Coal is still the biggest source of energy for electricity production and the biggest contributor to carbon dioxide (CO_2) emissions and CO_2 is the most important representative of the greenhouse gases (GHGs) responsible for the warming effect [5].

Developed countries are in particular faced with the problems of air pollution in big cities and industrialized areas and sometimes they try to solve this problem by the relocation of theirs industrial plants in less developed countries. Apparently, this is a bad example of temporary problem solving and transferring the problem to another location. Currently, some politicians and special interest groups in the

more developed parts of the world still refuse to act in order to reduce environmental pollution on the global level, mainly due to significant economic impacts of
particular industries or some segments of the economy. On the other hand, in the
developing countries and countries in transition, development is usually based on
"dirty" industry and unsustainable technologies. The consequence is pollution on a
global scale, due to the fact that air pollutants can be transported over long
distances from their source. The amount of time that an air pollutant, like ozone,
greenhouse gases and others, will stay in the atmosphere is determined by its
reactivity and removal rate by natural mechanism (dry or wet deposition, natural
reservoirs or sinks). Some air pollutants remain in the atmosphere from a few
hours, days to several thousands of years. Therefore, protecting human health
and the environment from the impacts of air pollution and climate changes become
the critical challenges of the twenty-first century. These challenges are additionally
complicated by the interrelation between the air quality, climate changes as well as
existing and advanced energy options [6].

Although there is significant progress in the last half-century because of numerous
efforts to reduce air pollution and greatly improve air quality, many people are still
living in counties that do not meet current air quality standards for the air pollutants
regulated by the environmental agencies, such as European Environment Agency (EEA)
and many national environmental agencies worldwide. Thus, continued improvements
and understanding of emissions, atmospheric processes, exposure and effects are still
critical tasks to ensuring continued improvement of the air quality. Reducing emissions
of critical pollutants into the atmosphere is technically difficulty and very expensive.
Although there are some natural air self-cleaning mechanisms, efficiency of such
natural mechanisms is insufficient to meet the regulatory requirements.

Air quality management is definitely one of the most demanding and complex
activities that involve a high degree of organization of all parts of human society
and require effective implementation of institutional system of air protection and
the whole society. The overall air protection strategy is based on the basic principles of the environmental protection, the principles of international politics as well
as environmental laws, regulations, decisions and directives with respect to the
latest scientific knowledge and the world's best practices. The most important
factors influencing air pollution control system are: policy or political commitment
at international, regional, national and local level, legislation, resources (human
and financial) and available equipment or technical devices. An effective air quality
management system is based on environmental policy that, besides other, includes
the continuous improvement of the installation by the efficient management, adop-
tion and implementation of adequate legislation and regulations to limit emissions
of air pollutants from all kind of sources and recommended emission limit values,
monitoring and measurement of emissions to air, processing the collected data and
presentation of obtained results to the wider professional community, development
and application of appropriate educational and promotional activities aimed at

raising public awareness of environmental protection, application of economic incentives and privileges, checking performance and taking corrective action with respect to the emission levels, product quality standards and in accordance with the conditions and mode of legal entities responsible to perform professional activities in the field of the environmental protection. Regulations are very important, but they are effective only after implementation. Since 1972 many countries published a lot of acts and rules to protect air and environment, but the effectiveness of the implementation of legislations is still questionable. The regional and global environmental problems related to atmospheric pollution cannot be solved only by the local regulations used for some types of air pollutants. In some cases the only solution is worldwide action and application of different approaches. Environmental policy in the European Union (EU) has a long history and in line with other international instruments and conventions. During the 1970s and 1980s, a series of problems and scandals, involving waste treatment, gave the necessary impetus for the creation of the common environmental policy and appropriate strategies. Since then several policy strategies have been adopted, but probably the most famous are *The Lisbon Strategy* (2001) and *The Europe 2020 Strategy* (2010). The Lisbon Strategy established a strategic goal for the EU with aim to transform EU into a competitive knowledge-based economy, with particular emphasis on environmental protection and sustainability. Term *sustainable development* was defined in 1987 in a report of the United Nations as "development which meets the needs of the present without compromising the ability of future generations to meet their own needs". Although according to some opinions, *The Lisbon Strategy* was not entirely successful, one of the most important goals of this strategy was consideration of an economy with low CO_2 emissions. In 2010 the European Commission inaugurated *The Europe 2020 Strategy*. The well-known objectives of this new strategy for viable development, the so-called 20-20-20 include the following: reducing the greenhouse gas emissions by 20 % against the 1990 baseline before the year 2020, raising the share of renewable energy sources to 20 % measured in terms of actual final consumption and increasing energy efficiency by 20 %. This strategy is based on the new technologies with low carbon emissions. Although scientists and politicians do not always agree on the best course of action and cooperation at global, European, national and local levels, most of them agree that some changes need to be made to slow global climate changes associated with air pollution. The current European policy includes implementation of the legal limits for ambient concentrations of air pollutants and the emission mitigation controls to national and specific sources or sectors. Over the last three decades these policies have resulted in significantly decreased emissions of some air pollutants (such as SO_2, CO, NO_x, lead and some particulates) and noticeable improvements in air quality, but there is still huge room for improvement of existing technical processes, particularly those for reducing greenhouse gas emissions and persistent organic compounds.

7.1.2 Basic definitions and classifications

Air is a mixture of gases forming the Earth's atmosphere, which cannot be seen, but which exists all around us. This layer of gases surrounding our planet is retained the Earth's gravity. The atmosphere is a very complex system in which physical and chemical reactions are constantly taking place, but dynamic of these processes can be disrupt due to the presence of some unwanted and harmful compounds.

The term **air pollution** can be defined in different ways (Figure 7.1). Generally, air pollution is the release or emission of different chemicals, particulates, biological materials or other harmful materials into the Earth's atmosphere, both from natural and from anthropogenic sources. Generally speaking, an emission refers to the discharge of pollutants into the atmosphere, including not only gaseous, vaporous and solid substances, but also heat, noise and radiation. In the context of air pollution extremely important are all stationary sources, such as smokestacks, vents, surface areas of domestic, commercial or industrial facilities as well as numerous mobile sources. Generally, the air pollution refers to any kind of the air contamination, regardless of the atmosphere, which can be indoors or outdoors atmosphere. Thus, general classification of air pollution involves both *indoor and outdoor air pollution.*

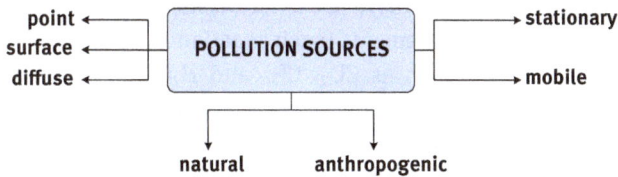

Figure 7.1: Classification of air pollution sources.

When a building is not properly ventilated, pollutants can accumulate and reach concentrations greater than those typically found outside. This problem has received media attention as the so-called *Sick Building Syndrome* (SBS). Tobacco smoke is one of the main contributors to indoor pollution as well as carbon monoxide, nitrogen monoxide and sulfur dioxide, which can be emitted from furnaces and stoves. Cleaning or remodeling a house contributes to elevated concentrations of harmful chemicals such as volatile organic compounds (VOCs) emitted from household cleaners, paint and varnishes. On the other hand, when bacteria die, they release endotoxins into the air, which can cause adverse health effects, too. More information on indoor air pollution can be found in the literature [7]. Indoor air pollution and urban air quality are often considered as two of the world's worst environmental pollution problems. According to data of the World Health Organization (WHO) due to air pollution each year die more than 3 million people, which represent about 3 % of deaths annually [8]. Surprisingly, the higher mortality rate refers to the indoor air pollution. Indoor air pollution is a very dangerous problem, because indoor air is far more concentrated with pollutants than outdoor one. According to some estimation, the level of pollutants indoors can reach values that are 2–5 times, sometimes even up to 100 times larger in comparison to those outside the buildings [5]. In addition to the different industrial emission sources and other sources connected with various human activities, mostly related to the heat and electricity production, the air pollution from mobile sources and transportation is becoming increasingly dominant in recent years, due to the high level of motorization in economically developing countries. According to some estimates, the air pollution from the mobile sources is growing faster than from industrial sources. Copper and Alley [5] have pointed out that motor vehicles account for 10–60 % of total air pollution emissions.

Air pollution sources can be classified as *natural (biogenic)* and *man-made (anthropogenic)* as a function of the origin of emissions or precursors. There are many *natural sources of air pollutant emissions* such as: volcanic activities (which produce sulfur, chlorine, ash, particulates), anaerobic microbial degradation in soil (which produce oxides of nitrogen), atmospheric electrical discharges (source of ozone), methane emitted by the digestion of food by animals, forest fires (which produce smoke and carbon monoxide), evaporation of droplets of sea water resulting in sea salt crystals being suspended in the air, some VOCs emitted from vegetation, natural radioactivity (which increases with altitudes and also depends on type of soil, e. g. marl soil and often is connected with intensive use of fertilizers), new buildings connected with the presence of radioactive elements (such as radon in concrete and building materials). Some air pollutant such as CO_2, CH_4 and N_2O can occur in significant amounts from the natural sources, but there are also some natural mechanisms of their removal from the atmosphere (or sinks).

Most air pollutants originate from *human-made sources or anthropogenic activities*, including: (a) *stationary sources* (all activities connected with the combustion of fossil fuels in the production of heat, electricity or other forms of energy, i.e. production processes and power plants, households and waste incinerators, furnaces and other types of fuel-burning heating devices, traditional biomass burning): different industrial plants (the so-called non-energy source like refining and petrochemical processing of oil, ceramic and glass industry, production of cement, fertilizers, steel industries); production and use of organic chemicals and solvents (including paints, adhesives, aerosols, metal cleaning, printing); activities such as mining operations, extraction and distribution of fossil fuels (dust, minerals and some chemicals); agricultural activities and farming (including crop growing, silage manufacture, sludge spreading, fertilizer application); waste treatment and disposal; produced or artificial radioactivity (nuclear reactors and accelerators, nuclear medicine, radiotherapy and (b) *mobile sources and all kind of transport vehicles* (motor vehicles, marine vessels, aircraft).

Another possible way of characterizing air pollutants emission sources according defined locality of emission and type of emission involve: point, line, surface (or area) and volume sources. *Point sources* of emission (like chimneys, stacks, vents and other functional openings) belong to the independent single sources, which typically emitted contaminants on a continuous basis. Because of the small exit area, the emission from such source will always be very large. *Line sources*, like vehicle emission from roadways, surface or area sources (e. g. volatile emissions from lagoons, forest fires) and *volume sources* (e. g. diffuse air pollutant emissions from oil refineries) are considered as *diffuse or fugitive sources*. Contrary to emissions from point sources, fugitive emissions are emissions that cannot reasonably be collected and pass through a stack, vent or similar opening. This difference also determines the technical devices adopted for the abatement of such emissions.

What is an air pollutant? **Air pollutant** is any kind of solid, liquid or gaseous substance present in the air in such concentration that can have adverse effects on human beings or other living organisms, natural biodiversity, construction, building and other types of common materials and the whole ecosystem. The atmosphere contains a number of air pollutants generated or emitted from different either natural or anthropogenic sources. The *common air pollutants*, their sources and some effects on human heath are summarized in Table 7.1.

Table 7.1: Common air pollutants, sources and theirs effects on human health.

Common air pollutants	Source of emission	Effects
Carbon monoxide (CO)	Fuel combustion from vehicles and engines (transportation, residential heating, industrial processes).	Reduces the amount of oxygen reaching the body's organs and tissues; aggravates heart disease, resulting in chest pain and other symptoms.
Volatile organic compounds (VOCs): non-methane hydrocarbons (NMHC) and oxygenated NMHC (e. g., alcohols, aldehydes, organic acids)	Photosynthesis, vegetation and oceans, biomass burning, agriculture; the use of paints, coatings, printing; surface cleaning; fuel production, distribution and combustion.	The outdoor air pollutant that contribute to the emergence of the Sick Building Syndrome (SBS); the aromatic NMVOCs (e. g. benzene, toluene and xylene) are suspected carcinogens and may lead to leukemia with prolonged exposure.
Nitrogen oxides (NO_x, $x = 1,2$)	Any kind of fuel combustion in air including mobile sources and stationary combustion sources (vehicles, electric utilities, big industrial boilers, wood burning, etc.).	Impacts on respiratory conditions causing inflammation of the airways at high levels; long term exposure can decrease lung function, increase the risk of respiratory conditions and increases the response to allergens.

(*continued*)

Table 7.1: (*continued*)

Common air pollutants	Source of emission	Effects
Sulfur dioxide (SO_2)	Fossil fuel combustion for electric power generation, industrial processes (petroleum refining, non-ferrous metal smelting, etc.).	Irritates the skin and mucous membranes of the eyes, nose, throat and lungs; inflammation and irritation of the respiratory system; people with asthma or chronic lung or heart disease are the most sensitive to SO_2.
Ground-level ozone (O_3)	Secondary pollutant formed by chemical reaction of VOCs and NO_x in the presence of sunlight.	Decreases lung function and causes respiratory symptoms (coughing and shortness of breath); makes asthma and other lung diseases get worse.
Lead (Pb) and other heavy metals such as arsenic (As), cadmium (Cd), chromium (Cr), nickel (Ni), mercury (Hg)	Mining and smelting operation, industrial production and use, combustion, metal corrosion and leaching of heavy metals, waste incinerators, agriculture	Pb attacks the brain and central nervous system to cause coma, convulsions and even death. Arsenic exposure affects virtually all organ systems (including the cardiovascular, dermatologic, nervous, hepatobilliary, renal, gastro-intestinal and respiratory systems). A short-term exposure to a high level of Ni showed damage to the lungs and kidneys, gastrointestinal distress, neurological effects. Cd, Pb and Hg are recognized as being toxic to biota.
Particulate matter (PM) or particulates	Materials handling processes (crushing or grinding ores, loading dry materials in bulk), combustion processes, industrial processes, farming (ploughing, field burning), unpaved roads and road constructions, etc.	Increased levels of fine particles in the air are linked to health hazards, such as heart disease, respiratory problems and lung cancer.

According to the last report of the EEA from 2016 [9] emissions of many air pollutants have decreased substantially over the past decades, resulting in improved air quality across the European region. However, a large proportion of European populations and ecosystems are still exposed to air pollution in exceedance of European standards and WHO Air Quality Guidelines (AQGs). Europe's most problematic pollutants in terms of harm to human health are PM, ground-level O_3 and NO_2. In addition, benzo(a)pyrene (BaP), an carcinogenic pollutant and indicator for polycyclic aromatic hydrocarbons (PAHs), is still of increasing concern, particularly in Central and Eastern Europe.

There are several classifications of air pollutants. The United States Environmental Protection Agency (US EPA) and most countries in the world recognized the *criteria air pollutants* including: ozone (O_3), atmospheric particulate matter (PM_{10}, $PM_{2.5}$), lead (Pb), carbon monoxide (CO), sulfur oxide (SO_2) and nitrogen oxides (NO_x). *Heavy metals* adhered on atmospheric particles are very dangerous for human health and impose a long-term burden on environmental quality. Among all the heavy metals, lead (Pb), arsenic (As), mercury (Hg), cadmium (Cd) and chromium (Cr) are of the greatest concern for public health because of their high toxicity. Heavy metals in atmospheric particles originate from some natural sources (volcanic eruptions, forest fire, etc.), but most environmental pollution and human exposures are caused by anthropogenic activities such as mining, smelting, industrial production and use, traffic, fossil fuel combustion, domestic and agricultural use of metals and metal-containing compounds. Other important groups of air pollutants are: greenhouse gases, minor air pollutants, persistent organic pollutants (POPs) and radioactive pollutants.

Among all air pollutants that are commonly associated with global environmental problems the most important are GHGs, most frequently CO_2, CH_4 and N_2O. These gases contribute directly to the climate change owing to their positive radiative forcing effect. Many countries that have signed and ratified the Kyoto Protocol are obligatory to reduce their GHGs by an agreed amount. The GHG inventory covers the seven direct greenhouse gases: carbon dioxide (CO_2), methane (CH_4), nitrous oxide (N_2O), hydrofluorocarbons (HFCs), perfluorocarbons (PFCs), sulfur hexafluoride (SF_6) and nitrogen trifluoride (NF_3). Compounds such as HFCs, PFCs, SF_6 and NF_3 are known as the synthetic gases. In addition, NO_x, CO, NMVOC and SO_2 are also sometimes considered as the indirect greenhouse gases.

Minor air pollutants include hazardous air pollutants (HAPS) and POPs. HAPS, also known as toxic air pollutants or air toxics, are pollutants that are known or suspected to cause cancer or other serious health problems (such as reproductive effects or birth defects and other adverse environmental effects). Almost 200 pollutants are recognized as toxic air pollutants and most representative examples include: benzene, perchloroethylene, methylene chloride, dioxin, asbestos, toluene as well as some metals (e. g. cadmium, mercury, chromium and lead). The US EPA is required to control 188 HAPS. POPs are halogenated organic compounds that are resistant to chemical, biological and photolytic degradation processes and also exhibit high lipid solubility. Thus, they bioaccumulate in fatty tissues and have negative impacts on human health as well as on the environment. Those compounds are also classified as PBTs compounds (**p**ersistent, **b**ioaccumulative and **t**oxic). Many POPs are currently used or were used in the past as pesticides, solvents, pharmaceutical and industrial chemicals. Some POPs arise naturally (volcanoes, various biosynthetic pathways, etc.), but mostly they are man-made. At the beginning, the so-called *Stockholm Convention list* (2001) recognized only 12 POPs, such as: aldrin, chlordane, dieldrin, endrin, heptachlor, hexachlorobenzene (HCB), mirex, toxaphene, polychlorinated biphenyls (PCBs), dichlorodiphenyltrichloro ethane (DDT), dioxins and polychlorinated dibenzofurans). Recently, new compounds such as PAHs, brominated flame retardants and others are added on this list. PAHs are composed of two or more condensed aromatic rings and contain only carbon or hydrogen (naphthalene, anthracene, pyrene, BaP). Heteroatoms such as nitrogen, oxygen and sulfur can be also substituted into PAHs (carbazole, dibenzofuran, dibenzo-*p*-dioxin, etc.).

Radioactive pollutants in the air are generated from geogenic and anthropogenic sources. Geogenic radioactivity results from the presence of radionuclides, which originate either from radioactive minerals in the Earth's crust or from the interaction of cosmic radiation with atmospheric gases. Anthropogenic radioactive emissions originate from nuclear reactors, the atomic energy industry (mining and processing of reactor fuel), nuclear weapon explosions and plants that reprocess spent reactor fuel. Since coal contains small quantities of uranium and thorium, these radioactive elements can be emitted into the atmosphere from the coal-fired power plants and other sources.

Waste heat and light pollution also belong to specific forms of atmospheric pollution, but they are not considered in details in this chapter. Briefly, *waste heat* contained in the flue gases or vapor streams is usually generated in process of the fuel combustion or by chemical reaction. A part of the waste heat can recovered and losses of the heat/energy minimized using the waste heat recovery devices, such as recuperators, regenerators, heat wheels, heat pipe exchanger, heat pumps and similar technical devices [10]. The strategy of the waste heat recovery mostly depends on temperature of waste heat gases and economics involved. *Light pollution (or photo pollution)* is a side effect of industrial civilization. This term refers to multiple problems, all of which are caused by inefficient or unnecessary using of the artificial light. Some categories of light pollution include over-illumination, glare and light clutter. Medical research shows that various harmful health effects, such as headache, worker fatigue, medically defined stress and increase in anxiety, may be caused by light pollution or excessive light exposure.

According to the aggregate state the air pollutants are classified into two phases:
- *gaseous air contaminants and vapors of volatile liquid substances (including odors)* and
- *PM or particulates (aerosols)* (Table 7.2).

The odors are sometimes distinguished as a separate category of air pollutants, although basically they belong to a group of easily VOCs [10].

Organic and inorganic gaseous and vaporous air pollutants include compounds such as SO_2, CO, NO_x, NH_3, different hydrocarbons, VOCs, PAHs, halogen derivatives and odorous substances. In order to understand the basic principles of

Table 7.2: Anthropogenic pollution sources and their harmful influence [11].

Pollutant and the source	Influence*			
	A	B	C	D
Organic gases and vapors (hydrocarbons, HC)				
Paraffins: processing and transport of oil products; solvents usage; motor vehicles, etc.		+	+	
Olefins: processing and transport of gasoline; motor vehicles etc.	+	+	+	
Aromatics: processing and transport of oil products; solvents usage; motor vehicles etc.	+	+	+	Odor
Other:				
oxidized HC (e. g. aldehydes, ketones, alcohols)		+	+	Odor
halogenated HC (e. g. CCl$_4$)		+	+	Odor
Inorganic gasses				
Nitrogen oxides: fuel combustion; motor vehicles; electrical discharges, etc.	+	+	+	+
Sulfur oxides: fuel combustion; chemical industry, etc.	+	+	+	+
Carbon monoxide: motor vehicles; petroleum and metal industry, etc.				+
Particles and aerosols				
Solid particles (carbon, ash or soot); fuel combustion; motor vehicles, etc.			+	+
Metal oxides and salts: catalyst particles; fuel combustion; motor vehicles, etc.			+	
Silicates, minerals, metal foams and other: metal industry, etc.			+	

*Influence: A-plant detriment, B-eye irritation, C-reduced visibility, D-other.

air pollution control it is important to distinguish the terms gas and vapor. Sometimes confusion arises from the fact that under various conditions of temperature and pressure the differences between gas and vapor are very small. A gas is a single well-defined thermodynamic phase and one of the four fundamental states of matter (along with solid, liquid and plasma). Generally, a substance in the *gaseous state* is considered as a gas if the temperature of this substance is above its critical points (the highest temperature at which this substance can be condensed), while *vapor* refers to a gas phase at a temperature where the same substance can exist in the liquid or solid state, which is below the critical temperature of the substance. For example, water vapor can coexists with liquid water and with ice.

Gases can be organic or inorganic and typical examples of gases are: O_2, CO_2, SO_2, O_3, NO_x, CH_4 and similar gases. Vapor can be also organic and inorganic according to their chemical composition and typical examples include H_2O, Hg, but mostly different VOCs compounds. There are several definitions of VOCs. The European Union defines a VOC as any organic compound having an initial boiling point less than or equal to 250 °C measured at a standard atmospheric pressure of 101.3 kPa. This includes most of the organic compound with less than 12 carbon atoms and compounds which contain organic carbon, i.e. carbon bonded to the carbon, hydrogen, nitrogen or sulfur (excluded are CO, CO_2, hydrocarbon acid, metallic carbides or carbonates and ammonium carbonate). VOCs are numerous and include both human-made and naturally occurring chemical compounds (e. g. isoprene, terpenes, green leaf volatiles, molds). Most *odors* are also VOCs. The VOC Solvents Emissions Directive is the main policy instrument for the reduction of industrial emissions of VOCs in the European Union. It covers a wide range of solvent using activities (e. g. printing, surface cleaning, vehicle coating, dry clean-ing and manufacture of footwear and pharmaceutical products) [12].

The terms *particulates* (or particles) and *aerosols* can be regarded as synonyms, i.e. they include heterogeneous mixture or suspension of solid particles or liquid droplets in the air, with particle sizes ranging from 0.001 to over 100 μm (including dust, fumes, smoke, soot, spray and mists). According to chemical composition particles can be classified as: organic carbon compounds (carbon soot, condensable organic compounds, PAHs, etc.), inorganic compounds (metal oxides, salts in the form of nitrates, sulfates, carbonates, chlorides, etc.) and others (silicates, minerals, etc.). Emission of particulates can be generated from different sources including: mechanical (powdering, crushing, traffic), chemical or thermal (particles formed by the chemical reactions or high-temperature evaporation followed by condensation) and biological sources (pollen, fungi, bacteria). Some particles are large and can be seen as soot or smoke and some are very small. Solid particles are at least of the order of about 0.002 μm (i.e. 2 nm). For the comparison a typical gas molecules dimensions are from 0.0001 to 0.001 μm, a typical strand of human hair is 70 μm in diameter and a grain of salt is about 100 μm. Solid particles with typical dimensions between 1 and 100 μm in diameter are called dust particles, while solid particles less than 1 μm in diameter are called fumes or smoke.

Dusts are solid airborne particles, often formed by operations such as grinding, crushing, milling and sanding. The size of the dust particles is important as there is a difference between inhalable and respirable dusts and the nature of the hazards they present. The EPA defines *inhalable dust* as a fraction of airborne dust entering the body, but lags behind in the nose, throat and upper respiratory tracts (size about 10 μm). *Respirable dust* containing dust particles small enough (all below 10 μm in size and 50 % below 3.5 μm) to penetrate deep into the lungs through the nose and upper respiratory system and as a rule, they can not "be cleaned" by natural mechanisms of mucous membrane cleaning and will be kept in the respiratory system. *Fumes* are solid particles formed by condensation from the gaseous state (e. g. lead fume). *Smoke* is a substance made up of small particles of carbonaceous matter in the air, resulting mainly from the burning of organic

material, such as wood or coal. The largest suspended particles are of the order of about 100 µm (i.e. 0.1 mm). Individual suspended particles can be classified according their size on:
- *coarse particles* (2.5–10 µm),
- *fine particles* (<2.5 µm) and
- *ultrafine particles* (<0.1 µm)

and commonly are referred to as PM_{10}, $PM_{2.5}$ and $PM_{0.1}$, respectively. For regulatory purposes PM_{10} fraction is usually collected. PM_{10} fraction includes all particles smaller than 10 µm, including $PM_{2.5}$ fraction. The EPA currently has standard that measure $PM_{2.5}$ and PM_{10}. The WHO air quality standards for PM_{10} is 50 µg/m^3 (24 h average) and 20 µg/m^3 (annual average) and for $PM_{2.5}$ 25 µg/m^3 (24 h average) and 10 µg/m^3 (annual average).

7.2 Transport and transformation of air pollutants in the atmosphere

Possibly one of the most important properties of air pollutants is their transboundary nature, because they can easily travel and affect the areas far away from the point of theirs origins. During movement or transport in the atmosphere air pollutants pass through various changes in location and chemical forms, due to chemical, biochemical and photochemical processes. The physico-chemical properties determine how a specific compound will interact in the environment. These properties also affect how those compounds move in the environment and how efficiently they can be removed from the environment.

Air pollutants released into atmosphere are transported by the wind and dispersed as a function of many variables, including (Figure 7.2):

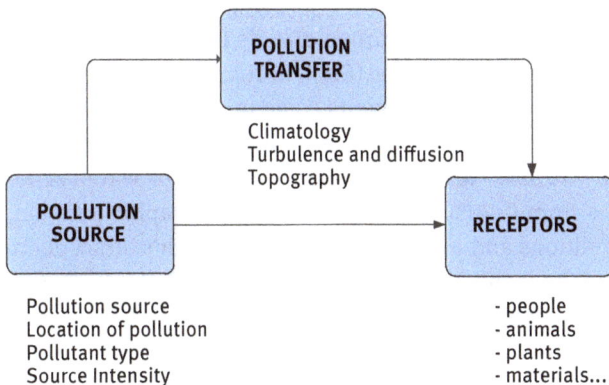

Figure 7.2: Air pollution transport and dispersion [13].

- properties of pollutant source (type of source, the concentration or quantity of pollutants, source location and intensity),
- climatology and meteorological parameters (e. g. wind speed and direction, atmospheric turbulence, ambient air temperature, relative humidity, solar radiation, atmospheric stability), local and regional geographic features (topography) and
- the receptor type and location [13].

Air quality is often modeled by air dispersion or diffusion models (e. g. the Gaussian plume models, the multiple cell models), represented by a set of mathematical equations describing physical and chemical processes in the air and solved by various numerical methods. They can be also used for air quality forecasting, which is important for emergency planning and operation of accidental chemical releases.

7.2.1 Primary and secondary air pollutants

Air pollutants may be additionally categorized as either primary or secondary. A **primary pollutant** is one that is emitted into the atmosphere directly from the source of the pollutant and retains the same chemical form, i.e. CO_2 or the ash produced by the burning of solid waste. The effect of primary pollutants is most pronounced in the immediate vicinity of their source. A **secondary pollutant** is formed by atmospheric reactions of precursor or primary emissions. Examples of the secondary air pollutants are ozone (O_3), secondary nitrogen dioxide (NO_2), peroxyacetyl nitrates (PANs), and secondary PM.

7.2.1.1 Formation of secondary air pollutants in the atmosphere

The emission of pollutants into the air as a result of various human activities may change the natural dynamics of processes in the atmosphere. On the dry volume base air consists almost entirely of the four gases: molecular nitrogen (N_2, 78 %), oxygen (O_2, 20.94 %), argon (Ar, 0.93 %) and carbon dioxide (CO_2, 0.04 %), while other gases like neon (Ne), helium (He), methane (CH_4), nitrous oxide (N_2O), ozone (O_3), hydroxyl radicals and some other constituents come in traces. Even though some gases such as O_3, N_2O, CH_4 are at trace concentration, they exert profound effects on the environment. Air also contains a variable amount of water vapor (H_2O), with value that varies from 0–4 %, depending on the geographic position, part of the day, weather conditions and other parameters. The Earth's atmosphere extends to a height of several thousand kilometers and can be divided into several layers: troposphere (0–12 km), stratosphere (12–50 km), mesosphere (50–85 km), thermosphere (85–500 km) and exosphere (>500 km). The most important parts of the atmosphere with respect to air pollution are **troposphere** and **stratosphere** (Figure 7.3).

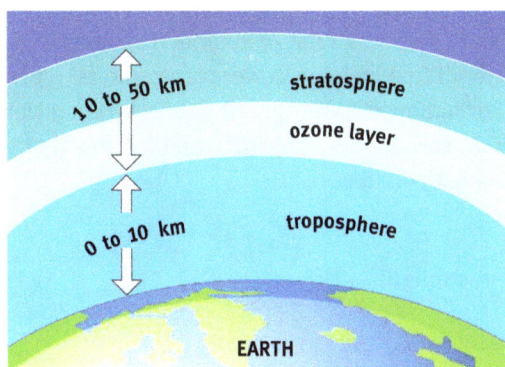

Figure 7.3: The most important layers of the atmosphere for the human life [14].

7.2.1.2 Tropospheric vs stratospheric ozone

Emissions of air pollutants contribute to local, regional (or transboundary) and global air pollution (see Figure 7.4.).

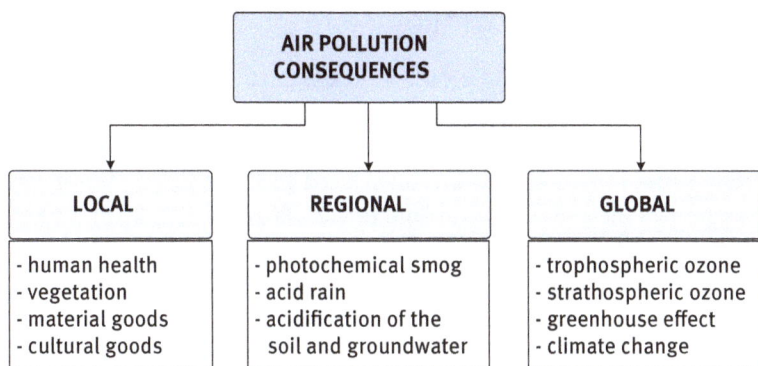

Figure 7.4: Consequences of air pollution – the influence of air pollution on different scales.

When we talk about the local and urban air pollution, the emphasis is mostly on the protection of the human health and preservation of natural biodiversity, protection of the materials and buildings, historical and cultural heritage materials. *Transboundary air pollution* refers to pollution transported in the atmosphere from one country or region to another and often undergoing chemical transformation in the process. Acidification, eutrophication and ground-level ozone are types of transboundary air pollution, arising from the emissions of the air pollutants such as: NO_x, SO_2, NMVOC, NH_3.

Ozone (O_3) is a secondary air pollutant which is found in two different layers of the atmosphere: troposphere and stratosphere. In the troposphere, the ground-level, **tropospheric or "bad" ozone** is a secondary air pollutant that damages human health, vegetation and many common materials. It is also a key ingredient of the urban photochemical smog and the third-most important greenhouse gas, after carbon dioxide and methane. The primary sources of the pollutants responsible for formation of photochemical smog are fossil fuel-burning power plants and all types of the internal combustion engines. Formation of tropospheric ozone by chemical reactions between nitrogen oxides (NO_x), carbon monoxide (CO) and hydrocarbons (HCs) in the presence of sunlight is summarized in eq. (7.1):

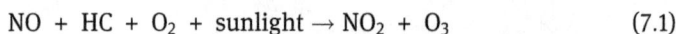

$$NO + HC + O_2 + \text{sunlight} \rightarrow NO_2 + O_3 \tag{7.1}$$

NO_x and O_3 interact in the atmosphere by the following reactions:

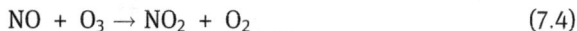

$$NO_2 + h\nu \rightarrow O + NO \ (\lambda < 420nm) \tag{7.2}$$

$$O + O_2 + M \rightarrow O_3 + M \tag{7.3}$$

$$NO + O_3 \rightarrow NO_2 + O_2 \tag{7.4}$$

where hv represents a photon of light of proper wavelength and M denotes any other molecule which must carry away some of the energy released in the reaction (usually N_2 or O_2). Instead of HCs the other terms like non-methane hydrocarbons (NMHC), VOCs and reactive organic gases (ROG) are often used in this context. The major anthropogenic sources of these chemicals are motor vehicle exhaust, industrial emissions and chemical solvents. Nitric oxide (NO) is also an important component of aircraft exhaust, which is formed by oxidation of atmospheric N_2 at the high operating temperatures of the aircraft engine. NO easily react with oxygen in the air to form nitrogen dioxide (NO_2), a foul-smelling brown gas – secondary air pollutant. Some of the NO_x reacts with hydroxyl radicals (OH˙) and produce HNO_3 which fall to the Earth in wet or dry forms, including acid rain, fog or dust that is acidic. Although O_3 is the main cause of the photochemical smog formation, some aldehydes (RCHO), peroxy and hydroperoxy radicals and peroxyacyl nitrates (PANs or $RC(O)OONO_2$) also contribute to this process through the similar mechanism [15]. PANs are powerful respiratory and eye irritants present in photochemical smog. As secondary air pollutants they are formed from other pollutants by chemical reactions in the atmosphere. For example, formation of the peroxyacetyl nitrate ($CH_3COOONO_2$) can be described by:

$$HC + O_2 + NO_2 + h\nu \rightarrow CH_3COOONO_2 \tag{7.5}$$

Since PANs dissociate quite slowly in the atmosphere into radicals and NO_2, they are able to transport these unstable compounds far away from the urban and industrial origin.

From a global point of view, the environmental problems which have to be solved or at least minimized involve the **stratospheric ozone depletion** (the so-called ozone holes), the greenhouse effect as well as the global climate changes and similar emerging environmental problems. In the stratosphere, we find **the stratospheric or "good" ozone** that protects life on the Earth from the harmful effects of the Sun's ultraviolet (UV) radiation. About 90 % of the ozone in the Earth's atmosphere lies in the stratosphere (between 16 and 48 km above the surface of the Earth), where it is a kind of layer (ozone layer) in the stratosphere. Although O_3 is concentrated in the ozone layer more than anywhere else, its concentrations are extremely small, typically only 1 to 10 parts of ozone per 1 million parts of air, compared with about 210,000 parts of oxygen per 1 million parts of air. The concentration of the stratospheric ozone is result of dynamic equilibrium between the chemical processes of its formation and processes of its destruction [16]. The first photochemical mechanism of the stratospheric ozone formation was proposed in 1930 by the famous British scientist Sydney Chapman, who assumed that ozone formation occurs by continuously cyclical processes that begin by photolysis of oxygen in the upper stratosphere. However, it was soon found out that Chapman mechanism predict much higher ozone concentrations than actually measured and it was assumed that ozone is consumed by additional chemical reactions. A large contribution to the understanding of the stratospheric chemistry was given by Paul Crutzen, who won the Nobel Prize for chemistry (in 1995 with Mario José Molina and Frank Sherwood). Paul Crutzen linked nitrogen oxides with the chemistry of stratospheric ozone depletion. He pointed out that emissions of nitrous oxide (N_2O), a stable, long-lived gas produced by soil bacteria or formed as result of the increasing use of fertilizers, could affect the amount of nitric oxide (NO) in the stratosphere. Mario José Molina and Frank Sherwood Rowland predicted the influence of chlorine released from chlorofluorocarbons degradation to stratospheric ozone depletion. It is worth mentioning that the ozone hole over Antarctica was discovered even in 1985, thanks to a British geophysicist Joe Farman.

According to the Chapman mechanism the ozone is produced by the photolysis of oxygen molecules into oxygen atoms, shown in eq. (7.6), followed by the reaction of one oxygen atom with an oxygen molecule, eq. (7.7):

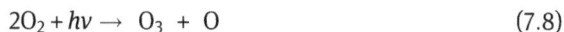

$$O_2 + hv \rightarrow O + O (\lambda < 242 nm) \tag{7.6}$$

$$O + O_2 + M \rightarrow O_3 + M \tag{7.7}$$

$$2O_2 + hv \rightarrow O_3 + O \tag{7.8}$$

The third molecule, M is needed to remove the excess energy and can be any other molecule in the atmosphere (usually N_2 or O_2). The overall reaction eq. (7.8) indicates that ozone and oxygen atoms are produced from oxygen molecules and light (hv). Ozone can be also destroyed by photodissociation of ozone in the presence of ultraviolet radiation (UV), eq. (7.9) or by reaction with an oxygen atom, eq. (7.10):

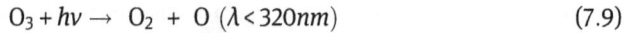

$$O_3 + hv \rightarrow O_2 + O \ (\lambda < 320nm) \tag{7.9}$$

However, reaction given by eq. (7.9) produces an oxygen atom and oxygen molecule, which can participate in the regeneration of oxygen molecule and the recombination of ozone as in eq. (7.7). The above reaction is the predominant mechanism by which ozone performs the function of shielding the Earth from harmful UV light. The formation of O_3 is balanced by its dissociation through several reactions and one of them is:

$$O + O_3 \rightarrow 2\ O_2 \tag{7.10}$$

There are many other ways in which ozone can be destroyed besides by the reaction given by eq. (7.10). Paul Crutzen assumed in the early 1970s that ozone can be catalytically destroyed in a cycle involving NO_x (x = 1, 2). He proposed the following catalytic cycle:

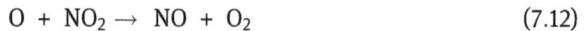

$$NO + O_3 \rightarrow NO_2 + O_2 \tag{7.11}$$

$$O + NO_2 \rightarrow NO + O_2 \tag{7.12}$$

The overall reaction is also given by the eq. (7.10), indicating that cycle of ozone destruction is only catalyzed by NO_x, because the NO consumed in reaction (11) is regenerated in reaction (12) and the NO_2 produced by reaction (11) is consumed in reaction (12). Obviously, this cycle can be repeat as long as NO_x is available for the reaction. NO_x in the stratosphere originate from natural and anthropogenic sources. A natural source of NO_x is the oxidation of N_2O released by bacteria in soil and oceanic microorganisms. An anthropogenic source of N_2O is fertilization. N_2O reacts with highly energized oxygen atoms which results in formation of NO molecules. Some other molecules can also participate in the chemistry of the stratospheric ozone destruction, such as chlorine or bromine molecule, eqs (7.13)–(7.15)

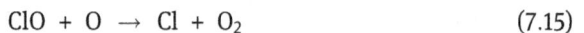

$$CCl_3F + hv \rightarrow CCl_2F + Cl \ (\lambda < 230nm) \tag{7.13}$$

$$Cl + O_3 \rightarrow ClO + O_2 \tag{7.14}$$

$$ClO + O \rightarrow Cl + O_2 \tag{7.15}$$

with the same overall reaction given by eq. (7.10). According to some estimation one chlorine atom can participate in the degradation of 10^4–10^6 ozone molecules. The main source of chlorine molecules is CFCs, known by the DuPont brand name *freons*. Many CFCs, like CF_2Cl_2 (Freon-12, CFC-12) and $CFCl_3$ (Freon 11, CFC-11), are chemically stable molecules in the troposphere, but they drift into the upper atmosphere where their chlorine components destroy ozone. In the past, CFCs were used as inert, non-toxic and easily liquefied chemicals in refrigeration, air-conditioning, as solvents, as well as aerosol propellants. Because of their recognized contribution to the depletion of stratospheric ozone, after the signing of the Montreal protocol in 1987, they are replaced with other compounds such as HFCs. HFCs are "safer" than CFCs for the ozone layer, because they react in the troposphere. Bromine species are 50–100 times more destructive for ozone than chlorine species [17]. The main source of bromine in the stratosphere is methyl bromide. It is emitted naturally by oceanic biological activity and from anthropogenic activities, such as soil fumigation and biomass burning. The well-known anthropogenic bromine sources are halons used in fire extinguishing systems, agriculture, dry cleaning and other applications. Fortunately, production of halons is currently being regulated for the same reasons like CFCs. Hydroxyl radicals ($OH^.$) can also catalyze the dissociation of ozone in the stratosphere; however, this kind of reaction is estimated to account for the degradation of only ca. 15 % of the stratospheric ozone. Many of the molecules released on Earth do not reach the stratosphere because they are soluble in water and return to the surface in precipitation or are broken down by chemical reactions in the troposphere.

7.2.1.3 Secondary particulate maters

PM can be directly emitted to the atmosphere (primary PMs) and formed in the atmosphere (secondary PMs). Primary PM originates from both natural and anthropogenic sources. Natural sources of PMs include sea salt, naturally suspended dust, pollen, volcanic ash. Anthropogenic sources of PMs, such as dust or soot (black carbon), include fuel combustion in thermal power generation, waste incineration, domestic heating for households and fuel combustion for vehicles, as well as vehicle (tyre and brake), road wear and other types of anthropogenic dust. The main precursor gases for secondary PMs are SO_2, NO_X, NH_3 and VOCs. The largest sources of NH_3 emissions are agricultural activities and NH_3-based fertilizer applications. Other sources of NH_3 emissions include some industrial processes, biomass burning (including forest fires) and to a lesser extent fossil fuel combustion and volatilization from soils and oceans. The result of the atmospheric reactions in which participate NH_3, SO_2 and NO_X is formation of ammonium (NH_4^+), sulfate (SO_4^{-2}) and nitrate (NO_3^-) compounds, which form new particles or condense onto existing ones and form *secondary inorganic aerosols*. Some VOCs also contribute to formation of the *secondary organic aerosols*.

7.2.2 Acid rain

Acid rain or acid deposition is considered as washout of air pollutants by precipitation. Any precipitation with a pH level less than 5.6 is considered to be acid rainfall. The dominant precursors of acid deposition are sulfur dioxide (SO_2) and nitrogen oxides (NO_x). The acidity of acid precipitation depends on emission level as well as on the chemical mixtures in which SO_2 and NO_x interact in the atmosphere through several chemical reactions (photo oxidation by means of ultraviolet light, reaction with moisture found in the atmosphere, reaction with ozone). A prerequisite for acid deposition is oxidation of the emitted SO_2 and NO_x either in the gas phase, after absorption into water droplets (sulfuric, nitric or nitrous acid) or after dry deposition on the ground (suspended sulfates and nitrates). These materials can be deposited on the ground unchanged (as primary pollutants) or in a transformed state (as secondary pollutants). Transformed pollutants can be deposited in wet form (rain, fog, snow) or in dry form (sulfates, nitrates).

7.2.3 Greenhouse effect

Many GHGs, such as water vapor, CO_2, CH_4 and N_2O occur naturally in the atmosphere. It is important to distinguish *the natural greenhouse effect* and *the enhanced or anthropogenic greenhouse effect*. The natural greenhouse effect is a phenomenon caused by gases naturally present in the atmosphere that affect the behavior of the heat energy radiated by the sun. Without this natural process the average temperature on Earth would be about −18 °C, instead of the current about 15 °C. On the other hand, the disruption to Earth's climate equilibrium caused by the increased concentrations of greenhouse gases has led to an increase in the global average surface temperatures and this process is called *the enhanced greenhouse effect*. It should be taken in mind that prior to the Industrial Revolution, the components of the greenhouse effect were in balance, but today's increasing amounts of greenhouse gas emissions are created mostly by the human activities, such as the burning of fossil fuels, new industrial processes, deforestation and extensive agriculture.

Concentrations of GHGs in the air vary daily, seasonally and annually. They have specific physical and chemical properties, which make them to interact with solar radiation and infrared light (heat) given off from the Earth. In such a way they affect the energy balance of the globe. As already mentioned, H_2O is also a greenhouse gas, since it traps heat in the atmosphere. However, the most abundant and the most problematic GHGs is CO_2. It is convenient to use carbon dioxide equivalent as a measure to compare the emissions from various greenhouse gases based upon their global warming potential. For example, the global warming potential for methane over 100 years is 21, which means that emissions of one million metric tons of CH_4 is equivalent to emissions of 21 million metric tons of CO_2. CH_4 is even more potent GHG than CO_2, but its atmospheric residence time (9 years) is far less than that of CO_2 (between 5 and 100 years). Since 1970, CO_2 emissions have increased by about 90 %, with emissions from fossil fuel combustion and industrial processes contributing about 78 % of the total greenhouse gas emissions increase from 1970 to 2011 [18]. On the global level, there are some signs of decreased global CO_2 emissions because of structural changes in the global economy, such as improvements in energy efficiency and the energy mix of the key global players (China, United States, India). However, small positive trends in this sense are probably coupled to a recent global economic recession [18].

7.2.4 Global climate change *vs* global warming

Sometimes is hard to recognize the difference between *global climate change* and *global warming*, because a term climate change is frequently used in reference to greenhouse gas emissions from human activities. The term global warming is used to describe the current increase in the Earth's average temperature. An indicator of global warming is the Earth's average global temperatures (AGT). The AGT for the Earth in the twentieth century was about 15 °C, but it increased significantly during the last three decades [5].

On the other hand, *climate change* refers not only to global changes in temperature but also to changes in wind, rising sea levels, changing precipitation, the expansion of deserts in the subtropics, the length of seasons as well as the strength and frequency of extreme weather events, including heat waves, tornados, droughts, heavy rainfall with floods and heavy snowfall.

7.3 Air pollution control

7.3.1 Selection of the air pollution control approach and control technique

From the engineering point of view, the good air quality can be achieved by using different approaches. Some of the possible approaches to the air pollution control include:
- *the primary or preventative approach* – all control measures applied to prevent pollutants formation directly on the potential source of emission,
- *the secondary approach* – technical devices applied to remove pollutants from the gas stream before releasing to the environment *and*
- *the integral approach (or process-integrated techniques)* – combination of different engineering approaches and methods of control, identification and application of specific measures for simultaneous removal of different air pollutants across various process plants in order to achieve the environmental targets at least costs.

Depending on simplicity or complexity of air pollution problem, the single or complex technical devices and even combination of different process equipment can be used for the purpose of reducing/eliminating emission of pollutants into the atmosphere. The economy (i.e. investment, operating and maintenance costs), regulation, space constraints and availability of control equipment in the market are some of the possible limitations which define choice and installation of the appropriate technical equipment (BREF). In the framework of the implementation of the *Industrial Emissions Directive* (2010/75/EU) [19] and as the result of the exchange of information between EU Member States, the European Commission's Joint Research Centre prepared *The best available techniques (BAT) reference document (BREF) for Common Waste Water and Waste Gas Treatment/Management Systems in the Chemical Sector (BREF)* [10]. This document provides general information on waste water and waste gases and describes techniques to prevent or to reduce their environmental impact. An overview of the available control techniques for common waste gas treatment in chemical sector is given in Figure 7.5.

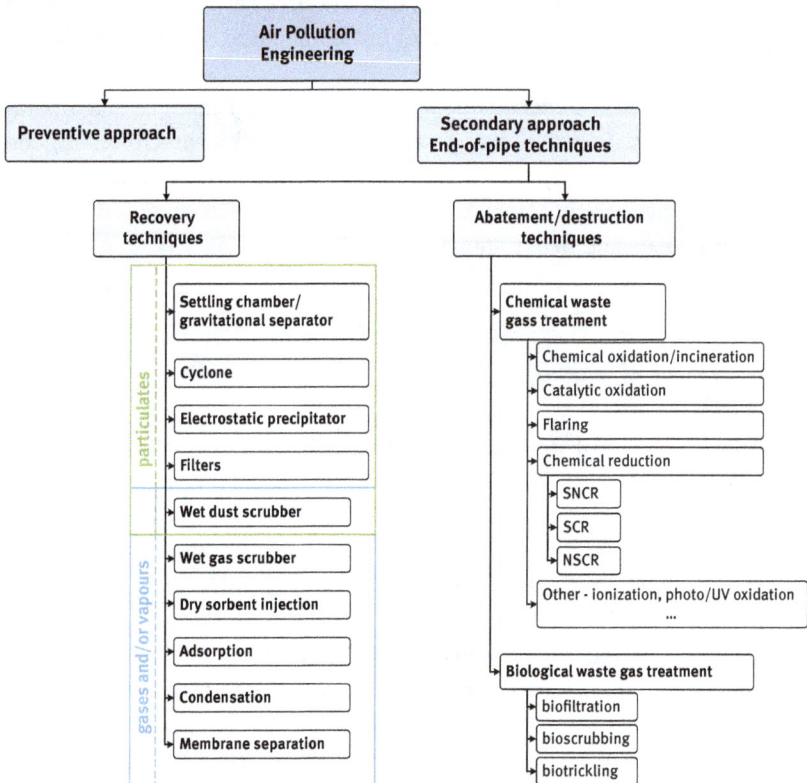

Figure 7.5: Air pollution control techniques for removing harmful gases, vapors and particulates from industrial process emissions (inspired by [10]).
*SNCR, selective non-catalytic reduction of NO_x; SCR, selective catalytic reduction of NO_x; NSCR, non-selective catalytic reduction of NO_x.

7.3.1.1 The primary or preventative approach to the air pollution control

An efficient air pollution control requires a complete knowledge of the air contaminants and the source of their emission. It is desirable to prevent the formation of pollutants whenever is possible. Changing or elimination of process steps that are basic sources of pollution is often one of the possible primary or preventative methods available for control of emissions of air pollutants discharged by the industries. Examples of such control measures to meet the emission standards are based on consideration of the source reduction opportunities, careful planning and process optimization to minimalize pollutants and carrier fluid generation at the source and include: operational changes, minimizing the volumetric flow rate of the stream to be treated, substitution of raw materials or fuels, use of efficient engines and more efficient burners design, delocalization of the plant, process equipment modification or replacement of old process equipment and other types of the process-oriented specific measures and control strategies. A modern nuclear power plants appears to be relatively pollution free solutions, compared to the more familiar fossil fuel-fired plants, which emit a lot of pollutants (carbon oxides, nitrogen oxides, sulfur dioxide, hydrocarbons and fly ash). However, waste and spent-fuel disposal problems may limit the apparent advantages [20, 21]. The biggest improvement in the air quality in most cities of the United States and European Union was achieved by replacement of coal with natural gas, by switching vehicles from gasoline to unleaded gasoline, compressed natural gas and other more environmentally friendly fuels, by addition of oxygenated compounds to motor fuels, by using of low-sulfur fuels, etc. [15]. There are also additional preventative ways to reduce air pollution including proper planning and location of the industrial areas, development of new technologies and legal regulations as well as using of administrative controls. The government plays a very important role in prevention of all kind of environmental pollution through government regulations that forced industries to reduce pollutant discharge and to promote new developments in technology.

7.3.1.2 The secondary approach to the air pollution control

The secondary air pollution control approach is relied on recovery or abatement of contaminants from the waste gas streams and application of the end-of-pipe treatment devices. This implies installation of control device (or equipment) between the source of pollutant generation and its release to the atmosphere. In that case control may consist of either removal/degradation of the pollutant or conversion to a less polluting form or recovery of economically valuable waste products. These solutions are focused on the specific air quality objectives or emission limits and primarily directed on the well-defined (or controllable) sources of emissions. *The end-of-pipe treatment devices* can be applied only for ducted emissions, which means that collection hoods and a ventilation system are required upstream of the end-of-pipe abatement system. *Abatement* is the term used for all technical and mechanical engineering devices, methods and technologies which may help in reducing quantity of pollutants emitted into atmosphere. Such devices are divided into two categories: devices (or equipment) for reducing particulates and devices for reducing emissions of gaseous and vaporous pollutants, although some of devices can be also used for simultaneous removal of the both groups of pollutants. In many cases, heating or cooling of the gaseous effluent is required before it enters the appropriate control device.

The environmental engineers must be aware of the gas laws, thermodynamic properties and all reactions ensuring a satisfactory design of the control device. The key design issues for air pollution control equipment are usually based on consideration: the process conditions, the physico-chemical and other important properties of pollutants and the added value of the emitted compound. *The process conditions* include: the total gas flow rate (or velocity) and volume of the waste gas to be treated, temperature limitations, allowable pressure drop, degree of variability depending on the operating and process conditions (i.e. variation in the pollutant concentration, gas flow rate, temperature), power/energy requirements, removal efficiency requirements, etc. Besides nature of the air pollutant and its concentration (including minimum and maximum values) *typical properties of pollutants* usually considered during design of air pollution control system are: solubility, corrosivity, flammability, toxicity or reactivity (if pollutants are gaseous) and size range and distribution, particle shape, agglomeration tendencies, corrosivity, abrasivity, hygroscopic tendencies, stickiness, flammability, toxicity, conductivity, electrical resistivity or reactivity (if pollutants are in the form of particulates). *The added value of emitted product* is also very important and can affect the selection of adequate air pollution control equipment or the chose between: *the recovery techniques* (e. g. fabric filter, cyclone, adsorption, condensation) and *the abatement or destruction techniques* (e. g. thermal or catalytic oxidation, chemical reduction, biofiltration). Typical air pollutants for which recovery technique is economically feasible are those which concentrations in the waste stream are high enough, e. g. VOCs from solvent vapors and vapors of low-boiling compounds, NH_3 (to recycle in the production process), SO_2 (converted into H_2SO_4, sulfur or gypsum), dust which contains higher amounts of solid raw products or end products [10]. Some of the mentioned recovery techniques can be used to minimize emissions of odors which also belong to the highly volatile compounds.

7.3.1.3 The integral approach (or process-integrated techniques)
In order to achieve all the stringent criteria of the air protection it is often necessary to combine different engineering approaches and methods of control, i.e. an integrated approach and comprehensive solution of the environmental pollution related to certain human activity. Some waste gas techniques (like scrubbing, adsorption, electrostatic precipitation) require further downstream treatment of the waste water or solid waste generated during the process of air purification.

Unfortunately, the cost of the waste disposal can be a significant part of the total cost of the air pollution control.

Engineers working in air pollution control have many responsibilities and one of them is the design of air pollution equipment in order to achieve emission standards and the requirements of legal regulations. The key challenges in using appropriate equipment for air pollution control are [10]:
- quantity and concentrations of pollutants in air streams (including physical and chemical properties of the effluent from the emission source),
- available space and equipment location,
- contribution of air pollution control system to wastewater and land pollution as well as disposal of waste recovered,
- economy (costs of investment and installation of the process equipment, costs of maintenance) and
- safety.

7.3.2 Appropriate use of units and some basics of air pollution control

To understand air pollution control, one first must understand some fundamental concepts and properties of the common air pollutants. Typical mass concentration units for ambient measurements of air pollutants are mass (µg) per unit volume (m^3). Concentration of air pollutant can be also expressed in parts per million (ppm), which is almost always ppm *by volume or by mole* if it is concentration in a gas and ppm *by mass or weight* when applied to liquids or solids. For example, concentration of 1 ppm can be visualized as 1 cubic centimeter (cm^3) of gas per cubic meter of air (or 1 molecule of the gas in question per 1,000,000 molecules of all gases present). For gaseous pollutants these concentration units are related to each other through the ideal gas law, taking into account that a gas may occupy different volumes depending on operating temperature and pressure. The ideal gas law is commonly written as

$$pV = nRT \qquad (7.16)$$

where p is the gas pressure (Pa), V is gas volume (m^3), n is number of moles (mol), R is universal gas law constant (8314 J mol^{-1}K^{-1}) and T is absolute temperature of the gas (K).

Particulates come in a variety of shapes and sizes and can be in the form of liquid droplets or dry dusts. Particles of concern to air pollution are typically measured in microns or micrometers (µm) or 1×10^{-6} m. Particles are frequently described by their diameter and size distributions. Many technical devices used for separation of particulates from the dirty gas streams exploit the difference in the physical size and/or mass (expressed by diameter and particles density) of particles and gas molecules. There are several diameters that are commonly used to define particle size, such as physical diameter, aerodynamic diameter, equivalent diameter, cut diameter and others. *The physical diameter* is the real diameter of a spherical or nearly spherical particle. To express the size of a non-spherical (irregular) particle as a diameter, several relationships are used, but most frequently aerodynamic and equivalent diameter. *The aerodynamic diameter, d_a* is the diameter of a spherical particle with the density of one gram per cubic centimeter that behaves aerodynamically in the same manner as the particle or aerosol of interest. That means that the actual particle and the spherical particle with density of $1 \, \text{g} \, \text{cm}^{-3}$ will have the same momentum and drag characteristics as well as the same terminal settling velocity, which is also an aerodynamic property. As known, the terminal settling velocity is result of the simple force balance between the force of gravity and resisting drag force. *The equivalent diameter, d_e* is the diameter of a sphere that has the same value of a physical property as that of the non-spherical particle and is given by

$$d_e = \left(\frac{6V}{\pi}\right)^{1/3} \tag{7.17}$$

where V is volume of the particle. Sometimes is also used term *cut diameter, d_c*, which represents the diameter of particles collected with 50 % efficiency. The non-spherical shape of a particle can be quantified by a single quantity such as the sphericity, ψ, using a perfect sphere as a reference.

Particle size distribution can greatly affect the efficiency of any collection device. The separation or collection efficiency of the control equipment is normally specified before the equipment is purchased depending on emission standards and regulations. *The overall separation or collection efficiency, η* for the control device depends on the operating conditions as well as on the particle-size distribution. It may be calculated as

$$\eta(\%) = \frac{amount \ collected}{amount \ input} = \frac{input - output}{input} \tag{7.18}$$

In the case of particulates removal it is important to distinguish fractional separation efficiency, η_j from the overall separation efficiency, due to different size fractions of particulates in the waste gases. Information on particle size distribution can be obtained based on total number of collected particles or, more often, based on total mass of collected pollutant or a volume-based distribution which is closely related to a mass distribution. Alternatively, a surface-based distribution can be also used, depending on the measurement technique that is applied. It is common practice to represent particle size distribution in the form of either a frequency distribution curve or a cumulative distribution curve. Generally, particle size distribution is represented by a list of values or a mathematical function that defines the relative amount, typically by mass, of particles present according to their size. If the particulate size distribution is known it is possible to calculate the overall separation efficiency based on *the fractional separation efficiency, η_j* as follows:

$$\eta = \sum \eta_j m_j \tag{7.19}$$

where η_j is fractional separation efficiency (e. g. separation efficiency for the j-th size range of particles) and m_j is mass fraction of particles in the j-th size range.

The control of PM is an important aspect of industrial air pollution engineering. In all air pollution control devices, particles are separated from the surrounding fluid by the application of one or more external forces. These external forces acting on a particle include gravitational, centrifugal, electrostatic and other forces (e. g. drag force, buoyant force, magnetic force) and separation mechanisms (e. g. inertial impaction, direct interception, diffusion). The consequence of acting forces on a particle is *the settling (or terminal) velocity.* The settling velocity is a constant value of velocity reached when all forces (gravity, drag, buoyancy, etc.) acting on a particle are balanced. In order to determine an unknown particle settling velocity in the appropriate control device, it is important to understand the movement of a particle in the fluid (gas) stream and to know the flow regime. The flow regime is characterized by the Reynolds number. The Reynolds number (Re) for a particle is defined by the following equation:

$$Re = \frac{d_p v_r \rho_f}{\mu} \tag{7.20}$$

where d_p is particle diameter (m), v_r is the relative velocity of the particle (m s^{-1}), ρ_f is density of fluid (kg m^{-3}) and μ is fluid viscosity (kg m^{-1}s^{-1}).

Depending on the values of Reynolds number, it is possible to distinguish three flow regimes with corresponding equations to calculate the settling velocity [22]:

–   laminar regime or Stokes' low range (Re<1):

$$v_t = \frac{g\, d_p^{\,2}\, (\rho_p - \rho_f)}{18\mu} \tag{7.21}$$

–   transitional regime or intermediate low range (2 < Re < 500):

$$v_t = 0.153 \frac{g^{0.71}\, d_p^{1.14}\, (\rho_p - \rho_f)^{0.71}}{\mu^{0.43}\, \rho_f^{0.29}} \tag{7.22}$$

–   turbulent regime or Newton's low regime (500 < Re < 200,000):

$$v_t = 1.74 \left(\frac{g d_p (\rho_p - \rho_f)}{\rho_f} \right)^{0.5} \tag{7.23}$$

In the Stokes range (Re≤1) the particle diameter, d_p must be smaller than

$$d_p \leq \frac{0.304}{\sqrt[3]{\rho_p}} [mm] \tag{7.24}$$

which means smaller than 0.03 mm (or 30 μm) in order to remain within Stokes range, which is often the case in technical processes [23]. In the previous equations, e. g. eqs (7.21)–(7.24), ρ_p refers to the particle density. However, if the particle diameter is smaller than a lower limit, the particle is able to "slip" between the gas molecules and this slippage can affects the values of the settling velocity predicted by Stokes law. In this case the Cunningham correction factor, C has to be used to correct the settling velocity for this effect [22]:

$$v_t = \frac{g \, d_p^{\,2} \, (\rho_p - \rho_f)}{18\mu} C \tag{7.25}$$

The correction factor C can be calculated as [24]:

$$C = 1 + 2.0 \frac{\lambda}{d_p} \left[1.257 + 0.40 \exp(-0.55 d_p / \lambda) \right] \tag{7.26}$$

$$\lambda = \frac{\mu}{0.499 P \sqrt{8M/\pi RT}} \tag{7.27}$$

where λ is the mean free path of the gas (m), P is absolute pressure (Pa), R is universal gas constant, M is molecular weight (g mol^{-1}), T is absolute temperature (K) and μ is absolute viscosity (kg m^{-1}s^{-1}).

7.3.3 Control of particulate air pollution

They are several *preventive measures* which are applicable specifically to PM emissions such as decreasing or elimination of PM production. This can be accomplished in several ways: by change to process that does not require operations such as crushing, grinding, milling, polishing, pulverizing, spraying, by change from solid to liquid or gaseous material, by change from dry to wet solid material, by change particle size of solid material or by change to process that does not require particulate material [10, 21]. *The secondary particulate control devices* may be divided into two large groups: dry dust removal equipment and wet dust removal equipment [23].

Dry dust removal equipment include: gravity settling chamber (or gravity settlers), centrifugal settlers (cyclones), electrostatic precipitators (ESP) and bag houses (fabric filters), while different kind of scrubbers (like packed column scrubber, vortex scrubber, Venturi scrubber) and some configurations of the so-called wet ESPs belong to the *wet dust removal equipment*. Selection of the most suitable device for control of particulate air pollution is always related to a specific particulates removal problem and depends on two groups of criteria: *a gas-stream specific criteria* (particles concentration, particles size distribution, gas temperature, tendency for agglomeration, chemical reactivity, flammability and explosiveness, toxicity, wettability, odors and foaming properties, etc.) and *equipment specific criteria* (fractional collection efficiency, security, availability, adaptability to various operating conditions, pressure drop, maintenance, sensitivity to erosion, corrosion and foam formation, water requirement, power requirement, operational and investment costs) [10].

It is important to emphasize that different separation devices operate in different particle size ranges. According to the additional classification these control devices can be divided in: process when an external force is applied to the particle (mechanical collectors and ESP) and processes where the gas stream is forced to go through a barrier that cannot be passed by the dispersed particles, in the form of holes smaller than the particles or a droplet cloud (different kind of filters and scrubbers). For larger particulates (>10 µm) gravity and centrifugal forces can be very effective means of the particles removal. For the fine particles (<2 µm) it is useful to apply an electrostatic force in combination with particle charging (ESP) or Venturi scrubbers. Venturi scrubbers are very efficient even for the separation of particles with size down to a few micrometers. Filters show very high efficiencies over wide size ranges and in that sense they are very flexible in comparison with other separation techniques, although they also have some disadvantages as will be explained later. Cost and energy requirement are very important in the selection of equipment as well as costs of the produced waste water and waste solid treatment.

7.3.3.1 Mechanical collectors

In the following text three types of control devices will be considered, including gravity settlers and cyclone separators, which belong to the mechanical collectors and electrostatic precipitators (ESPs). All these devices are function of driving the particles to a solid wall, where they adhere to each other to form agglomerates that can be removed from the collection device and disposed of. Therefore, these devices are sometimes called *the wall collection devices*. Instead of driving the particles to a wall, filters and scrubbers divide the flow into smaller parts in order to collect the particles. For this reason they are called *the dividing collection devices* [15].

7.3.3.1.1 Gravity settler

An gravity settler is a long chamber through which the polluted gases pass very slowly (less than about $0.5\,\mathrm{ms^{-1}}$ for good results), allowing particles enough time to settle out at the bottom of the chamber (in the hopper or collector) under the action of gravity (Figure 7.6). The chamber is cleaned manually to dispose the waste. Gravity settlers are more efficient in the laminar flow conditions. Under intense turbulent flow conditions deposition of particles is difficult and therefore the efficiency is lower. The separation effect is more pronounced by reducing the gas stream velocity by means of baffles and similar design elements (e. g. horizontal trays or shelves). Gravity settler is the simplest control device which is efficient only for large particles (>50 μm) and therefore it is useful for processing of very "dirty" gases (from cement industry, metallurgical processes, etc.). Even with horizontal trays the minimum particle size which can be removed is about 10 μm. The pressure drop through the gravity settler is very small and consists mostly of entrance and exit losses. Separated particles might be contaminated with toxic or hazardous contents which need to be considered for further treatment or disposal. Gravity settler is usually used as pre-cleaner, e. g. as preliminary step in various filter systems and scrubbers [10].

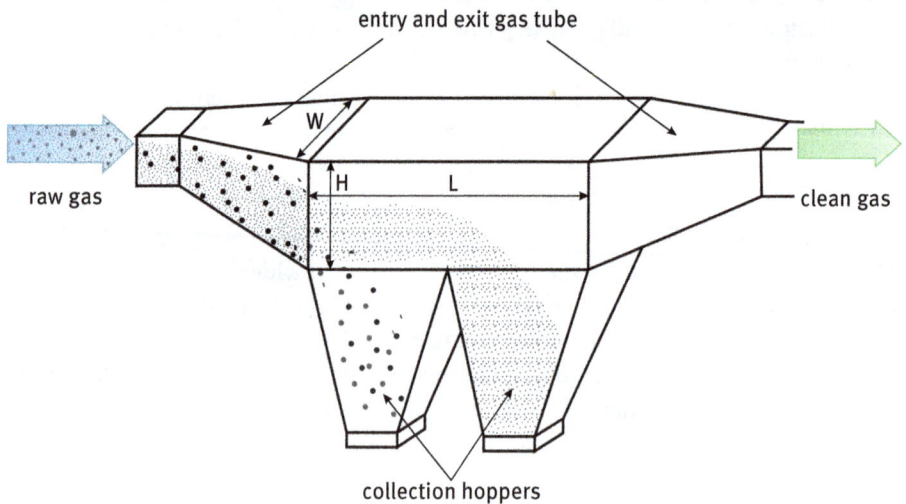

Figure 7.6: Schematic of gravity settler (inspired by [25]).

The mathematical analysis of the gravity settlers is very easy and with some exceptions it appears in similar form for cyclones and electrostatic precipitators. The separation efficiencies for laminar and turbulent settling chambers are given by the following equations [23]:

$$\eta_{la\,min\,ar} = \frac{v_t L}{uH} = \frac{g d_p^2 \rho_p}{18\mu} \frac{L}{uH} \tag{7.28}$$

$$\eta_{turbulent} = 1 - \exp\left(-\eta_{laminar}\right) \tag{7.29}$$

where v_t is the particle settling velocity (m s^{-1}) (Stokes low), L is chamber length (m), H is chamber height (m) and u is the average horizontal gas velocity in the chamber:

$$u = \frac{Q}{WH} \qquad (7.30)$$

In the eq. (7.30) Q is the volumetric air flow rate (m^{3}s^{-1}).

7.3.3.1.2 Cyclone

Cyclones are the most common centrifugal or inertial separators. Their efficiency is much better compared to gravity settlers and yields about 90–99 % for particles with diameter between 5 and 10 µm [10, 22]. At modest velocities and common diameters the centrifugal forces acting on particles can be two orders of magnitude larger than the gravity forces [15]. The cyclone consists of a vertically placed cylinder which has an inverted cone attached to its base. The gas stream containing particulates enters tangentially at the inlet point to the cylinder. The velocity of this inlet gas stream is then transformed into a confined vortex (called "external vortex"), from which centrifugal forces tend to drive the larger suspended particles to the walls of the cyclone. Due to the reduction of the flow cross section, the increase in pressure and the higher concentration of solid particles in cone part, a part of "external" vortex turn into "inner" directed upward at the bottom of the cone part. The clean gas (also with the smallest particles) is removed from a central cylindrical opening at the top (called "apex"), while the dust particles are collected at the bottom in a storage hopper by gravity. Schematic of a cyclone separator and the flow field inside a cyclone is given in Figure 7.7.

Cyclones are sized on the basis of centrifugal force, F_c

$$F_c = \frac{mv_c^2}{r} = m\omega_c r \qquad (7.31)$$

where m is mass, v_c is velocity, r is radius and ω_c is angular velocity for rotational motion of particle inside the cyclone. Using a centrifugal equivalent of Stokes' law, i.e. after substitution of the gravitational force by the centrifugal force in eq. (7.20) it is possible to calculate the settling velocity of particle for the centrifugal settling [15]:

$$v_t = \frac{v_c^2}{r} \frac{d_p^2 \rho_p}{18\mu} \qquad (7.32)$$

The cyclone diameter is important design parameter that has the highest effect on the cyclone separation efficiency. For a given pressure drop, smaller the diameter, greater is the efficiency, because centrifugal action also increases with decreasing

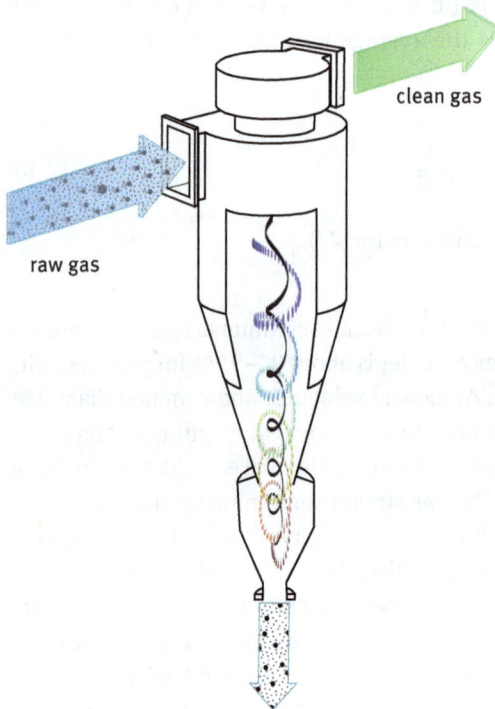

Figure 7.7: Schematic flow diagram of a standard centrifugal separator – cyclone (inspired by [25]).

radius of rotation. The efficiency of a cyclone can be increased by the use of a large number (up to seven thousand) or a battery of smaller cyclones (with a diameter ranging between 0.005 m and 0.3 m) operating either in parallel or serial. This kind of cyclone arrangements with a common inlet chamber, outlet plenum and collection system is known as multiple cyclones or multi-clones. The separation efficiency will also increase with increasing particle size, particle density, inlet particle loading, cyclone body length and ratio of body diameter to gas outlet tube diameter [10, 15, 26]. However, as efficiency increases the operating costs also increase because of the higher pressure drops. Common problems related to the work of the cyclone are erosion, corrosion and a dust cake builds up on the cyclone walls, especially if the dust is hygroscopic [26]. Cyclones are usually designed by set of procedures. More information on design considerations can be found out in the literature [15, 23, 26].

7.3.3.2 Electrostatic precipitator

An electrostatic precipitator (ESP) is a separation devices like gravity settler or cyclone, but it use electrostatic force to move particles entrained within a waste gas stream to the walls of the collector plates. It is very useful for the removal of fine dusts (between 1 and 10 μm, but mostly less than 5 μm) from different kind of waste gases with very high efficiency (approaching even 100 %) [10, 26]. The principle behind all electrostatic precipitators is to give electrostatic charge to particles in a given gas stream and then pass the particles through an electrostatic field that drives them to a collecting electrode of opposite charge. The overall process of dust separation in the ESP consists of the following steps: generation of electric field, generation of electric charges, transfer of electric charge to a dust particle, movement of a charged dust particle in an electric field to the collection electrode, adhesion of the charged dust particle to the surface of the collection electrode and mechanical removal of the dust layer from the collection electrode (by rapping, vibration or washing) to a hopper.

The electrostatic precipitators require maintenance of a high potential difference (20–100 kV, but mostly 40–60 kV) between the two electrodes: discharging electrode and collecting electrode. The applied voltage depends on the distance between electrodes. The high voltage between electrodes results in a strong electric field and electric discharge of ions from the discharging electrode called the corona. Besides on process conditions (flow rate, composition, temperature, dew point) and properties of particles such as concentration, size distribution, hygroscopy and tendency to agglomerate, the efficiency of ESP significantly depends on the particle resistivity. The particle resistivity is a measure of the resistance of the dust layer to the passage of a current. For efficient work of ESP, the resistivity should be between 10^7 and 10^{10} ohm-cm [10, 23]. Operation at the lower or higher resistivity leads to a decrease in removal efficiency. To improve efficiency it is possible to perform chemical conditioning of the flue gases (with moisture, by addition SO_3 gas, sodium or ammonium salts) or to work at the lower operating temperature (with caution because of the possible condensation and corrosion) [15].

The most important challenge in consideration of ESP design is calculation of collection efficiency or collection area. The collection efficiency and/or the collection area of an ESP can be estimated using several equations and one of them is known as the Deutch-Anderson equation [16, 26]:

$$\eta = 1 - \exp\left(-\frac{\omega A}{Q}\right) \tag{7.33}$$

where A is the effective collecting plate area of the precipitator (m^2), ω is the drift velocity or the particle migration in an electric force field ($m\ s^{-1}$) and Q is volumetric gas flow through the precipitator ($m^3 s^{-1}$). More accurate estimates of collection efficiency can be obtained by modifying the eq. (7.32) or by decreasing the calculation of collection efficiency by a factor of k. The resulting Matts–Ohnfeldt equation is given by [15, 26]:

$$\eta = 1 - \exp\left(-\frac{\omega A}{Q}\right)^k \tag{7.34}$$

where k is constant (usually 0.4 to 0.6). The particle migration velocity is the speed at which a particle, once charged, migrates toward the grounded collection electrode. Variables affecting ω are particle size, the strength of the electric field and the viscosity of the gas. The theoretical drift velocity of a spherical particle (or particle settling velocity in ESP) subjected to constant gas flows and electrostatic fields in an ESP is given by

$$\omega = \frac{Cd_p}{\mu} \varepsilon_0 \frac{\varepsilon}{(\varepsilon+2)} E^2 \tag{7.35}$$

C is the Cunningham correction factor, μ is gas viscosity (kg m^{-1} s^{-1}), ε_0 is permittivity of free space (8.85 10^{-12} C V^{-1} m^{-1}), ε is dielectric constant for the particle relative to free space and E is average field strength (V m^{-1}). The drift velocity is also related to the corona power, P_c (W):

$$\omega = \frac{kP_c}{A} \tag{7.36}$$

where k is an adjustable constant. The drift velocities are usually in the range 0.03 and 0.2 m s^{-1} [27]. After substitution of eq. (7.32) into **the Deutsch-Anderson equation**, eq. (7.32) the collection efficiency can be calculated as

$$\eta = 1 - \exp\left(-\frac{kP_c}{Q}\right) \tag{7.37}$$

Based on the shape of the collection electrode and depending on the mode of operation there are several types of ESPs: dry and wet wire-plate ESPs as well as dry and wet wire-pipe (or tubular) ESPs [10]. The discharge electrodes are commonly in the form of a wires (made from noncorrosive materials like tungsten or alloys of steel and copper), because of small radius of curvature. In the wire-plate ESP the high voltage electrodes are long wires (discharge electrodes) that are weighted and hang between the plates (collection electrodes). In this case the waste gas flows horizontally and parallel to vertical plates of ESP. Plates are usually 1–2 m wide, 3–6 m high and plates are at distance between 15 and 35 cm [10]. In the wire-pipe type precipitator the discharge electrodes are in the form of wires and the collecting electrodes consist of parallel pipes which may be round, square and octagonal. The pipe electrodes are 2–5 m high and about 30 cm (or less) in diameter. The space between the electrodes is in the range 8–20 cm. In dry ESPs the PM is removed from the collecting electrodes by rapping, while in the wet precipitator water or any other fluid can be used for removal of the solid particulates. Wet precipitators are more efficient, but dry type plate precipitators are mostly used in the practice. The most important fields of their application are: cement factories, steel plants, chemical industry, petroleum industry.

7.3.3.3 Fabric filters

Fabric filtration is a well-known physical technique in which a gas stream containing **dry particles** passes mostly through a porous fabric medium, which retains the solids. During this process, the contaminated gas flows into and passes through a number of filter bags placed in parallel, leaving the solid particles retained by the fabrics. The air pollution control equipment where this process takes place is called bag house. The fabric is a filter medium; however it is also a support for the layer of dust (dust cake or filter cake) that accumulates on it. This dust layer can significantly increase the efficiency of filtering, especially for very small particles. However, these filter bags must be periodically cleaned, because with the increase in the thickness of the filter cake also increases the pressure drop or the air resistance through the filter. The filter bags are usually cleaned by rapping, shaking or vibration as well as by the reverse air flow, causing falling of the filter cake into the hopper. Generally, the economy of filtration depends on separation efficiency, pressure drop and lifetime. According to some opinions, fabric filters belong to the most effective dust removal devices. Therefore, they are widely used in industrial application (steel industry, foundry, cement industry, power stations, etc.), although depending on the applied filter medium, they may be quite sensitive to high temperatures and humidity.

According to the chemical composition filter medium can be natural or man-made fibers, plastic, ceramic, glass or metals. The choice of fabrics is determined by the following conditions: temperature, gas composition and/or presence of acids or alkalis in a gas streams, gas flow rate as well as size and shape of dust particles and its concentration. The filter medium can be fibrous (e. g. cloth), granular (e. g. sand), a rigid solid (e. g. screen) and a mat (e. g. a field pad). It can be made in different forms, i.e. in the shape of a bag (the most common form), tube or sheet with a number of the individual filter units housed together in a bag house as well as in the form of bed or fluidized bed. Generally, filters can be divided into the following types: fabric or cloth filters (surface filters) and fibrous or deep bed filters (depth filters). *Surface filters* usually compete with cyclones and ESPs for treatment of gas stream contaminated with particulates. In the cloth filters, the filter is in the form of tubular bags, while in the case of deep-bed filters, a fibrous material (cotton, wool, cellulose, etc.) has the role of separator and the collection of particulates takes place in the interstices of the bed. The number of filter bags may vary between a few hundred and a few thousand. The diameter of a bag filter ranges from 0.1 to 0.4 m and the height may be up to 10 m. The velocities at which a gas stream is passed through the bags are very low 0.4–1 m min^{-1} [26]. **The total pressure drop through a filter bag**, ΔP (N m^{-2}) is given as the sum of the pressure drop owing to the fabric, ΔP_f and the drop owing to the caked or adhered particles, ΔP_p [22]:

$$\Delta P = \Delta P_f + \Delta P_p = u\mu \left(\frac{x_f}{60K_f} + \frac{x_p}{60K_p} \right) \tag{7.38}$$

where u is superficial filtering velocity (m min^{-1}), μ is gas viscosity (kg m^{-1}s^{-1}), x_f and x_p are thicknesses of the filter and the particulate layer (m), K_f and K_p are permeabilities of the filter and the particulate layer (m^2) (values of K_f and K_p must be determined experimentally for each type of fabric and type of dust removed) and 60 is conversion factor (s min^{-1}). The superficial filtering velocity, u is also known as **the air/cloth ratio** and can be determined as the ratio of the volumetric gas flow rate, Q (m^3 min^{-1}) and the cloth area, A (m^2). Excessive air/cloth ratio lowers particulate removal efficiency.

Ceramic and metal filters also belong to the surface filters. There are more resistant at much higher temperature than fabric filters (up to 1000 °C in the case of metal filters) and also they have higher chemical resistance. The process of filtering and removal of particles from the contaminated gas is comparable to that on fabric filters and only significant difference is in the filtering material. There are some special designs applicable for removal of compounds such as HCl, NO$_x$, SO$_x$ and dioxins, but the process of filtering in these cases involve the presence of catalysts and the injection of some reagents [10].

Deep-bed filters can be filled with granular materials as well as with fibers. They are usually composed of mats of wood, cellulose, glass or iron fibers. The most important filed of their application is in air conditioning, heating and ventilating systems as well as for the cleaning under the conditions of light dust loads. The advantage of a granular bed over a fiber filter bed can be the greater structural strength of a packed bed. Deep-bed fiber filters have the potential to remove the particles that are too small to be removed efficiently by any other type of surface filters. The low capital costs of small fiber filters make them very attractive for small gas streams (e. g. household furnace filters, air filters for automobiles).

7.3.3.4 Wet dust scrubbers

Wet collectors or scrubbers have several important advantages in comparison to other air pollution control devices. The most important advantages of scrubbers are: low initial cost, moderately high collection efficiency for small particles and applicability for the high temperature installations. They can handle large volume of gases and collect dust particulates like flammable, sticky and explosive dusts with a little risk. Additionally, wet scrubbing (or absorption) is process which can be used for the removal of gaseous pollutants (such as SO$_2$, NH$_3$, H$_2$S, hydrogen halides, VOCs) as well as for the removal of mists, fumes, heavy metals and suspended dusts [28]. They also provide cooling hot gases and neutralization of corrosive gases and dust. Based on the main purpose of their using, scrubbers can be divided on the wet gas scrubbers and the wet dust scrubbers. Special configurations of the so called dry scrubbers belong to the additional group scrubbers, which are very efficient for the elimination of the problematic liquid waste stream, acid and odorous gases by means of a dry sorbent injection (mostly alkaline material) into a gas stream. The dry scrubber is usually followed by a bag house to clean the effluent to acceptable levels.

Wet dust scrubbers are devices that remove PM by direct contact of the dirty gas stream with liquid drops (mostly water) (Figure 7.8). Scrubbing is usually carried out by intensively mixing of the incoming gas stream with water in combination with the removal of particles by means of centrifugal force. The absorption of particles into the liquid phase can be both physical and chemical absorption, depending on the particle and liquid phase and gaseous properties. By replacement of water with other aqueous solutions (lime, potassium carbonate or slurry of MgO) it is possible to perform simultaneous removal of gaseous and solid pollutants from a gas streams. Different types of scrubbers can be found on the market. Most of them can be grouped according to the contacting mechanism with liquid. The most popular scrubber designs which vary in terms of both function and performance are: spray tower scrubber, Venturi scrubbers, cyclone scrubbers, packed scrubbers and mechanically aided scrubber. According to energy requirements they can be classified as low-, moderate- and high-energy scrubber units. The simpler types of scrubbers with low energy requirements are efficient in collecting particles above 5–10 μm in diameter, while the more efficient, high energy input scrubbers are favorable for separation of very small particles (1–2 μm). Energy requirements are

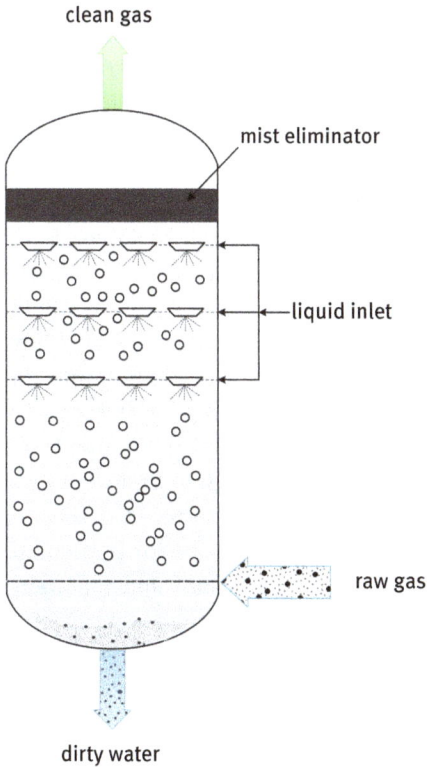

Figure 7.8: Spray tower with counter-current flow (inspired by [29]).

related to the pressure drop across the scrubber or to the level of contacting power. Obviously, the disadvantages of scrubbers are previously mentioned high power consumption for higher efficiency, moderate to high maintenance costs (corrosion and abrasion) as well as disposal of waste sludge which can be very expensive.

A key parameter in the design of scrubbers is the liquid-to-gas ratio (L/G). This ratio is determined by the solubility of the gas pollutants, concentrations of pollutants and particulates in the gas stream and the mass transfer properties of the scrubber. The increase in the (L/G) increases the separation efficiency of the system, so finding the optimum ratio is very important for the balancing of performance with operation costs. Besides energy needs and requirements for performance efficiency, the other important parameters for the selection of scrubber for a particular application include compatibility of materials used in the construction of the scrubber with the chemical (acidity, reactivity), physical properties of the gas stream (abrasivity), optimal operating temperature, need for addition reagents, etc.

In the *spray tower scrubber* the washing liquid is sprayed by spray nozzles, fast-spinning nebulizer disk or rotating sprays, creating a large contact surface for the drops and the inlet gas. Droplet size is controlled to optimize particle-liquid contact and to provide easy separation of droplets from the air stream. Very high (L/G) are required ($>3 \, 1 \, m^{-3}$) to capture fine particles. Power consumption for spray tower scrubber is low and they operate with a low pressure drop. Spray tower are capable for removal of particles $<8 \, \mu m$ with 90 % efficiency [22]. *Cyclone spray scrubbers* combine the capture techniques of cyclones and spray towers. Gas streams enter into the device tangentially at high speeds. The high speeds in combination with the centrifugal force promote droplets separation, allowing the use of a smaller droplet size, which also increases collection efficiency (95 % for particles $>5 \, \mu m$). They are more efficient than spray towers and have lower liquid requirements, but require more power, due to higher pressure drop. They are preferred over spray towers for gas streams with heavier particulate loads. *Venturi wet scrubbers* are applied for removal of fine particulate from corrosive, high temperature, hazardous or volatile gas streams. These are scrubbers with a Venturi-shaped chamber including converging and diverging sections. Water is injected at low pressure into the throat of the Venturi, through which the gas stream passes at high velocities. The energy from the gas atomizes the liquid, allowing particles and pollutants to be entrained in droplets. Venturi jet scrubbers use a modified design in which liquid at high velocity is jetted (injected) into the throat of a Venturi, rather than the gas stream. Venturi scrubbers have very high separation efficiencies for particulate pollution ($>98 \, \%$ for particles $>0.5 \, \mu m$) and are simple to install and maintain. However, they require large pressure drops leading to higher power requirements than other scrubber designs. Other often used designs involve packed scrubbers and mechanically aided scrubber. Packed scrubbers are occasionally used for particulate control, but they are restricted to separation of fine and/or soluble particulates, aerosol and mists, because of possible blockage of nozzles and plugging of bed. In line with this, they are limited to applications with low particle content ($<0.5\,g\ m^{-3}$). Scrubber can also use mechanically enhanced gas-liquid contact. However, the main disadvantage of mechanical scrubber is high maintenance cost.

7.3.4 Control of gaseous and vapor pollutants

This section gives an overview of several control devices for handling gaseous pollutants. Gaseous and vaporous pollutants may be removed from the waste gases by changing the process that produces the pollutants or by using the secondary treatment devices with the aim of trapping the pollutants from the waste stream (absorption, adsorption) or changing the physical state of matter from gas phase into liquid phase by cooling (condensation) or changing them chemically or biologically. A large number of gases and vapors which can be present in a gas streams are potentially explosive. Therefore, one of the main factors that must be considered in designing a gaseous contaminant control system is the concentration range at which one or more of the contaminants can be ignited.

7.3.4.1 Absorption

Absorption (also called gas absorption, gas scrubbing or gas washing) is the standard method of removing any compounds from a gas stream, which can be used if is possible to find a liquid solvent in which the gaseous pollutant is more soluble than the other compounds in the gas stream [10, 25]. The species transferred to the liquid phase are referred to as solutes or absorbate. This is probably one of the most important techniques in the control of gaseous pollutant emission, although the most important disadvantage of this process is the generation of waste water. The most important fields of application this technique include control of SO_2 and NO_x from combustion sources, removal and recovery of NH_3 in fertilizer production, removal of HF from glass furnace exhaust, control of odorous gases from rendering plants, recovery of water-soluble solvents such as acetone and methyl alcohol, etc. Absorption takes place in the absorption unit (absorber or scrubber) and basic principle of this process is the selective mass transfer of gaseous substance from the gas to liquid phase, owing to the preferential solubility of a gaseous component in the liquid. During this process the substance can be only dissolved in the liquid phase (physical absorption) or may react with the liquid or with a specific substance contained in the liquid (chemical absorption). As mentioned previously, water is commonly used liquid or absorbent. However, gases of very little solubility such as SO_x, H_2S and NO_x, can be absorbed readily in an alkaline solution such as NaOH [10]. Thus, when water is used as the solvent, it may contain added species, such as acids, alkaline, oxidants or reducing reagents to react with the gas being absorbed and to enhance its solubility. Absorption is favorable process for the recovery of products or for purification of a gas streams that have high concentration of organic compounds, otherwise the process is not economically acceptable. Generally, absorber can achieve separation efficiency higher than 95 %.

Basic principles of the process operation are described previously (see Section 7.3.3.4). The packed column absorber/scrubber is most commonly used technical device for separation of gaseous pollutants. The packed absorbers/scrubbers are usually filled with an inert material such as plastic or ceramic with the aim to increase the gas-liquid interface. The adsorbent can be in a wide variety of physical forms such as pellets in a thick bed, small beads in a fluidized bed or fibers pressed onto a flat surface. There are many types of packing and some of the commercially used are Raschig rings, Berl saddles, Pall rings and Intalox saddles. Generally, it is required for the packing to have large surface area per unit volume and low pressure drop of gas through it. Bubble caps and sieve trays can be also used in the interior of the absorption/scrubbing device to enhance counter-current contact between the liquid and gas stream. The main elements of the absorber design are: material balance to determine a liquid circulation rate, calculation of a packed absorber high or number of plates (depending on the type of absorber selected for the process) and determination of the absorber diameter sufficient to handle the required liquid and gas flow rates. The process opposite to the gas absorption is the stripping (Figure 7.9). After the solubility equilibrium is achieved during the process of absorption, the liquid which contains most of the gaseous compound removed from the waste gas, passes to the stripper – another device which works at much higher temperature and/or at much lower pressure than the absorber. Because of this the solubility of the gas in liquid is reduced, the gas comes out from the liquid and after the cooling it can be sent to storage. At the same time, the stripped or recovered liquid can be sent back to the absorber unit.

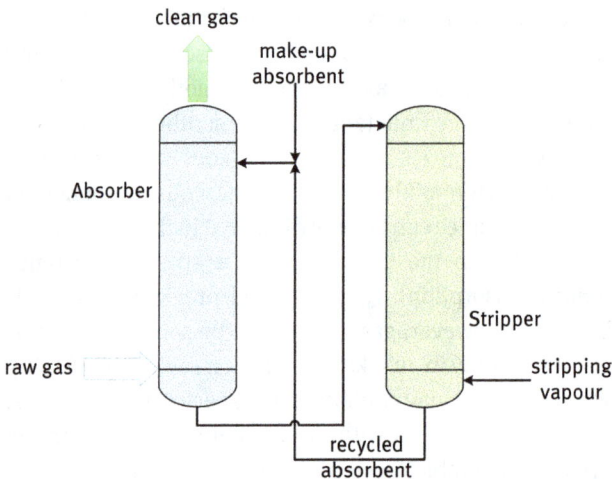

Figure 7.9: Absorption-stripping system.

7.3.4.2 Adsorption

Adsorption is another type of equilibrium separation process; in this case it involves a gas-solid equilibrium. Adsorption means the attachment or the binding of gaseous and vaporous molecules to the surface of a solid adsorbent or in the pores or interstices of adsorbent material. The adsorbing solids are usually called *adsorbent*, while the gas which is adsorbed is referred to as the *adsorptive* (Figure 7.10).

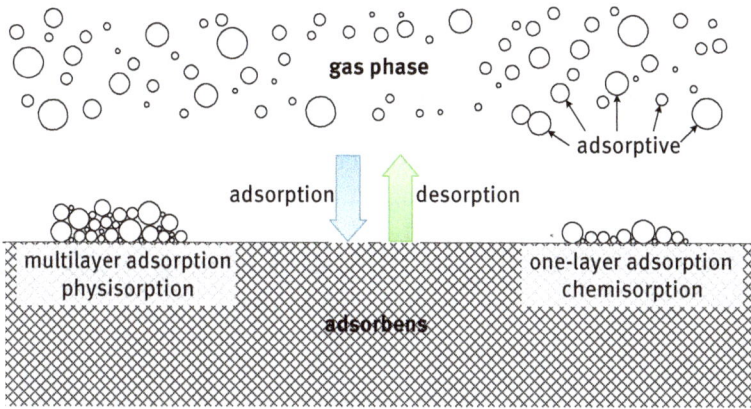

Figure 7.10: Simple illustration of gas adsorption (adopted from [23]).

Adsorption may be physical or chemical (chemisorption), depending on the strength of molecules binding to a surface of adsorbent. *Physical adsorption* (reversible process) is used primarily for the control of organic compounds, while *chemical adsorption* (irreversible process) is frequently used for the control of acid gases (hydrogen chloride, hydrogen fluoride, hydrogen sulfide) and mercury vapor. Depending on the flow rate of the polluted air, the adsorption process involves one or several adsorbers that operate in parallel. Adsorption is an exothermic process, because in the adsorption process the molecule loses kinetic energy. Contrary, the reverse process of adsorption is desorption and this process is endothermic. Conventional adsorption plant usually consists of two devices: one for the adsorption cycle and another for desorption, because the adsorbent must be regenerated after adsorption in order to be available for another process. Adsorption is mostly used to capture, concentrate and recovery a pollutant that is present in dilute form in waste stream. This means that adsorption may result in the recovery of economically valuable final product or in disposal of pollutants after previous treatment if there is a need. Separation efficiency of 95–99 % can be achieved using the adsorption process based on activated carbon as adsorbent [26].

The most often used adsorbent are activated carbon, silica gel, alumina, molecular sieve, polymer-based materials and zeolites. The common types of *adsorbent equipment* are thin bed absorbers and deep bed adsorbers. As the names of devices imply, the thin bed adsorber consists of a thin layer of adsorbed solid, which is preferable if adsorption is very quick process, while another type of adsorber is usually based on a deep bed of adsorbent to adsorb contaminants from the air that is heavily polluted.

The adsorbers usually work in the temperature range between 30 and 60 °C. Besides on temperature, the rate of adsorption is depended on the concentration of the adsorptive in the gas phase, the surface area and pore volume of the adsorbent and other properties of adsorbent (e. g. porosity, adsorption capacity, selectivity) and adsorptive. Regeneration of the adsorbent is usually carried out by increasing the temperature, by means of water vapor desorption at temperatures of 100 °C (and higher), by contact with a hot inert gas or by decreasing the total pressure (under vacuum). Reactivation of adsorbent in the case of chemisorption can be performed at the high temperatures (ca. 700 °C, depending on adsorbent nature). Design and analysis of adsorption system involves heat and mass transfer, fluid dynamics, process control and chemical analysis. More information on design of adsorbers can be found in [22, 23].

7.3.4.3 Condensation

Condensation is a separation process used to convert one or more volatile components of a vapor mixture to a liquid through saturation process. Condensation can be accomplished by reducing the temperature or by increasing the pressure. The first approach is more often used in the practice, because an increase in the vapor pressure can be expensive [30]. Condensers are typically used as a pre-treatment device to reduce the volume of the effluent gas before treatment by other devices. They can be also used for the reducing emissions from the high concentration gas streams to recovery of economically valuable products in a waste stream or to remove corrosive components and components that can damage other part of the system. Two most important representatives of condensers are contact condensers and surface condensers. In contact condenser the contaminated gas stream comes into contact with cold liquid by spraying ambient or chilled liquid directly into the gas stream. The direct mixing of the coolant and contaminant causes separation or extraction before coolant reuse. This additional separation process may lead to a disposal problem or secondary emissions. In the surface condensers, the coolant does not mix with the gas stream (which is advantage comparing to contact condensers), but contacts a cooled surface in which cooled liquid or gas is circulated. Surface condensers require less water and generate 10–20 times less condensation than contact condensers do. Removal efficiency of condensers range from 50 % to 95 %, depending on theirs design and application.

7.3.4.4 Chemical waste gas treatment

Chemical waste gas treatment methods involve process of *incineration* (thermal oxidation and catalytic oxidation) and flaring, in which an organic pollutant can be converted to CO_2, water and some inorganic gases. Incineration (or combustion) refers to the rapid oxidation of pollutants through the combination of oxygen from the air with supplementary combustible fuel in the presence of heat. The chemical conversion of pollutants (mostly gaseous but also solid and liquid particles) contained in waste gases is the most efficient process in the air pollution control which results in the complete degradation of almost all pollutants, independent on their chemical properties. Equipment for control waste gases by incineration can be categorized as: thermal oxidizers or afterburners, catalytic oxidizers and direct combustor or flare.

   The thermal oxidizer involves specifying a temperature (650–825 °C) of operation along with a desired residence time (0.5–1.0 s) and then optimum sizing the device to achieve the desired residence time and temperature with proper flow velocity. Efficiency of thermal oxidation can be higher than 99 % when operated correctly. However, if the process of ignition is not complete, some pollutants can be released into the atmosphere, which is undesirable. In *the catalytic oxidation* the presence of catalyst is responsible for achievement of high reaction rate at much lower temperatures (315–485 °C), which results in considerable savings in the fuel costs. The space requirement for operation of the catalytic oxidation is also much smaller. However, the costs of the catalytic systems are higher than those of the thermal oxidation systems, because of expensive catalysts and associates systems. Anyway, application of catalytic afterburners is favorable because of possibility to avoid formation of the thermal NO_x, dioxins and other undesirable products. Thermal NO_x are formed at the high temperatures (>1300 °C) due to complex reaction of oxygen and nitrogen with high energetic free radicals (O, N, OH, H and hydrocarbons that have lost one or more hydrogen). The catalysts are usually based of combination of a noble and other metals (Pd, Pt, Cr, Mn, Cu, Co, and Ni) deposited on an alumina monolithic support to minimize the pressure drop, which is one of the critical parameters of the incinerator designs. Limitation of the catalytic systems is catalyst deactivation, due to the presence of contaminants, attrition or catalyst sintering. Catalytic systems are suited for a gas stream with low VOC content. Depending on the types of heat recovery, both thermal and catalytic incinerators can be classified into two categories: recuperative and regenerative. The recuperative incinerator can recover about 70 % of the waste heat from the waste gases, while the regenerative incinerator can recover up to 95 % of the energy. *Flare gas systems* are used in a wide variety of applications and are generally used to burn off excess gas, usually hydrocarbons. A flare gas system may contain open flame flares or combustors (flame enclosed) with small differences between them. A flare can be used to control almost any emission stream containing VOCs [23, 28]

7.3.4.5 Biological waste gas treatment

In the biological control systems gaseous pollutants are converted into less harmful or harmless substances by microorganisms. Almost all kind of organic (hydrocarbons, chlorinated hydrocarbons, ketones, esters, aldehydes, odors) and inorganic pollutants (NH_3, NO_x, H_2S, CS_2) can be treated by microorganisms. There are two groups of microorganisms that may be used for this purpose: autotrophic bacteria which are particularly suitable for the conversion of inorganic substances and heterotrophic bacteria which are well suited for the conversion of organic pollutants. The microorganisms oxidize the pollutants and produce cell substrate, CO_2, water and inorganic salts. This process takes place in the presence of water and it is possible only for the very low pollutant concentration. Biological systems are especially suitable for degradation of VOCs, odors and toxic compounds. Obviously, they are similar to biological systems used for treatment of the wastewater and can be divided into three categories: biofilters, biotrickling filters and bioscrubbers (Table 7.3).

Table 7.3: Types of biological air pollution control systems [22].

Application	Biomass	Water phase
Biofilter	fixed	fixed
Biotrickling filter	fixed	flowing
Bioscrubber	flowing (suspended)	flowing

Biofilms of microorganisms are grown on porous media in biofilters and biotrickling filter, while a contaminated gas stream is passed through a porous packed bed at a low velocity. The main difference between those two systems is that liquid phase is stationary in a biofilter and moves through the porous media in a biotrickling system. Bioscrubbing combines wet gas scrubbing (absorption) and biodegradation. The scrubbing water with suspended microorganisms is recirculated through the reactor and the absorbed pollutants are degraded in aeration sludge tank. Bioscrubbing is suitable for removing of readily biodegradable compounds, such as NH_3, amines, hydrocarbons, H_2S and odorous pollutants. Biofiltration mainly use compost as the porous media is cheapest and thus widely used biological waste gas treatment for the removal of low concentration of pollutants that are easily soluble in water [10, 22].

Readers interested in the additional information on design and scale-up principles and operational guidance regarding to the control devices for cleaning gaseous pollutants can refer to available literature [5, 10, 13, 15, 23, 31].

References

[1]   https://www.theatlantic.com/photo/2012/12/2012-the-year-in-volcanic-activity/100423/. Accessed: 15 May 2017.

[2]   Trends in global CO$_2$ emissions: 2016 Report. The Hague: PBL Netherlands Environmental
       Assessment Agency and European Commission, Joint Research Centre, 2016. Available at:
       http://www.pbl.nl/en/publications/trends-in-global-co2-emissions-2016-report. Accessed:
       18 May 2017.

[3]   Stocker TF, Qin D, Plattner GK, Alexander SK, Allen NL, Bindoff FM, et al. Technical summary, In:
       The physical science basis. Contribution of Working Group I to the Fifth Assessment Report of
       the Intergovernmental Panel on Climate Change, IPCC, 2013:33.

[4]   Kremer M. Population growth and technological change: one million B. C. to 1990*. Q J Econ
       1993;108:681–716.

[5]   Cooper CD, Alley FC. Air pollution control: a design approach. Long Grove: Waveland Press, Inc.,
       2002.

[6]   https://www.epa.gov/research/strategic-research-action-plans-2016-2019. Accessed: 24 May
       2017.

[7]   http://www.epa.gov/iaq/pubs/hpguide.html. Accessed: 18 May 2017.

[8]   http://www.who.int/mediacentre/news/releases/2016/air-pollution-estimates/en/.
       Accessed: 20 July 2017.

[9]   Air quality in Europe: 2016 report. Luxembourg: European Environment Agency, 2016. Available
       at: https://www.eea.europa.eu/publications/air-quality-in-europe-2016. Accessed: 20 May
       2017.

[10]   Best available techniques (BAT) reference document for common waste water and waste gas
       treatment/management systems in the chemical sector. Luxembourg: Publication office of the
       European Union, 2016. Available at: http://eippcb.jrc.ec.europa.eu/reference/BREF/
       CWW_Bref_2016_published.pdf. Accessed: 18 May 2017

[11]   Petrović M, Tomašić V, Macan J. Environmental pollution. In: Macan MK, Petrović M, editors.
       Environmental analytics. Zagreb, Croatia: HINUS and Faculty of Chemical Engineering and
       Technology University of Zagreb, 2013:65.

[12]   http://ec.europa.eu/environment/archives/air/stationary/solvents/index.html. Accessed: 21
       May 2017.

[13]   Schnelle KB, Brown CA. Air pollution control technology handbook. Boca Raton: CRC Press, 2002.

[14]   http://schooltutoring.com/help/biology-review-of-the-ozone-hole-and-global-warming/.
       Accessed: 15 May 2017.

[15]   De Nevers N. Air pollution control engineering. New York: McGraw Hill, 1995.

[16]   https://chem.libretexts.org/Core/Physical_and_Theoretical_Chemistry/Kinetics/
       Case_Sudies%3A_Kinetics/Depletion_of_the_Ozone_Layer. Accessed: 22 May 2017.

[17]   Middlebrook AM, Tolbert MA. Stratospheric ozone depletion. Sausalito: University Science
       Books, 2000.

[18]   http://infographics.pbl.nl/website/globalco2-2016/. Accessed: 18 May 2017.

[19]   May 2017 http://eur-lex.europa.eu/legal-content/EN/TXT/PDF/?
       uri=CELEX:32010L0075&qid=1496146818076&from=en. Accessed: 23 May 2017.

[20]   May 2017. Accessed: 20 May 2017 https://beeindia.gov.in/sites/default/files/2Ch8.pdf.

[21]   Boubel RW, Fox DL, Turner DB, Stern AC. Fundamentals of air pollution control. San Diego:
       Academic Press, 1994.

[22]   Cooper CD, Theodore L, Buonicore AJ. Industrial air pollution control equipment for particu-
       lates. Cleveland: CRC Press, 1976.

[23]   Bauer H, Varma YB. Air pollution control equipment. Berlin: Springer Verlag, 1981.

[24]   Wark K, Warmer CF. Air pollution – its origin and control. New York: Harper & Row, 1981.

[25]   Kumar KS, Chenchaiah S. Controlling industrial air pollution emissions. In: Proceedings of
       International Conference on Impact of Industrialization on Environmental Pollution – Its Control
       and Abatement, Orissa, India, 2006:86–92.

[26] Meikap BC, Swar AK, Mohanty C, Sahu JN, Hung YT. Air pollution and its control. In: Hung YT, Wang LK, Shammas NK, editors. Handbook of environment and waste management: air and water pollution control. Singapore: World Scientific Publishing Co., 2012.

[27] Weiner RF, Matthews R. Environmental engineering. Amsterdam: Butterworth Heinemann, 2003.

[28] Vallero DA. Air pollution. Kirk-Othmer encyclopedia of chemical technology. John Wiley & Sons, 2015.

[29] Chandrappa R, Kulshrestha UC. Sustainable air pollution management- theory and practice. Switzerland: Springer Int. Pub., Environmental Science and Engineering, 2016.

[30] Pham TD, Lee BK, Lee CH, Nguyen MV. Emission control technology. In: Nejadkoorki F, editor(s). Current air quality issues. Rijeka, Croatia: InTech, 2015.

[31] Reynolds JP, Jeris JS, Theodore L. Handbook of chemical and environmental engineering calculations. New York: Wiley InterScience, 2002.

Questions and exercises to get you thinking

- What is the difference between primary and secondary air pollutants. Give some examples!
- Explain the greenhouse effect and specify some greenhouse gases!
- What is "sick building syndrome" and what does it link to?
- What are you doing to improve the air quality in your household?
- What are the major parameters that influence dispersion of air pollutants in the atmosphere?
- Describe typical sources and types of air emissions resulting from the oil refineries. What are examples of prevention pollution practice in this industry which kind of and which kind of the emission control equipment are used for the same purpose?
- Compare the major types of air pollution control devices used for particulates removal. Visit the Web and make a list of suppliers for these devices.
- Explain the better efficiency of multiple cyclones in comparison to a single cyclone.
- Describe the processes of absorption and desorption. Explain the similarities and differences of these processes.
- Adsorption isotherms can be described by different equations. List some of them!
- Describe the characteristic types of scrubbers that can be used for the simultaneous removal of particulates and gaseous pollutants.
- Does the Henry's law absorption limit influence the amount of air pollutant that can be absorbed in a wet scrubbing type system?
- Compare the most important advantages and disadvantages of thermal and catalytic oxidizers!
- Explain some heat recovery options during the incineration process.

Danijela Ašperger, Davor Dolar, Krešimir Košutić, Hrvoje Kušić,
Ana Lončarić Božić and Marija Vuković Domanovac

8 Water and wastewater treatment engineering

Abstract: This chapter introduces water and wastewater characterization and gives
an overview of the most important instrumental techniques used in water analysis.
Pollution sources, origin of pollutants (inorganic, artificial organic), and modern
removal techniques are particularly emphasized. The chapter describes the basic
principles of biological water treatment, as well as modern processes involving
membrane separation processes and advanced oxidation processes.

Keywords: water, wastewater, characterization, membrane separation processes,
biological treatment, advanced oxidation processes

8.1 Water and wastewater characterization and analysis

Water is in fact the most widespread mineral substance on the earth's surface. A
casual observation of the world map would suggest that the supply of water is
endless since it covers over two-third of the earth's surface. However, we cannot
use it directly since 90 % is in the salty oceans, 2 % is tied up in the polar ice
caps, and most of the remainder is beneath the earth's surface. Therefore only a
small fraction of the water in the world is available to manage for human use. It
is self-evidently an essential for life and many chemical analyses are carried out
to ensure that water of appropriate quality is supplied for human consumption.
Water has a wide variety of end users associated with it, like for recreation
(swimming, boating and fishing), and drinking are common water usages for
families, just as cooling, washing, and steam generation are common usages of
water for industry. Each of these usages has different water quality criteria.
Commonly there is some confusion between criteria and standards. Criteria are
the scientific requirements which a water source must meet in order to support a
designated use. Thus water quality criteria govern the input of water to a
particular use and will be different for each intended use. Standards, on the
other hand, govern the quality of the water after the user is through with the
water and before he discharges it back to the environment [1, 2].

Water quality criteria depend upon the use of the water and vary considerably in
the number and levels of the parameters to be considered. The water quality criteria for
streams has resulted in the development of four primary classes of water, where it is
probably the single most important criterion used in classifying a water as "polluted" is
the microbial count:

1. Class A: *Potable water supply* – quality criteria: microbiological counts, color,
   turbidity, pH, dissolved oxygen, toxic materials, taste, odor, temperature;

https://doi.org/10.1515/9783110468038-008

2. Class B: *Bathing, primary contact recreation, fishing* – quality criteria: same as A, but less stringent levels;
3. Class C: *Industrial, agricultural, fishing, navigation* – quality criteria: dissolved oxygen, pH, suspended solid, temperature;
4. Class D: *Cooling, sailing* – quality criteria: floating material, pH, suspended solids [2].

Quality criteria for wastewater uses are varied and generalizations are difficult. The bridge between the discharge of the water and the standard involves water treatment and purification. Since most industrial and domestic application use water for solids transport or for dissolving solids, it is imperative that many of these impurities be removed before the water is discharged to the environment. The water use cycle is a closed loop since water is conserved on our planet and consisting various stages: criteria-use-purification-standards-environment-criteria. We can no longer rely solely on the environment to absorb and treat the waste quantities of materials in our wastewater. To comply with effluent standards, the wastewater must be purified prior to its discharge [2].

Governmental bodies establish regulations setting standards for water discharge into the environment so that the criteria which have been set for reuse of that water can be met. Generally, those regulations are defined through parameters like soluble degradable organics [such as biochemical oxygen demand (BOD), chemical oxygen demand (COD), total organic carbon (TOC), and total oxygen demand] (Table 8.1) control the utilization or depletion of dissolved oxygen (DO) by the aerobic bacteria present in receiving streams or lakes. In setting standards

Table 8.1: Application of classical methods in analysis of water samples [3].

Parameters	Methods	Procedure
Volatile suspended solids	Gravimetry	Drying in an oven
Total solids	Gravimetry	Calcine in muffle furnace
Suspended solids	Gravimetry	Filtration, drying, calcine
Acidity	Titrymetry	Acid-base titration
Alkalinity	Titrymetry	Acid-base titration
Total hardness	Titrymetry	Complexometric titration
Carbonate hardness	Titrymetry	Acid-base titration
Chemical oxygen demand (COD)	Titrymetry	Redox titration
Biochemical oxygen demand (BOD)	Titrymetry	Redox titration
Oil and grease	Gravimetry	Extraction, distillation
Residue chlorine	Titrymetry	Redox titration
Chlorides	Titrymetry	Precipitation titration
Sulphates	Gravimetry	Filtration, calcine, precipitation titration
Sulphides	Titrymetry	Distillation, redox titration
Cyanides	Titrymetry	Titration of the complex formation
Ammonia	Titrymetry	Distillation, acid-base titration

for discharges, it is important to consider the concentration level of the pollutant and the flow rate of wastewater. A high flow rate can impose an unacceptable amount of pollutant on a receiving stream even if the concentration of pollutant is low. Percent removal is not an environmentally acceptable basis for setting standards. The total quantity as well as concentration of pollutant in the discharge should be controlled by the regulations. The water quality standards have legal backing and specify key characteristics for the water, dependent upon its intendent use. In conjunction with the specified tests and analyses a formal sampling protocol is also documented [2, 3].

Pollution of natural surface waters and groundwater by entering of unknown organic (pharmaceutics, pesticides and other organic substances) and inorganic constituents are increasing, how the number of inhabitants is increasing. At the same time, the consumption of potable water increases what caused requirements for better treatment of wastewaters. Namely, many wastewater treatment plants so far do not effectively remove all pollutants and become the main source of drinking water contamination. By recognizing the harmful effects of organic and inorganic pollutants, the requirements or standards to be met by quality drinking water are also changed. The increasingly stringent standards are also seeking to constantly improve the methods of processing but also the introduction of new modern procedures. For this reason, membrane processes have been increasingly used in practice, not only for the treatment of drinking water (sea, brackish and groundwater), but also for the treatment of wastewaters. In addition to the above-mentioned membrane processes, advanced oxidation processes (AOPs) are increasingly used, and classical water treatment methods are improved in accordance with established criteria and/or standards by scientific demand and government [2].

Drinking water treatment implies the most common pre-oxidation (aeration, chemical oxidation), polishing (removal of organic matter and micro-pollutants), disinfection, membrane processes, eliminating planktons and algal metabolites. Depending on the characteristics of raw water, specific treatments for removal of iron, manganese, ammonia, nitrates, hydrogen sulfide, arsenic, salinity, other metalloids and heavy metals, as well as organics should be applied. On the other hands, treatment of municipal or industrial wastewater is more and more demanding because the conventional methods of water treatment do not effectively remove so-called new pollutants like very small molecules of pharmaceuticals, pesticides and hormones. Therefore, it is essential to develop new methods of wastewater treatment (membrane processes, AOPs), to improve traditional or conventional methods, and to determine their effectiveness by monitoring pollutants removing after wastewater treatment with modern and highly sophisticated analytical techniques. In general terms the analytical methods used in the assessment of water and wastewater characteristics are very much the same irrespective of the ultimate purpose of the analysis. The choice of analytical method is often dictated by validation requirements such as detection limits, quantification limits, selectivity and sensitivity of methods

[2, 3]. In the same time industry is the greatest user of water, so future industrial growth will be restricted to regions having adequate water supplies. The major industrial users of water are the primary manufacturers of metals, chemicals, papers, petroleum and food products. The pollutant concentrations in wastewater are often characterized by solid content and by BOD, which is a measure of the DO used by microorganisms in biological oxidation of organic matter. The total BOD of aqueous industrial wastes is three times the total BOD of wastes entering municipal wastewater treatment plants. Over 90 % of the industrial BOD is generated by the chemical, paper, food, and petroleum industries. The primary metals industry together with these four industries contributes 90 % of the solids entering industrial wastewater. The total solid entering sewage treatment plants from domestic wastes are less than one-half of the total solid in industrial wastes [2].

How life standards are progressing, demand for water accelerates. We have to meet the needs of an increasing world population by irrigating more of the unproductive areas and fulfilling the demands for an even greater industrial output. These needs require a stewardship of our water resources to preserve water quality through waste treatment and ensure adequate quantities through recycle. This means there is a pressing need to protect water and treat it, whether for the purpose of producing fresh water for human consumption, for specific industrial uses, or for limiting pollution discharged into the environment [1, 2].

The classification of water and wastewater depends on the type of measurements test that has to be performed. Thus, water characterization is carried out on the basis of biological, physical and chemical measurements [4].

8.1.1 Biological characteristics

Biological tests on water and wastewater determine whether pathogenic organisms are present by testing for certain indicator organisms. Biological information is needed to measure water quality for such uses as drinking and swimming, and to assess the degree of treatment of the wastewater before its discharge to the environment. Biological tests include testing on microorganisms which are classified as follows: animal (rotifers, crustaceans, worm, and larvae), plant (mosses and ferns) and protista (bacteria, algae, fungi and protozoa) [4].

8.1.2 Physical characteristics

The many properties of water which appeal to the natural senses are termed the physical characteristics. The most important physical characteristic of water is its solids content as it affects the esthetic, clarity, and color of the water. Other physical parameters are temperature and odors which are largely the result of baseline levels for that geographical area and are not commonly altered in a wastewater treatment plant [4].

8.1.3 Chemical characteristics

Nowadays environmental contamination is a recognized worldwide problem. Significant part of the environmental pollution is caused by application of pesticides in agriculture, horticulture and forestry, by emissions of hazardous chemical substances formed during industrial processes and combustion reactions. A large number of these compounds and/or their degradation products are highly toxic and they have negative effects not only on the ecosystem, but also on human health. In this context it is necessary to control the presence of organic pollutants in the environment, and at the same time, to assess the risk to human health due to the presence of the chemicals at workplace or in the environment in order to prevent adverse effects. The chemical characteristics of wastewater can adversely affect the environment in many different ways. Soluble organics can deplete oxygen levels in stream, and cause an unpleasant odor and taste of water supplies. Toxic substances affect on to the food chain and human health. Nutrients can cause eutrophication of lakes. Although some chemical tests are specific, many determine broad classifications due to the variety of compound found in wastewater [4].

Therefore, prior to processing of potable water and wastewater it is necessary to chemically characterize it, which implies determination of pH-values, temperature, conductivity, cation and anion analysis, DO and oxygen demand, TOC, determination inorganic and organic of pollutants. Determination of water parameters shown in the Table 8.1 are determined by classical analytical methods and these methods are the routine methods of each laboratory for water analysis. However, the determination of inorganic and organic pollutants may vary from water to water, especially in wastewater of industrial and municipal origin. Therefore, require the application of sophisticated analytical methods for the determination of inorganic and organic pollutants, especially if they are in trace amounts in water [3, 4].

Where water quality and characteristics are defined by regulation, the questions of what measurements to make, the methods to use, and the sampling protocol are clearly recorded. When carrying out surveys of environmental water for investigational or research purposes the situation is less clear cut. Clearly defined aims for the survey or investigation must be established before the appropriate methods can be selected.

As stated earlier, chemical characterization also depends on the type of contamination whether it is inorganic or organic, and depending on it, different methods of analysis are also applied. Today, they are the biggest trace contamination problem because they require highly sophisticated analytical techniques that are often time-consuming and costly for preparing analytical samples. Also, trace contaminations are difficult to remove by conventional methods of processing any water and wastewater.

Before analyzing any sample, it should be in mind that each analysis is proceeded by sampling and sample preparation. Sampling, sample preparation and analysis procedures must be customized depending on whether the analyte is the main, minor component or trace component. The following section provides an overview about general analytical procedure for analysis of inorganic and organic trace pollutants in the environment.

8.1.3.1 Inorganic trace elements in water and wastewater

A wide range of human activities contributes to the trace element pollution of the aquatic environment. The major activities include mining and ore processing, coal and fuel combustion, industrial processing (chemical, metal alloys, chloro-alkali and petroleum), agricultural (fertilizers, pesticides and herbicides) domestic and agricultural effluents or sewage, and nuclear activity. Elemental or heavy metal input can be from atmospheric fallout, leaching or dumping from the lithosphere, or directly into the aquatic environment (including ground, surface, river, lakes, estuarine, oceans, etc.). The impact of water pollution depends upon the value of trace element input, duration of input, physical and chemical form and associated ligands or chemicals. All of these inter-related factors will determine the elemental concentrations in the water system, and their relative availability, transport, and toxicity. The most important factor is the chemical form in which the element exists in solution. This depends upon pH, solubility, temperature, the nature of other chemical species, and many basic factors of solution chemistry. Many trace element contaminants entering the aquatic environment can have dramatic effects on the bioavailability and toxicity of biological processes.

Prior to analyzing inorganic contaminants, especially the trace substances, representative sampling and good sample preparation are required. In general water sampling requires a very close inspection of the required procedures because most trace elements to be measured in water samples will be at very low or ultratrace levels ($\mu g\ L^{-1}$ or $ng\ L^{-1}$). Sample contamination and analyte losses are potential problems. Water samples can be collected from numerous sites from domestic faucet outlets to the effluents associated with industrial processes, and sewage or water treatment works. The frequency and duration of sampling is important in order to obtain both a representative and reproducible sample. It is common to take composite samples, which are composed of several individual samples taken at regular intervals.

Various container pre-treatment procedures should be undertaken before sample collection. Only polyethylene or teflon vessels should be used. Prior to use all new containers should be washed and stored in ultra-pure grade 10 % HNO_3 for 2 days, and rinsed with double distilled deionized water. All cleaned containers should be sealed in clean polypropylene bags until required. Acidification of the sample will reduce or eliminate trace element adsorption and hydrolysis. All water samples should be in theory analyzed immediately after collection. Most water samples require filtration immediately following collection with 0.45 or 0.50 μm membrane filters. All water samples should be stored in the dark, either by refrigeration (4 °C) or by deep-freezing (−20 °C). Long-term storage of frozen water samples may also be a problem for some trace metal species. Prior each water sampling, it is necessary to be well acquainted with sampling procedures depending on the requirements of the analyte which will be determined by one of the analytical methods.

The ideal analytical technique for measuring trace elements generally in environmental samples must offer a very low detection limits, a wide linear dynamic range, simple interference-free data, qualitative, semiquantitative and quantitative analysis, possible simultaneous multielement capability, simple sample preparation, and high throughput and low cost per determination. In practice and in reality there is no ideal or universal analytical technique for environmental sample analysis. All the main contenders, including atomic absorption spectrometry (flame or electrothermal), atomic fluorescence spectrometry, inductively coupled plasma optical (or atomic) emission spectrometry, neutron activation analysis, X-ray fluorescence, proton-induced X-ray emission, spark source or isotope dilution mass spectrometry, electrochemical (anodic stripping voltammetry and polarography) or inductively coupled plasma-mass spectrometry have their advantages and disadvantages. The main limitations for most of these analytical methods are sensitivity or precision problems due to interferences and sample matrix effects. In conclusion, it is always important to remember that the trace element analysis of samples must be associated with an appropriate validation or quality control scheme. For the measurement of inorganic trace elements in water or wastewater samples the use of certified reference materials, or interlaboratory test comparison must always be undertaken and published along with the results of the study.

8.1.3.2 Organic trace elements in water and wastewater

Environmental organic pollutants include aliphatic hydrocarbons, aromatic hydrocarbons, alcohols and phenols, anilines and benzidines, chlorinated solvents, polychlorinated biphenyls, pesticides (including insecticides and herbicides), and products used in the polymer industry such as phthalates and adipates, pharmaceuticals, pesticides etc. A detailed account of the analytical protocols used for the analysis of compound from each category is beyond the scope of this chapter; however, the importance of analytical procedures in the analysis of trace organic pollutants in the environment is emphasized here. The analysis of organic materials can be divided into two major classifications, the first being the identification of bulk materials, and the second the analysis of trace organic materials. The analysis of bulk materials is usually conducted on a single homogeneous component of known composition, and the procedure is one of identification, quantitation not usually being required. Techniques such as nuclear magnetic resonance and infrared (IR) spectroscopy which have been traditionally used as aids in identification rather than for quantitation are most applicable [3, 4].

Trace organic analysis is required where the components of interest are present at very low concentrations and contained within a bulk matrix that is often of an undetermined and heterogeneous composition. Consequently the methods used for the analysis of trace analytes are not as easily defined as for bulk organic materials. Physical separation (extraction and partition) of the analyte from the matrix is usually required before analysis can begin. Once achieved, further physical separation may be needed to remove interfering components from the extract that may have been co-extracted with the analytes. The presence of co-extractives interferes not only with analyte determination, but contributes to the fouling of chromatographic columns and detectors, and hence adversely affects the sensitivity and precision of analysis. Only after the analytes have been isolated can analysis proper begin. Where determination of a single analyte is required, direct spectroscopic methods of detection are favored as they can simply and quickly provide a means of positively identifying the analyte as well as enabling quantitation [3, 4].

Where several analytes have been isolated as a mixture, spectrometric and non-spectrometric physical detectors are used in conjunction with a variety of chromatographic techniques to enable quantitation. The combination of detectors and chromatographic techniques such as those commonly used for high performance liquid chromatography (HPLC) or gas chromatography (GC), are particularly suited to the analyses of multiresidue samples. Method commonly used for trace organic analysis can be assigned to one of three subdivisions: sample preparation, chromatography and detection [4].

Sample preparation is a necessary prerequisite for almost all trace organic analysis. The method of sample preparation is defined by the physical characteristics of the sample matrix and analytes, and phrase "like dissolves like" best describes the approach to successful analyte preparation. The method of sample preparation must also be compatible with chromatographic and associated detection systems that will be used later for analysis. In general, most protocols for sample preparation include several common steps which can be summarized as follows: maceration, dissolution, extraction, partition and concentration. Following sample preparation, isolation of individual analytes from a mixture can be achieved by highly efficient column chromatography [3, 4].

The diversity of trace organic compounds that have been found to contaminate water supplies has necessitated the development of many procedures for analysis. Statutory regulation of drinking water quality has meant that the ability to identify contaminants in $\mu g\ L^{-1}$ or even $ng\ L^{-1}$ concentration is required. Organic materials found in water can be placed into one of three classifications: volatile, semi-volatile and non-volatile compound. GC has been used for the analysis of volatile and some semi-volatile compounds. HPLC is ideally suited to the analysis of water samples, as a minimum of sample clean-up is required, and samples may be injected directly on column as aqueous solutions. HPLC is the method of choice for the analysis of semi-volatile and non-volatile compounds. The Environmental Protection Agency (EPA) has published a series of bulletins for chromatographic analysis of trace organic priority pollutants in wastewater [4].

Direct analysis for trace components in water is often not practical, as the detection of species at very low concentrations would require extremely sensitive means detection, and would most certainly yield ambiguous results due to the presence of interfering compounds. Prior any analysis a concentration of analyte is generally required, where the concentrate must be enriched by at least a 100- and preferably a 1000-fold. For semi and non-volatile organic compound, liquid/liquid partition or solid phase adsorbents (SPE) are used extensively, especially SPE techniques are frequently used to trap and concentrate semi- and non-volatile organic materials from water and wastewater samples. Also, where GC is to be used for final analysis, analytes concentrated by SPE can be eluted by heating the extraction cartridge. If the cartridge has been mounted in line with the gas inlet to a chromatograph, then eluted analytes are swept directly onto the GC column for separation. Combined with mass spectrometry for identification, this purge and trap technique is a very powerful tool for the analysis of trace volatiles and semi-volatiles organic materials in water. When the HPLC system is combined with UV or fluorescent (FL) detector, transfer of the extract to solvents such as methanol and acetonitrile affords solvent compatibility with the mobile phase and detectors. Recently, for the purpose of analyzing semi-volatile and volatile organic pollutants in traces, mostly related chromatographic techniques are used with sophisticated detectors based on mass spectrometry (MS) [3, 4].

8.2 Water and wastewater treatment by membrane processes

Membrane phenomena can be traced to the eighteenth century. French philosopher Abbé Nolet devised the word "osmosis" to describe permeation of water through a diaphragm in 1748 [5]. Osmosis is a natural process involving fluid flows across a semipermeable membrane. A breakthrough, as far as industrial membrane application were concerned, was achieved by the development of asymmetric membranes by Loeb and Sourirajan in 1962 [6]. In last few decades membrane processes are one of the fastest growing and fascinating fields in separation technology.

These processes are capable of performing certain separation by use of a membrane. The main concept of the membrane processes is relatively simple but until now often unknown. Membrane (*lat.: membrane = thin skin*) might be described as a conventional filter but with much smaller pores and different in size to enable the separation of tiny particles, even molecules and ions [7]. A specific separation is accomplished by use of a membrane with the ability of transporting one component more readily than another. This means that the membrane is more permeable to certain components than other components because of differences in physical or chemical properties between membrane and the components that are transported through the membrane.

Membrane processes are characterized by the fact that the feed stream, i. e. bulk solution, is divided into two streams, retentate and permeate, with different compositions as presented in Figure 8.1. *Retentate* stream (concentrate, brine) is the stream in which most particles, molecules or ions are retained by the membrane, while permeate is the one that passes through the membrane, containing much less solutes than in the feed stream. Both streams, retentate and permeate could be the product, i. e. if the aim of the separation is concentration, the retentate will be the product stream. However, in the case of purification mainly permeate is the desired product.

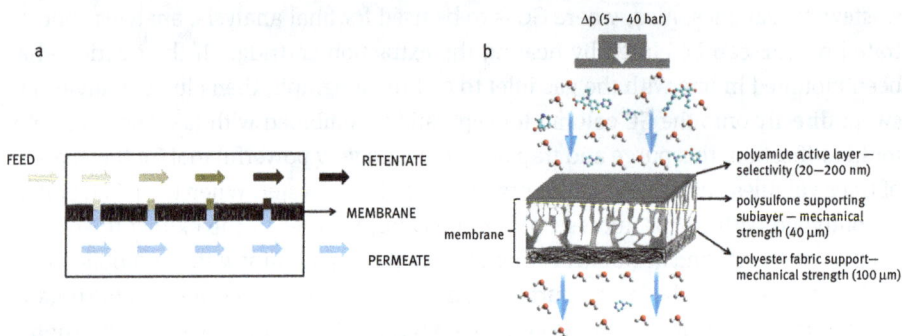

Figure 8.1: Schematic presentation of (1) membrane cross flow process and (2) presentation of mass transport through thin film composite (TFC) membrane under pressure driving force [8].

Due to their multidisciplinary character membrane processes can be use in a large number of separation processes. Comparison between membrane processes and other separation processes is difficult but the main benefits of membrane processes are: separation can be carried out continuously, energy consumption is generally low, they can easily be combined with other separation processes (hybrid processes), separation can be carried out under mild conditions, up-scaling is easy, and they belong to the nondestructive and environmentally friendly processes. However, following drawbacks should be mentioned: membrane fouling, concentration polarization, and up-scaling factor that is more or less linear [6]. It is important to mention that membrane fouling and concentration polarization should in fact not be considered as a disadvantage since these phenomena are inherently part of the separation process. Measures should be taken to reduce these phenomena as much as possible and this requires sufficient basic and application know-how.

There are many membrane processes, based on different separation principles or mechanisms and specific problems can cover the broad size range from particles to molecules and ions. In spite of these various differences, all membrane processes have one thing in common, the membrane. Although it is difficult to give an exact definition of a membrane, a general definition could be: *a selective barrier between two phases, the term "selective" being inherent to a membrane or a membrane process.* It should be noted that this is a macroscopic definition while separation should be considered at the microscopic level. The definition says nothing about membrane structure or membrane function. The membrane is the heart of every membrane process and can be considered as a permselective barrier or interphase between two phases [6].

To accomplish particular separation and transport through the membrane a certain driving force on the feed stream has to be applied. In many cases the permeation rate through the membrane is proportional to the driving force, i. e. the flux-force relationship can be described by a linear phenomenological eq. (8.1):

$$J = -A \cdot \frac{dX}{dx},$$ (8.1)

where A is called the phenomenological coefficient and dX/dx is the driving force expressed as the gradient of X (temperature, concentration, pressure and electric potential) along a coordinate x perpendicular to the transport barrier.

According to the used driving force, pressure, activity (concentration), electrical potential, and temperature differences, membrane processes can be divided into pressure driven separations, concentration gradient membrane processes, thermally driven membrane processes, and electrically driven membrane processes, respectively. A general classification of membrane processes and technically relevant main membrane processes in water treatment are presented in Table 8.2.

The next section will be focused on the pressure driven membrane processes, in which the flux through the membrane is induced by hydrostatic pressure difference between the feed and permaete stream (Figure 8.1) and depending of the value of applied pressure microfiltration (MF), ultrafiltration (UF), nanofiltration (NF) and reverse osmosis (RO) can be distinguish. The biggest difference between these processes, as go from MF to RO, is in the membrane pore size, and consequently, the size (or molecular weight) of the separated solutes diminishes. Due to membrane pore size decrease the driving force (pressure) has to be increased (0.1–2.0 bar, 1.0–5.0 bar, 5.0–20 bar, 10–60 bar for MF, UF, NF and RO, respectively), and on the other side permeability (L m^{-2} h^{-1} bar^{-1}) decrease (>50, 10–50, 1.4–12, and 0.05–1.4, respectively). Pore size decrease, from MF to RO, effects on retention of various solutes in the water. Namely, viruses, proteins and bacteria are removed >99 % with UF, NF and RO, while small organic molecule and salts (ionic solids) are most completely (>99.9 %) retained by RO membranes which pore size are below 1 nm. The biggest differences can be seen between tight and loose NF membranes. For tight NF membrane monovalent and divalent ions are removed between 70% and 98 %, while for loose NF membranes monovalent ions are removed between 0 and 70 %, and divalent ions between 0 and 99 %. Due to its nondestructive character, membrane processes, in particular RO and NF, play a very important role in removing emerging pollutants

Table 8.2: Technically relevant main membrane operations in water treatment [9].

Membrane process	Driving force	Mechanism of separation	Membrane structure	Phase	
				1	2
Microfiltration	Pressure	Sieve	Macropores	L	L
Ultrafiltration	Pressure	Sieve	Mesopores	L	L
Nanofiltration	Pressure	Sieve + (solution/diffusion + exclusion)	Micropores	L	L
Reverse osmosis	Pressure	Solution/diffusion + exclusion	Dense	L	L
Pervaporation	Activity (partial pressure)	Solution-diffusion	Dense	L	G
Membrane stripping	Activity (partial pressure)	Evaporation	Macropores	L	G
Membrane distillation	Activity (temperature)	Evaporation	Macropores	L	L
Dialysis	Activity (concentration)	Diffusion	Mesopores	L	L
Electrodialysis	Electrical potential	Ion exchange	Ion exchange	L	L

from water and wastewater. Emerging contaminants are organic compounds, commercial products, used in large quantities in everyday life, such as pharmaceutically active compounds, endocrine disrupting compounds, pesticides, human and veterinary drugs, surfactants, textile dyes etc.

The fundamental components of pressure membrane processes are the membrane chemistry (material) and the configuration into which the membrane is embedded. Membrane chemistry can be divided into organic (polymeric) and inorganic. Polymeric membranes are, at the moment, the most used membranes all over the world and they can be made from various polymers. First membrane, from 1960s, was made from cellulose (diacetate, triacetate). Significant number of polymers could be used for membranes but most frequent are aromatic polyamide (PA), polyacrylonitrile (PAN), polysulfone (PSf), polyethersulfone (PES), polytetrafluoroethylene (PTFE), polyvinylidenefluoride (PVDF), polyethylene (PE), polycarbonate (PC), and polypropylene (PP). In last few years great effort is invested in development of new generation membrane materials [10]. This includes nanocomposite, aquaporin, nanotube, and graphene-based membranes. The inorganic membranes are mainly ceramic made from Al, Zr and Ti oxide, nitrite, and carbides. These membranes possess high porosity, high permeability, defined pore size, and good mechanical properties, which make them suitable for use in wide range of application, such as water and wastewater treatment, medical and pharmaceutical, industrial processing, and food and beverage [11].

There are two main parameters that describe the performance of efficiency of a given membrane; *selectivity* and the flux (J), or *flow* rate (Q) through the membrane. Flux is defined as the volume flowing through the membrane per unit area and time ($L\,m^{-2}\,h^{-1}$) while for flow rate the unit is $m^3\,d^{-1}$. The selectivity of a membrane toward a mixture can be expresses by the rejection (R) or the separation factor (α). The rejection factor is given by eq. (8.2):

$$R = 1 - \frac{y_p}{y_f} = \frac{y_f - y_p}{y_f},\qquad(8.2)$$

where y_f and y_p are concentrations of solute in feed and permeate, respectively. The value of R varies between 100 % (complete retention of solute; in this case we have an "ideal" semipermeable membrane) and 0 % (solute and solvent pass through the membrane freely) [6].

The basic membrane configurations (modules) that are used in commercial applications are hollow fiber, spiral wound, tubular, and plate-and-frame. Nevertheless, spiral wound and hollow fiber are widely used in modern water treatment applications. In industrial practice, it is possible to connect up to seven spiral modules in the one pressure vessel without causing a significant drop in pressure (see Example 8.1).

The main application of pressure driven membrane processes, especially RO, is sea water desalination and brackish water treatment. The largest RO desalination plants are installed in Israel (Sorek, Hadera, Ashkelon) with a capacity of 420,000, 355,000 and 325,000 $m^3\,d^{-1}$ respectively, and in Europe (Barcelona, 200,000 $m^3\,d^{-1}$). The price of so-obtained drinking water is 0.5–0.6 $ m^{-3}. If the other processes are included, together with development of new membranes, application is involved in almost all industries (textile, food and beverage, pulp and paper, power industries, etc.) [12]. It is assumed that application of pressure membrane processes will rise since the global membranes market for the 5-year period 2016–2021 is projected to reach 11.95 billion USD by 2021 [11].

Example 8.1. Calculation of the rejection factor and recovery.
In the diagram are given fluxes and concentrations in feed, retentate, and permeate. According to the presented data calculate rejection and recovery.

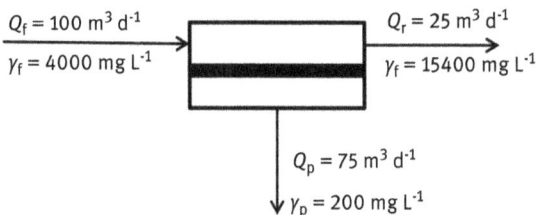

$Q_f = 100\ m^3\,d^{-1}$
$y_f = 4000\ mg\ L^{-1}$

$Q_r = 25\ m^3\,d^{-1}$
$y_f = 15400\ mg\ L^{-1}$

$Q_p = 75\ m^3\,d^{-1}$
$y_p = 200\ mg\ L^{-1}$

Recovery:

It can be calculated through flow rate (or flux) and concentrations: $Y = \dfrac{Q_p}{Q_f} = \dfrac{J_p}{J_f} = \dfrac{(\gamma_r - \gamma_f)}{(\gamma_r - \gamma_p)}$

Flow rate: $Y = \dfrac{Q_p}{Q_f} = \dfrac{75 \ m^3 \ d^{-1}}{100 \ m^3 \ d^{-1}} = 0.75 \cdot 100\% = 75\%$

Concentrations: $Y = \dfrac{(\gamma_r - \gamma_f)}{(\gamma_r - \gamma_p)} = \dfrac{(15400 - 4000) \ mg \ L^{-1}}{(15400 - 200) \ mg \ L^{-1}} = \dfrac{11400}{15200} = 0.75 \cdot 100\% = 75\%$

Rejection:

$$R = \dfrac{\gamma_f - \gamma_p}{\gamma_f} = \dfrac{4000 \ mg \ L^{-1} - 200 \ mg \ L^{-1}}{4000 \ mg \ L^{-1}} = 0.95 \cdot 100\% = 95\%$$

Example 8.2. *Selection of the module and calculation of technical conditions of membrane process for treatment of surface water to produce drinking water.*

The goal is to find optimal membrane treatment to decrease concentrations of dissolved organic carbon (DOC) and hardness of water (γ(Ca)) using 7 membrane modules in a vessel. Concentration of DOC in surface water was 7 mg L^{-1} and γ(Ca) = 140 mg L^{-1}. After the treatment DOC has to be 4.2 mg L^{-1} and γ(Ca) = 60 mg L^{-1}. In Table 8.3 the manufacturers' data for typical RO and NF modules are given.

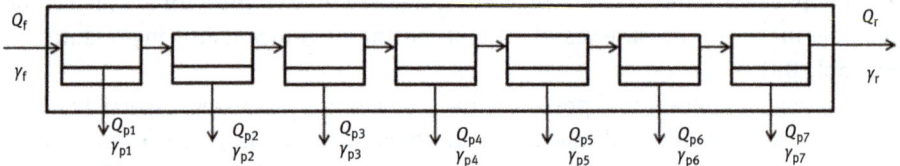

Table 8.3: The manufacturers' data for typical RO and NF modules.

	RO1	RO2	NF1	NF2
$Q_p / [m^3 \ d^{-1}]$	44.0	41.0	33.0	27.7
$p / [bar]$	10.2	10.2	7.0	5.5
R_{DOC}	0.96	0.96	0.946	0.95
R_{Ca}	0.99	0.99	0.946	0.45
Y	0.15	0.10	0.15	0.15
$A / [m^2]$	37.0	30.7	37.0	37.0

For membrane water treatment, depending on quality and quantity of feed water, it is necessary to choose membrane module. From technical data of membrane modules it is possible to calculate number of modules for treatment of certain amount of water.

According to the Table 8.3 differences between RO and NF membranes are visible. For the first calculation NF1 membrane will be used due to lower pressure compared to RO membranes and higher R_{Ca} than NF2.

Main parameters for calculation of module characteristics are recovery (Y), working pressure p, rejection factors of dissolved solutes R, permeate flow rate Q_p, and concentration of solutes in the feed which is necessary to decrease. Recovery for one module is 0.15 while concentration of feed and working pressure changes in pressure vessel from first to last module. This is due to increase of feed concentration and low decrease of pressure in each module.

Permeate and retentate flow rate can be calculated:

1. module – permeate flow rate: $Q_{p1} = 0.15 \cdot 220 = 33.0$ m³d^{-1}

retentate flow rate: $Q_{r1} = Q_{f1} - Q_{p1} = 220.0 - 33.0 = 187.0$ m³d^{-1}

5. module – permeate flow rate: $Q_{p5} = 0.14 \cdot 116.3 = 16.3$ m³d^{-1}

retentate flow rate: $Q_{r5} = 116.3 - 16.3 = 100.0$ m³d^{-1}

The same way is for calculating for all other modules and results are presented in Table 8.4.

$$\Sigma Q_p = 146 \text{ m}^3\text{d}^{-1}$$

Table 8.4: Calculation of flow rate in the pressure vessel of the process containing seven NF1 modules. $p = 7$ bar, $Q_p = 33.0$ m³ d^{-1}, $R = 0.946$.

Module	$Q_f/[\text{m}^3 \text{ d}^{-1}]$	Y	$Q_r/[\text{m}^3 \text{ d}^{-1}]$	$Q_p/[\text{m}^3 \text{ d}^{-1}]$
1	220.0	0.15	187.0	33.0
2	187.0	0.15	159.0	28.0
3	159.0	0.15	135.2	23.8
4	135.2	0.14	116.3	18.9
5	116.3	0.14	100.0	16.3
6	100.0	0.14	86.0	14.0
7	86.0	0.14	74.0	12.0

Next step is calculation of DOC and Ca concentrations in permeates and retentate for each module.

The water mass balance for a membrane element is given by: $Q_f = Q_p + Q_r$.

Also, the dissolved material of mass balance can be derived by: $Q_f \gamma_f = Q_p \gamma_p + Q_r \gamma_r$.

Therefore, concentration of DOC in module 1 retentate is:

$$\gamma_{r1} = \frac{Q_f \gamma_f - Q_p \gamma_p}{Q_r} = \frac{220 \cdot 7.0 - 33 \cdot 0.38}{187} = 8.17 \text{ mg L}^{-1}$$

All concentration of DOC and Ca are calculated on the same basic and concentrations are presented in Table 8.5.

Table 8.5: Calculation of concentrations in the pressure vessel of the process containing seven NF1 modules. $p = 7$ bar, $\gamma_{f,DOC} = 7$ mg L^{-1}, $\gamma_{f,Ca} = 140$ mg L^{-1}, $R = 0.946$.

Module	$\gamma_{f,DOC}$ [mg L^{-1}]	$\gamma_{r,DOC}$ [mg L^{-1}]	$\gamma_{p,DOC}$ [mg L^{-1}]	$\gamma_{f,Ca}$ [mg L^{-1}]	$\gamma_{r,Ca}$ [mg L^{-1}]	$\gamma_{p,Ca}$ [mg L^{-1}]
1	7.00	8.17	0.38	140.0	163.6	7.6
2	8.17	9.53	0.44	163.6	190.9	8.8
3	9.53	11.12	0.51	190.9	222.8	10.3
4	11.12	12.83	0.60	222.8	260.0	12.0
5	12.83	14.81	0.69	260.0	300.0	14.0
6	14.81	17.09	0.80	300.0	346.3	16.2
7	17.09	**19.72**	0.92	346.3	**399.6**	18.7

As seen in Table 8.5, a concentrations of DOC and Ca in all permeates are different. If permeates are collected together average concentrations of DOC and Ca can be calculated. Results are: $\gamma_{p,DOC} = 0.62$ mg L^{-1} DOC and $\gamma_{p,Ca} = 12.5$ mg L^{-1}Ca.

Overall recovery can be calculated: $Y = \dfrac{Q_p}{Q_f} = \dfrac{146 \text{ m}^3 \text{ d}^{-1}}{220 \text{ m}^3 \text{ d}^{-1}} = 0.664 \cdot 100\% = 66.4\%.$

8.3 Biological wastewater treatment

Biological treatment is one of the most widely used methods for organic matter removal as well as for partial or complete stabilization of biologically degradable substances in wastewaters. Biological unit processes involve microbial activity, which is responsible for organic matter degradation and removal of nutrients. Microorganisms require energy for respiration and organic carbon food sources (COHNS) in order to synthesize new cells ($C_5H_7NO_2$). Metabolic reactions, occurring within a biological treatment process, can be divided into three phases: organic matter oxidation, cell material synthesis and endogenous respiration or cell material oxidation. The three phases are [13, 14]:

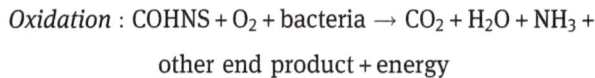

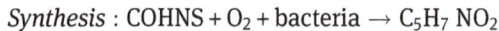

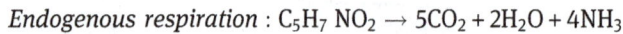

$$Oxidation : COHNS + O_2 + bacteria \rightarrow CO_2 + H_2O + NH_3 +$$
$$\text{other end product} + energy$$

$$Synthesis : COHNS + O_2 + bacteria \rightarrow C_5H_7 NO_2$$
$$Endogenous\ respiration : C_5H_7 NO_2 \rightarrow 5CO_2 + 2H_2O + 4NH_3$$

The primary objective of biological wastewater treatment processes is conversion of biodegradable organics into a microbial biomass. Organic pollutants are defined by three common parameters: BOD, COD and TOC. Table 8.6 shows the typical concentrations (mg L^{-1}) of organics found in untreated domestic wastewater [13].

Table 8.6: Typical composition of untreated domestic wastewater.

Constituents	Typical concentration		
	Low	Medium	High
BOD/[mg L^{-1}]	110	190	350
COD/[mg L^{-1}]	250	430	800
TOC/[mg L^{-1}]	80	140	260

BOD is used commonly as an indirect measure of organic matter and it represents the amount of oxygen utilized by microorganisms to biologically oxidize organic compounds in aerobic conditions in the dark at 20 °C over a period of 5 days. Five days as the representative period is assumed to determine the characteristic of BOD. The COD is the amount of oxygen necessary to oxidize organic compounds chemically. The COD test is used to determine the oxygen equivalent of the organic matter that can be oxidized by a strong chemical oxidizing agent (potassium dichromate) in an acid medium. Typical values for the ratio of BOD/COD for untreated municipal waste-water, as reported in Table 8.7, are in the range from 0.3 to 0.8 [13]. This ratio has been commonly used as an indicator for biodegradation capacity and is called biodegrad-ability index. If the COD value is much higher than the BOD value, the sample contains large amounts of organic compounds that are not easily degradable. If the BOD/COD ratio for untreated wastewater is 0.5 or greater, the waste is considered to be easily treatable by biological means. If the ratio is below about 0.3, either the waste may have some toxic components or acclimated microorganisms may be required in its stabilization. TOC represents the TOC in organic compounds and independent of oxidation state of organic matter. The corresponding BOD/TOC ratio for untreated wastewater varies from 1.2 to 2.0. In using these ratios it important to remember that they will change significantly with the degree of treatment the waste has undergone [13–15].

Table 8.7: Comparison of ratios of various parameters used to characterize wastewater.

Type of wastewater	BOD/COD	BOD/TOC
Untreated	0.3–0.8	1.2–2.0
After primary settling	0.4–0.6	0.8–1.2
Final effluent	0.1–0.3	0.2–0.5

Example 8.3. What is the relationship between the BOD and COD values in wastewater?

To establish the BOD/COD ratio for your wastewater, simply have both COD and BOD run on several wastewater samples. Divide the BOD concentration by the COD concentration for each sample and average the results. For example, below is the BOD/COD ratio developed using three wastewater samples:

Sample 1 : BOD = 183 mg L^{-1}; COD = 352 mg L^{-1} : BOD/COD = 183/352 = 0.52
Sample 2 : BOD = 172 mg L^{-1}; COD = 358 mg L^{-1} : BOD/COD = 173/342 = 0.48
Sample 1 : BOD = 165 mg L^{-1}; COD = 340 mg L^{-1} : BOD/COD = 175/350 = 0.49

$$\text{BOD/COD} = (0.52 + 0.48 + 0.49)/3 = 0.50$$

Explanation: The BOD/COD ratio has been used as one of the well-adopted surrogates for biodegradability of TOC. This ratio varies with the wastewater and degree of treatment. If BOD/COD ratio is > 0.5 than easily biodegradable 0.4–0.5 average biodegradable, 0.2–0.4 slowly biodegradable, and < 0.2 not biodegradable. If the BOD/COD ratio is small, the reason can be toxicity, low biodegradability or both together, which results in inhibition of metabolic activity of microorganisms. COD test for both (biodegradable and non-biodegradable) using chemical oxidizing agent, while BOD test is only for (biodegradable) using bacteria. For that reason COD is always higher than BOD.

8.3.1 Activated sludge process

Biological wastewater treatment has been utilized globally for decades. There are many different processes that vary depending on the origin of wastewater, i. e. the existing contamination. Still, one of the most frequently used process is biological wastewater treatment with activated sludge, because of its effectiveness in treatment of wastewater. The activated sludge process is a complex biological system in which organic matter is removed from wastewater by aerobic microorganisms. The microorganisms are incorporated within or on floc of activated sludge and aeration of wastewater suspension ensures the exchange of products of metabolism and substrates, as well as the oxygen required to maintain aerobic respiration and catabolic reactions. In the activated sludge process the incoming wastewater is mixed and aerated with existing activated sludge (microorganisms). Aerobic oxidation of organic matter is carried out in aeration tank. The content of the aeration tank in an activated sludge system is called mixed liquor. The mixed liquor suspended solids (MLSS) is the total amount of organic and mineral suspended solids, including microorganisms, in the mixed liquor. Organics in the wastewater come into contact with the microorganisms and are utilized as food and oxidized to CO_2, H_2O, NH_4, and new cell biomass. As the microorganisms use the organics as food they reproduce, grow, and die. The organic portion of MLSS is represented by mixed liquor volatile suspended solids (MLVSS), which comprises non-microbial organic matter as well as dead and live microorganisms. The MLVSS/MLSS ratio is typically in the range 0.65–0.85. Primary effluent is introduced and mixed with return activated sludge to form the mixed liquor, which contains 1500–3000 mg L^{-1} of suspended solids. The aeration tank is provided with fine bubble diffused aeration pipework at the bottom to transfer required oxygen to the biomass and also ensure completely mixed reactor. An important characteristic of the activated sludge process is the recycling of a portion of the biomass. This practice helps to maintain an activity of microorganisms in effectiveness of oxidizing organic compounds in a relatively short time. The detention time in the aeration basin varies between 4 and 8 h. The aeration tank is a completely mixed bioreactor where specific concentration of biomass is maintained along with sufficient DO concentration (typically 2 mg L^{-1}) to effect biodegradation of soluble organic impurities measured as BOD or COD. Excess biomass is wasted to the sludge handling and dewatering facility [14–16]. The purification process occurs in two steps: transformation of the solved and particular wastewater contents to activated sludge and separation of the activated sludge from cleaned wastewater through sedimentation. Separation of the activated sludge is possible because of the formation of the activated sludge flocs [15]. The characteristic schematic layout of activated sludge process is shown in Figure 8.2.

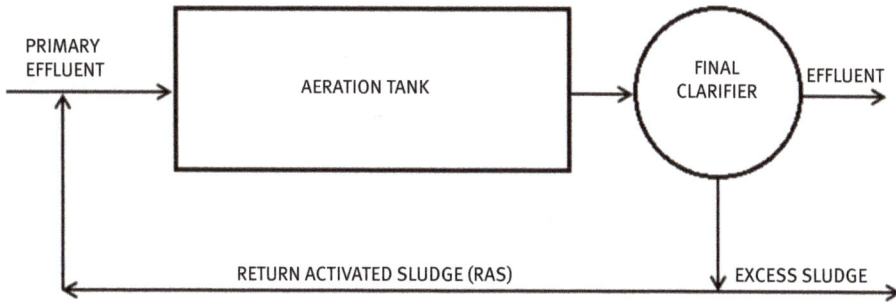

Figure 8.2: Conventional activated sludge system.

8.3.2 Environmental factors

Environmental factors which influence biological growth include temperature, pH, mixing intensity, the presence of DO, and the food/microorganism ratio, F/M [14, 16]. Temperature is an important factor that can affect biological activity. Both temperature and pH have an important effect on the survival and growth of microorganisms. If the temperature gets higher or lower, then the process is functioning less efficiently. Because of these increasing temperatures, the possible amounts of DO in water decrease, and this occurrence has a negative influence on the efficiency of the aerobic treatment system. The pH-level is an important factor of biological wastewater treatment. Low pH-values may inhibit the growth of nitrifying organisms, while the growth of filamentous organisms is encouraged. Optimal pH range for microbial growth is between 6 and 9. It is the sudden change in pH that the most upsets microbial metabolism. **Mixing** is important to ensure effective contact between the active microorganisms and the organic matter, and to prevent accumulation of products of microbial decomposition. During oxidative biodegradation, microorganisms consume oxygen. DO concentration in aerobically treated wastewater is maintained at a certain adjusted level. Concentrations lower than $2\,mg\,L^{-1}$ must be prevented since such a low concentrations may induce simultaneous anaerobic processes. The F/M ratio is a measure of the food entering the activated sludge process and the microorganisms in the aeration tank and it is expressed in kilogram BOD per kilogram of MLSS per day. The F/M ratio tells us something about growth and cell condition. If the F/M ratio is high, the growth is normally quite rapid (because this means there is a lot of "food" available in comparison to the amount of microorganism); if the F/M ratio is low, the growth is normally very slow (because little food is available for growth).

8.3.3 Membrane bioreactor

Activated sludge process is characterized by reliable operation and flexible capacity. However, when requirements for a higher effluent quality are imposed, activated sludge process must be expanded and additional purification procedures added. In recent years, new membranes that have been specially developed for use in wastewater treatment, made the membrane bioreactors application a promising alternative to the aerobic processes. The combination of an aeration tank and membrane filtration unit for the separation of activated sludge is defined as the membrane-coupled activated sludge process. Consequently, the difference from the conventional activated sludge process lies only in the separation of the activated sludge. Ultra- or micro-filtration membrane possesses advantage in possible replacing conventional sedimentation for the separation of the treated water from the sludge. The use of submerged membranes has reduced the power consumption of membrane bioreactors and hence increased their potential for the application of membranes in wastewater treatment. The complete retention of activated sludge allows operation at much higher biomass concentrations [16, 17].

8.3.4 Biology of activated sludge

Activated sludge system is an aerobic biological wastewater treatment process which contains mixed culture of microorganisms. It is a complex and variable heterogeneous suspension containing both feedwater components, metabolites produced during the biological reactions and biomass. In the biological oxidation of wastewater, both synthesis and oxidation occur. Many groups of activated sludge microorganisms take part in carrying out the process. In an activated sludge system that is properly operated, biological flocs are produced that incorporate most of the important groups of microorganisms. Bacteria, fungi, algae, protozoa, and metazoan constitute the biological component, or biological mass, of activated sludge. The population active in a biological wastewater treatment is mixed, complex and interrelated. In general, the bacteria in the activated sludge process seemed to be gram-negative and include members of the genera *Pseudomonas*, *Zoogloea*, *Achromobacter*, *Flavobacterium*, *Nocardia*, *Bdellovibrio*, *Mycobacterium*, and the two nitrifying bacteria, *Nitrosomonas* and *Nitrobacter*. Additionally, various filamentous forms, such as *Sphaerotilus*, *Beggiatoa*, *Thiothrix*, *Geotrichum*, may also be present. In the normal operation of the activated sludge plants, the bacteria are shaped in a floc form. In fact not all microorganisms in activated sludge mixed-liquor are beneficial. Filamentous organisms cause swelling of floc by increasing the water content of the bacterial cell. Under this condition the floc particle will have similar density as the water itself and particles will not settle efficiently. This type of sludge is called bulking sludge. The species of microorganism that dominates in a system depends on environmental conditions, process design, the mode of treatment plant operation, and the characteristics of the wastewater. Floc structure, like size and morphology, plays an important role in determining the efficiency and economics of the activated sludge process [14]. Figure 8.3 present microphotographs of aerobic flocs without/ with and filamentous bacteria.

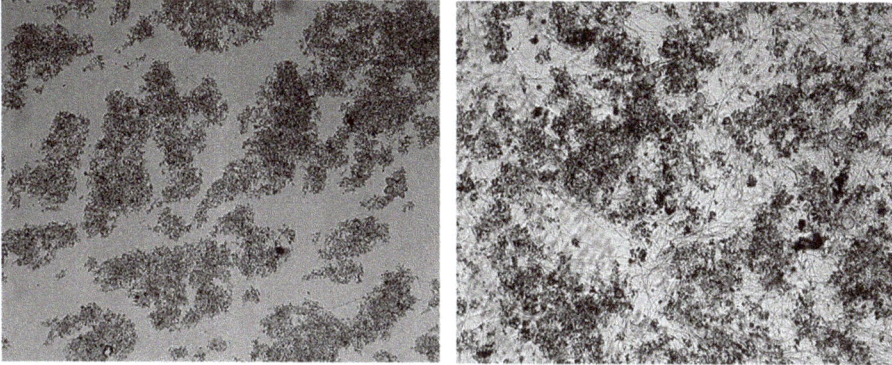

Figure 8.3: Microphotographs of flocs, and flocs with filamentous bacteria, M = 100×.

Protozoa are significant predators of bacteria in activated sludge as well as in natural aquatic environments and indicators of biomass health and effluent quality. Protozoa have an important role in maintaining a good balance in the biological ecosystem: they eliminate the excess bacteria and stimulate their growth and they promote flocculation (Figure 8.4).

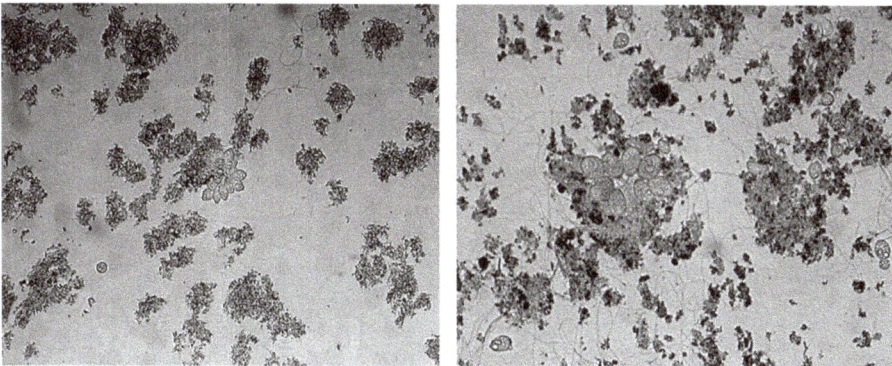

Figure 8.4: Microphotographs of flocs and protozoa, M = 100×.

Microphotographs were taken at Faculty of Chemical Engineering and Technology, University of Zagreb, Department of Industrial Ecology.

8.3.5 The Monod equation

The Monod equation [15, 16] assumes that the rate of biomass production is limited by the rate of enzyme reactions involving utilization of the substrate compound that is in shortest supply relative to its need. Equation (8.3) shows this relationship,

$$\mu = \mu_{max} \cdot \frac{S}{K_s + S} \tag{8.3}$$

where μ represents specific growth rate, 1/d; μ_{max} represents the maximum specific growth rate, 1/d S represents substrate concentration, g L^{-1}; K_s is the substrate saturation constant, g L^{-1}, defined as the concentration of substrate at half the maximum specific growth rate. The biodegradation process results in microbial growth with the removal of substrate. Substrate concentration decreases with the growth of microorganisms. Therefore, the following equation can be developed:

$$r_s = \frac{1}{Y} \frac{\mu_{max} X_v S}{K_s + S} \tag{8.4}$$

where r_s is substrate consumption rate, g $L^{-1} d^{-1}$; Y is the growth yield coefficient, g/g; X_v is biomass concentration, mg L^{-1}. The carbon and energy source, measured as COD, is usually considered to be the growth-limiting substrate in biological wastewater treatment processes. MLVSS represents the concentration of biomass in activated sludge. Microbial growth rate can be described by eq. (8.5):

$$r_x = \frac{\mu_{max} X_v S}{K_s + S} - k_d X_v \tag{8.5}$$

where r_x is biomass growth rate, g $L^{-1} d^{-1}$; and k_d is the decay coefficient, 1/d. Cell concentration reduction is known as endogenous respiration stage.

Example 8.4. Write the linearized Monod's equation and show its graph. How would you calculate the μ_{max} and K_s?

Monod's equation can be linearized using the Lineweaver–Burke equation:

$$\frac{1}{\mu} = \frac{K_S}{\mu_{max}} \cdot \frac{1}{\mu_{max}}$$

Graph shows a plot of $1/\mu$ vs. $1/S$. The slope, y-intercept, and x-intercept are (K_S/μ_{max}), $(1/\mu_{max})$, and $(-1/K_S)$, respectively. This plot allows the computation of K_S and μ_{max}.

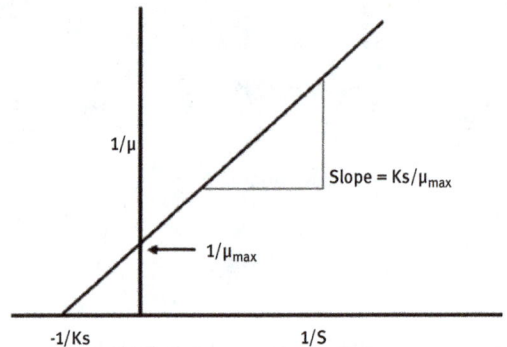

8.4 Advanced oxidation processes

There are varieties of (waste)water chemical treatment methods based on the application of strong oxidants or *in situ* generation of oxidizing species. However, their constrains related to their chemistry, which may lead to the production of harmful by-products, as well as the reactivity and selectivity toward targeted pollutants, may provide significant implications at larger scale and should be taken into account when considering appropriate treatment method for the removal of organic, particularly recalcitrant, pollutants [18]. Accordingly, in last two – three decades, advances in chemical treatment of water and wastewater resulted with the development of a numerous alternative chemical technologies; so-called AOPs. The interest of scientific community and water professionals rapidly increased for the development and application of AOPs as (waste)water treatment methods, yielding numerous scientific publications as well as designs and applications on full-scale treatment facilities [18]. The technologies based on AOPs may offer viable solution for the treatment of (waste)waters of different origin. One of the main advantage of AOPs over conventional biological, physical-mechanical and chemical (waste)water treatment processes is capability to degrade wide range of organic pollutants into the less harmful or biodegradable by-products, and eventually convert them into CO_2, H_2O and inorganic ions [19]. Besides, AOPs are considered as low- or even none-waste generation technologies [20], thus secondary treatment is avoided. The principle mechanism of AOPs is based on the *in situ* generation of highly reactive transient radical species that readily attack various recalcitrant organic pollutants, kill waterborne pathogens, and break down disinfection by-products. Hydroxyl radicals ($E^o(HO\bullet) = 2.80$ V), are considered to be the main reactive species possessing high reactivity (Table 8.8) and non-selectivity toward organics [21].

Table 8.8: Standard redox potential of some oxidant species [22].

Oxidant	Redox potential, E^o, V
Flour	3.03
Hydroxyl radical	2.80
Sulphate radical	2.60
Atomic oxygen	2.42
Ozone	2.07
Persulfates	2.01
Hydrogen peroxide	1.77
Permanganate ion	1.67
Chlorine	1.36
Chlorine dioxide	1.27

AOPs demonstrated high flexibility in practical application; thus can be used either separately or in combination, or in combination with other traditional wastewater treatment methods. Although the total mineralization of organic structures might be achieved by AOPs, such task could be sometimes very costly. Hence, AOPs may serve as an effective tool for converting non- or low-biodegradable organics to more readily biodegradable intermediates [23]. Accordingly, AOPs might precede secondary (biological) treatment in to increase increasing biodegradability and lowering toxicity, i.e. to destruct recalcitrant chemical structures, or can be used as tertiary treatment in order to remove remained organics upon secondary treatment. AOPs possess great advantage over common traditional (waste)water treatment methods; they can be conducted at ambient conditions, i.e. atmospheric pressure and room temperature. AOPs can be effectively used for the treatment of low to moderately loaded wastewaters; with organic content expressed by the means of TOC in range of 100–1000 mgC L^{-1} [24]. For highly loaded wastewaters, with COD values over 3–5 g L^{-1}, other wastewater treatment processes also generating OH radicals can be used. Although some authors consider such processes as AOPs, these processes, including wet air oxidation, supercritical wet air oxidation and other hydrothermal oxidation processes requiring pressure and temperature highly over ambient conditions (even above 1 MPa and 150 °C), can be classified as non-ambient AOPs [22, 24].

Generally, HO• can be generated, by chemical, electrical, mechanical or radiation energy. Thus, AOPs can be classified into chemical, electrical, mechanical and photo-assisted processes. *Chemical AOPs* include Fenton-based processes, involving the application of transitions ions (mostly iron) combined with strong oxidants (hydrogen peroxide or persulfate salts), as well as ozone-based processes, involving various combinations of catalytic ozonations and peroxone processes combining two or more oxidants. *Photo-assisted AOPs* include various photochemical processes, involving the application of UV irradiation sources with powerful oxidant (ozone, hydrogen peroxide and/or persulfate salts), and photocatalytic processes, involving the application of UV or solar energy and various semiconducting materials as photocatalyst. These two groups of AOPs are mostly investigated and are operating at larger scales to treat various types of wastewater, thus are discussed further in more details.

8.4.1 Chemical AOPs

8.4.1.1 Fenton-based processes

The processes involving catalytic reaction between transition metal ions and hydrogen peroxide (H_2O_2) can be classified under Fenton-based processes. The reactivity of such system was firstly observed by Fenton in 1876 [25, 26], applying iron in a form Fe^{2+} ions and H_2O_2. Upon more detailed research focused on the undergoing mechanism generating HO• within such system in 1930s, the application of Fenton process for the degradation of various classes and types of organic pollutants in water is continuously growing and is expanded into large-scale applications. The reactivity of Fenton process is based on reaction that involves oxidation of Fe^{2+} ions to Fe^{3+} ions with H_2O_2 and subsequent generation of $HO\bullet$, as demonstrated by eq. (8.6):

$$Fe^{2+} + H_2O_2 \rightarrow Fe^{3+} + HO\bullet + HO^- \tag{8.6}$$

The generated $HO\bullet$ would undergo reaction with present organics within water matrix. The above given reaction (6) is often considered as Fenton reaction [27], however it represents only first, but key step in the Fenton process. The so-called Fenton catalytic cycle involves following reactions eqs. (8.7)–(8.13) as well, describing iron recovery and quenching within the system and reacting mechanism responsible for degradations of present organics (RH).

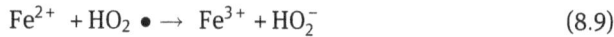

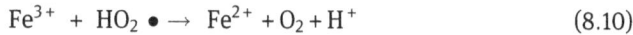

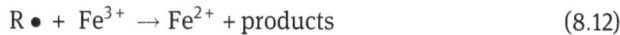

$$Fe^{2+} + HO\bullet \rightarrow Fe^{3+} + HO^- \tag{8.7}$$

$$Fe^{3+} + H_2O_2 \rightarrow Fe^{2+} + HO_2\bullet + H^+ \tag{8.8}$$

$$Fe^{2+} + HO_2\bullet \rightarrow Fe^{3+} + HO_2^- \tag{8.9}$$

$$Fe^{3+} + HO_2\bullet \rightarrow Fe^{2+} + O_2 + H^+ \tag{8.10}$$

$$RH + HO\bullet \rightarrow H_2O + R\bullet \tag{8.11}$$

$$R\bullet + Fe^{3+} \rightarrow Fe^{2+} + products \tag{8.12}$$

$$R-COO^- + Fe^{3+} \rightarrow Fe^{3+} - complex \tag{8.13}$$

The formed by-products through reaction eq. (8.12) undergo further degradation through radical mechanism up to complete mineralization forming eventually CO_2 and H_2O [28]. The reaction eq. (8.13) describing the tendency of ferric ions (Fe^{3+}) to form organic complexes with carboxylates [29], demonstrates quenching of iron from Fenton catalytic cycle, and is rate limiting for the overall process effectiveness due to restraining the iron recovery and causing lack of $HO\bullet$ within the system.

Fenton-based processes include systems with similar reaction chemistry that apply alternative catalysts instead of homogeneous ferrous iron for the incitation of Fenton reaction; (i) ferric salts, (ii) heterogeneous iron-based materials including: iron powder, iron-oxides, iron-ligands, or iron ions doped in zeolites, pillared clays or resins, and (iii) other metal ions, e. g. copper, manganese or cobalt [21, 27, 30].

Lately, besides alternative catalyst in Fenton-based processes, the alternative oxidant is also investigated. Thus, persulfates (usually in form of Na^+, K^+ and NH_4^+

salts), characterized by high solubility and stability at the ambient temperature, was found to effective alternative to H_2O_2. The persulfate anion, $S_2O_8^{2-}$ is a strong oxidant (Table 8.8) that can be easily activated by transition metals (M) (or by heat or UV irradiation) to generate highly reactive sulfate radical, $SO_4 \bullet$ eq. (8.14) that is also unselective, while the sulfate ions, as major products of persulfate reduction, can be considered as relatively harmless and environmentally friendly [31].

$$S_2O_8^{2-} + M^{2+} \rightarrow M^{3+} + SO \bullet_4^- + SO_4^{2-} \tag{8.14}$$

However, it should be noted that iron (or transition metal) in persulfate reduction, described by above reaction, does not have catalyst role, but it acts as activator; i. e. it can not be recovered as in Fenton catalytic cycle [31].

The primary benefits of Fenton-based processes, besides the ability to convert a vast array of organics into harmless and/or biodegradable products, are relatively cheap reagents that are ease and safe to handle, and are moreover environmentally benign [27]. However, the effectiveness is constrained by following disadvantages: (i) the necessity to remove remaining iron ions after the treatment, and (ii) afore-mentioned quenching of ferric ions into stable complexes [27], that can be overcome by the usage of alternative heterogeneous catalyst mentioned above and UV- or solar-assisted systems enabling the destruction of formed Fe^{3+}-complexes and continuous involvement of iron ions into the Fenton cycle [29]. One should be aware that Fenton-based processes are multi-factor systems, and that effectiveness strongly depends on key operating parameters: (i) iron concentration, (ii) source of (iron) catalysts, H_2O_2 (or $S_2O_8^{2-}$) concentration, catalysts/oxidant ratio (so-called Fenton reagent), temperature, pH and treatment time [21, 27]. Accordingly, the minimal threshold concentration of ferrous ions allowing Fenton reaction to occur ranges between 3 and 15 mg L^{-1}, however, iron levels <25–50 mg L^{-1} can significantly prolong treatment period (even up to 10–24 h). Typical catalyst/oxi-dant ratios range between 1:5 and 1:25, while it is markedly proven that the optimal pH range for homogeneous Fenton process is between 2.8 and 3.0, but application of heterogeneous catalyst may alter operating pH toward neutral, similar as usage of alternative oxidant persulfate salts [21, 27].

8.4.1.2 Ozone-based processes

Ozonation in fact pertains to classical chemical methods for water/wastewater treat-ment. However, combination within various catalytic, bi-/multi oxidant, ultrasound and/or UV-systems results with $HO\bullet$ chemistry.

Generally, ozone is naturally present in the upper atmosphere where its molecule made of three oxygen atoms is formed through diatomic oxygen photolysis. However, it can be easily artificially generated, either by aforementioned photolysis or under the influence of strong electrical field, as mostly applied in commercial ozone generators.

The history of water purification by ozone is quite rich; it is used in second half of nineteenth century for drinking water preparation through Germany and in Monaco, while its application for such purposes was even extended in 1970s across Europe (France, Switzerland and Germany) due to documented generation of production of organohalogenated hydrocarbons during water chlorination [22]. Ozone reactivity is based on its electronic configuration, providing characteristics to acts as electrophilic, dipolar or even nucleophilic agent. Accordingly, ozone can react with organics through direct (i. e. "molecular") and indirect (i. e. "radical") mechanisms. The later can be easy accomplished by increasing pH of treated water toward basic values, providing ozone to generate HO• radicals eq. (8.15):

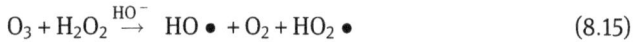

$$O_3 + H_2O_2 \xrightarrow{HO^-} HO\bullet + O_2 + HO_2\bullet \tag{8.15}$$

Such process is also called catalytic ozonation. Other catalytic variations of ozonation include homogeneous or heterogeneous catalysts containing the reduced transition metals such as iron, manganese, and copper. The undergoing mechanism differs from applied catalyst. For instance, iron application is followed by similar chemistry to above mentioned Fenton process, yielding HO• formation eqs (8.16) and (8.17):

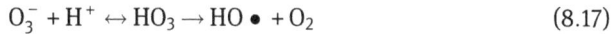

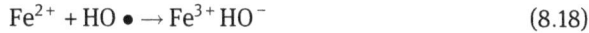

$$Fe^{2+} + O_3 \rightarrow Fe^{3+} + O_3^- \tag{8.16}$$

$$O_3^- + H^+ \leftrightarrow HO_3 \rightarrow HO\bullet + O_2 \tag{8.17}$$

$$Fe^{2+} + HO\bullet \rightarrow Fe^{3+} HO^- \tag{8.18}$$

On the other hand, the usage of manganese does not consider HO• formation; the degradation of organics (RH) occur through following complex mechanism eqs (19)–(23) [32]:

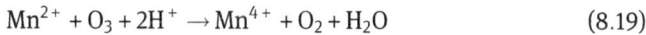

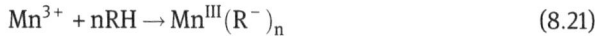

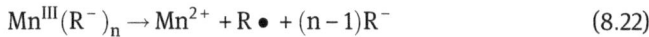

$$Mn^{2+} + O_3 + 2H^+ \rightarrow Mn^{4+} + O_2 + H_2O \tag{8.19}$$

$$Mn^{2+} + Mn^{4+} \rightarrow 2Mn^{3+} \tag{8.20}$$

$$Mn^{3+} + nRH \rightarrow Mn^{III}(R^-)_n \tag{8.21}$$

$$Mn^{III}(R^-)_n \rightarrow Mn^{2+} + R\bullet + (n-1)R^- \tag{8.22}$$

$$R\bullet + O_3 + H^+ \rightarrow products \tag{8.23}$$

Another widely investigated ozone process modification is peroxone process; a combination of ozone with another strong oxidant, H_2O_2, yielding HO• formation eq. (8.24):

$$H_2O_2 + 2O_3 \rightarrow 2HO\bullet + 3O_2 \tag{8.24}$$

Regardless the modification applied; catalytic, oxidant, UV or ultrasound driven, such process is more powerful than its classic analogue, possessing faster kinetics due to the presence of highly unselective radical species. The main limiting factor of ozonation process is efficiency of ozone mass transfer from gas to liquid phase; the aforementioned combined applications of ozonation may diminish such effect, as well as modifications of process equipment such as installment of porous electrode for *in situ* ozone generation.

8.4.2 Photo-assisted AOPs

8.4.2.1 Photochemical processes
The class of photochemical processes considers the application of UV irradiation and some powerful oxidant; the most common is H_2O_2, while alternatives might be ozone, persulfate and lately even chlorine. However, not any UV irradiation, that comprises energies from about 300 kJ Einstein^{-1} (UV-A radiation, 1 Einstein = 1 mol of photons), up to 1200 kJ Einstein^{-1} (vacuum UV), can be used to generate radical species through photolysis of mentioned oxidants. The mostly often used is UV-C, monochromatic, so-called germicidal, irradiation at 254 nm, that can be achieved by low-pressure vapor mercury lamps, while lately is application of LED devices also investigated. Considering the kinetics of direct photolysis, that depends on parameters such as: molar absorption coefficient (ε, dm^3 mol^{-1} cm^{-1}), quantum yield (Φ, mol Einstein^{-1}), intensity of emitted irradiation which is directly related to the incident photon flux, the irradiation wavelength, photoreactor design determining the optical length path, it is not surprising that UV-C$_{254}$ lamps are mostly used. This kinetic relationship can be represented by modified version of a semiempirical "LL model" based on the Lambert's law, eq. (25) [22]:

$$r_{UV} = -\frac{dc_i}{dt} = \Phi_i \times F_i \times I_0 \times \left[1 - \exp\left(-2.303 \times l \times \sum \varepsilon_j \times c_j\right)\right] \tag{8.25}$$

being $F_i = \dfrac{\varepsilon_i \times c_i}{\sum \varepsilon_j \times c_j}$

where Φ_i, ε_i, I_0 and l, stand for the physical properties of specie i, its quantum yield and the molar absorption coefficient, as well as the incident photon flux by reactor volume unit and the effective optical path in the reactor, respectively.

One of the simplest way to generate HO$\bullet$ is by the cleavage of H_2O_2 under UV-C irradiation yielding the formation of two HO$\bullet$ per photon absorbed eq. (8.26) [18]. However, the quantum yield of below reaction, when UV-C irradiation at 254 nm is applied, equals 0.5, with the regard to consumed oxidant, or 1, in respect toward formed radicals; another radical is lost in water matrix cage undergoing radical chain reactions. Nevertheless, the effectiveness of such photolytic reaction is still high. Another advantage of using H_2O_2 is its rather harmless nature and low environmental risk.

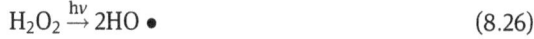

$$H_2O_2 \xrightarrow{h\nu} 2HO \bullet \qquad (8.26)$$

The persulfate ($S_2O_8{}^{2-}$), already mentioned as alternative oxidant within Fenton-based processes, can be used as alternative to H_2O_2 in photochemical processes as well [33]. As mentioned, persulfate can be activated by UV energy as well; similarly to H_2O_2, persulfate also forms two radicals per photon absorbed under UV activation eq. (8.27). It should be noted that effectiveness of persulfate system is higher than that of UV-C/H_2O_2; the quantum yield is 0.9 regarding consumed oxidant, i. e. 1.8 regarding the radicals. Besides, depending of process conditions HO$\bullet$ may be generated as well eq. (27) [31].

$$S_2O_8^- \xrightarrow{h\nu} 2SO_4^- \bullet \qquad (8.27)$$

$$SO_4^- \bullet + H_2O \rightarrow HSO_4^- + HO \bullet \qquad (8.28)$$

Ozone can also be used as an oxidant in photochemical processes; it has significantly higher molar absorption coefficient at 254 nm comparing to that of H_2O_2 (3300 dm^3 mol^{-1} cm^{-1} $>>>$ 18.6 dm^3 mol^{-1} cm^{-11}), resulting with almost 1000 times higher rate of photolysis yielding H_2O_2 eq. (8.29) formation, that undergo further decomposition according to eq. (26):

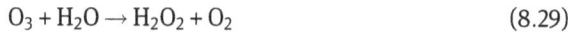

$$O_3 + H_2O \rightarrow H_2O_2 + O_2 \qquad (8.29)$$

Such oxidative system can be considered as multi-beneficiary regarding mechanisms involved in HO$\bullet$ generation; (i) through H_2O_2 photolysis eq. (8.26) and (ii) through peroxone chemistry eq. (8.24), and degradation of organics; (i) direct photolysis, (ii) HO$\bullet$ attack and (iii) direct ozonation [21].

As most of AOPs, photochemical processes are multi-factor systems; the effectiveness strongly depends on the process parameters related to both reactor design and characteristics of reaction mixtures. Hence, parameters that need to be optimized through reactor geometry are effective length path and flow regime, while characteristics of irradiation source are directly related to the radiant power and irradiation wavelength. The operating parameters such as pH, oxidant type and concentration would dictate the quantity of formed radical species and need to be optimized as well. Besides, constrains of photochemical processes are related to the presence of suspended matter, lowering solution transparence, and various species (e. g. counter ions), providing scavenging effect toward generated radical species.

8.4.2.2 Photocatalytic processes

Heterogeneous photocatalysis, employing semiconductor (nano)particles such as TiO_2, ZnO, Fe_2O_3, CdS, ZnS, SnS_2, and many others, is based on the photonic excitation of a solid. Underlying mechanism includes following phenomena, and is presented through eqs (8.30)–(8.34). Upon illumination of semiconductor (nano)particle by efficient energy electrons from valence band would cross the gap to be transferred into the conductance band, creating the electron-hole (e^-h^+) pair at the semiconductor (nano) particle surface. Such photogenerated electrons could react with adsorbed electron acceptor species, while the holes react with electron donor species, creating reduced and oxidized products, respectively. Accordingly, in the case of polluted water, electron acceptor specie is DO, resulting with superoxide radical $(O_2\bullet^-)$ formation or its protonated form $(HO_2\bullet)$. The electron donor specie could be either (i) water (or hydroxyl ion), resulting with the formation of hydroxyl radicals $(HO\bullet)$ capable to degrade organics, or (ii) pollutant molecule itself (usually organic specie; commonly marked as RH), which can be directly degraded through the formation of its radical by-product [34]. The most common semiconducting material applied in photocatalytic water treatment titanium dioxide (TiO_2); it has been promoted due to the: (i) high photocatalytic activity under the incident photon wavelength of 300 nm < λ < 390 nm and stability in consecutive catalytic cycles, and (ii) multi-faceted functional properties of TiO_2 catalyst, such as chemical and thermal stability, resistance to chemical breakdown and attractive mechanical properties [35].

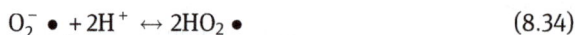

$$TiO_2 \xrightarrow{h\nu} h^+_{VB} + e^-_{CB} \tag{8.30}$$

$$H_2O + h^+_{VB} \rightarrow HO\bullet + H^+ \tag{8.31}$$

$$R + h^+_{VB} \rightarrow R^+ \tag{8.32}$$

$$O_2 + e^-_{CB} \rightarrow O^-_2\bullet \tag{8.33}$$

$$O^-_2\bullet + 2H^+ \leftrightarrow 2HO_2\bullet \tag{8.34}$$

However, the application of TiO_2 (nano)catalysts for water treatment is still experiencing a series of technical challenges. The separation of (nano)particles after water treatment remains the major disadvantage for commercial purposes. The fine particle size of the TiO_2 powders, and their large surface area-to-volume ratio and surface energy, creates a strong tendency for agglomeration during the operation, being highly detrimental in views of particle size preservation, surface-area reduction and its reusable lifespan. Other technical challenges include the limitation of TiO_2 to absorb irradiation energy in broader incident radiation range and its integration with feasible photocatalytic reactor systems [35, 36]. Hence, current research is directed to the immobilization of semiconducting materials on various supports in order to avoid in-treatment agglomeration and posttreatment separation issues and catalyst modification in order to enlarge its "narrow" wavelength spectrum for photoactivation, which can improve the overall light utilization properties. Such harvesting a broader spectrum of solar irradiation involves lowering of the band gap of semiconducting material, while inhibiting the recombination of photogenerated charges. Strategies including doping with non-metals, incorporation or deposition of noble metals (ions), and material engineering solutions based on composites formation using transition metals, carbon nanotubes, dye sensitizers, conductive polymers, graphene (oxide) and semiconducting materials, present viable solutions for set tasks [37].

8.4.3 Reactions and kinetics of hydroxyl radicals

According its mechanism, the reactions of HO• with organics can be classified within three groups [38]:
i. Electrophilic addition of HO• on double bond:

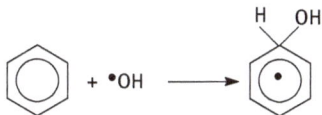

ii. Abstraction of hydrogen atom:

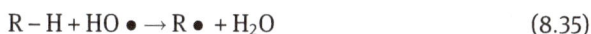

$$R - H + HO \bullet \rightarrow R \bullet + H_2O \tag{8.35}$$

iii. Electron transfer:

$$R - H + HO \bullet \rightarrow (R - H) \bullet + HO^- \tag{8.36}$$

All three mechanism yield initial radical product, while further reactions result in the formation of non-radical products; e. g. alcohols, aldehydes, carboxylic acids, ketones, dimers, and eventually final mineralization products CO_2 and H_2O.

Regardless the occurring mechanism presented above, the general reaction of HO• with organic pollutant (P) can be presented as:

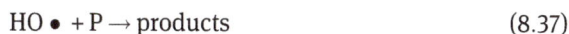

$$HO \bullet + P \rightarrow products \tag{8.37}$$

while the rate of reaction is given by:

$$r = -\frac{dc_P}{dt} = k_{HO\bullet, P} * c_{HO\bullet} * c_P \tag{8.38}$$

where $k_{HO\bullet, M}$ stands for reaction rate constant between HO• and targeted organic pollutant.

Considering the fact that the concentration of organic pollutant (P) is significantly higher than the concentration of HO• ($c_P \gg c_{HO•}$), the above eq. (8.38) can be simplified form second order kinetics to the "pseudo"-first order kinetics:

$$r = -\frac{dc_P}{dt} = k_0 * c_P \tag{8.39}$$

where k_0 stands for reaction rate constant of "pseudo"-first order:

$$k_0 = k_{HO•,P} * c_{HO•} \tag{8.40}$$

By integrating the above eq. (8.40) in the interval from t0 to t the following equation is obtained:

$$-\ln\left(\frac{c_{P,t}}{c_{P,t_0}}\right) = k_0 * t \Rightarrow c_{P,t} = c_{P,0} * e^{-k_0} \tag{8.41}$$

where $c_{M,0}$ and $c_{M,t}$ are concentrations of organic pollutant during the treatment.

The value of "pseudo"-first order reaction rate constant k_0 can be determined graphically when t is plotted in the relationship to c_P [22, 39].

As stated previously, HO• react with variety of organic compounds at rather high rates. Hence, in Table 8.9 the reaction rates with most common groups of organic pollutants are summarized [39].

Table 8.9: Reaction rate constants of HO• and common groups of organic compounds.

Organic compounds	$k/[dm^3\ mol^{-1}\ s^{-1}]$
Chlorinated alkenes	10^9-10^{11}
Phenols	10^9-10^{10}
Organic compounds with nitrogen	10^8-10^{10}
Aromatics	10^8-10^{10}
Ketones	10^9-10^{10}
Alcohols	10^8-10^9
Alkanes	10^6-10^9

Example 8.5. Different water contaminants and natural water constituents compete for reacting with hydroxyl radicals, thus affecting the overall efficiency of advanced oxidation treatment. The application of the kinetic data for the hydroxyl radicals can be used to determine which of several competing reactions will predominate in the aqueous solution in the presence of more compounds [40]. Second-

order rate constants for reaction of hydroxyl radicals with different compounds in aqueous solutions can be found in the literature [41, 42].

For example, when three solutes are present in water; phenol ($c(Ph) = 1{\times}10^{-3}$ mol dm^{-3}) as a target pollutant, hydrogen peroxide ($c(H_2O_2) = 1{\times}10^{-4}$ mol dm^{-3}) and sulphate ($c(SO_4{}^{2-}) = 6{\times}10^{-4}$ mol dm^{-3}), they compete for OH radicals according to the following reactions with the corresponding reaction rate constants [40–42]:

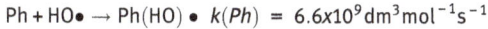

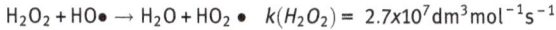

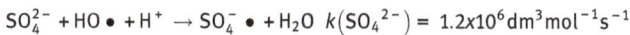

$$Ph + HO\bullet \rightarrow Ph(HO)\bullet \quad k(Ph) = 6.6x10^9 \, dm^3 mol^{-1} s^{-1}$$

$$H_2O_2 + HO\bullet \rightarrow H_2O + HO_2\bullet \quad k(H_2O_2) = 2.7x10^7 \, dm^3 mol^{-1} s^{-1}$$

$$SO_4^{2-} + HO\bullet + H^+ \rightarrow SO_4^-\bullet + H_2O \quad k(SO_4{}^{2-}) = 1.2x10^6 \, dm^3 mol^{-1} s^{-1}$$

The extent of reaction is proportional to the product of the reaction rate constant and the concentration of the solute:

$$k(Ph)'c(Ph) : k(H_2O_2)'c(H_2O_2) : k(SO_4{}^{2-})'c(SO_4{}^{2-}) = 70000 : 3 : 1$$

Accordingly, > 99.9 % of hydroxyl radicals reacting in the system would be effectively consumed in reaction with phenol, while with their loss in side reactions could be considered as negligible.

References

[1]   Chaussade J-L, Mestrallet G. Water treatment handbook. Volume 1. 7th ed. France: Degrémont, Rueil-Malmaison, Lavoisier SAS, 1995.
[2]   Sundstrom DW, Klei HE. Wastewater treatment. Englewood Cliffs, NY, USA: Prentice-Hall, Inc., Prentice-Hall, 1979.
[3]   Kaštelan-Macan M, Petrović M. Analitika okoliša (in Croatian language). Zagreb, Croatia, Denona: HINUS and Faculty of Chemical Engineering and Technology, 2013.
[4]   Fifield FW, Haines PJ. Environmental analytical chemistry. 2nd ed. Oxford, UK: Blackwell Science Ltd, Blackwell Science, 2000.
[5]   Baker RW. Membrane technology and applications, 2nd ed. Sussex: John Wiley & Sons, Ltd, 2004.
[6]   Mulder M. Basic principles of membrane technology. 2nd ed. Dordrecht: Kluwer Academic Publisher, 1996.
[7]   Cui ZF, Muralidhara HS. Membrane technology – A practical guide to membrane technology and application in food and bioprocessing. Oxford: Elsevier, 2010.
[8]   Dolar D, Košutić K. Removal of pharmaceuticals by ultrafiltration (UF), nanofiltration (NF), and reverse osmosis (RO). In: Petrović M, Barcelo D, Perez S, editors. Comprehensive analytical chemistry, Vol. 62. Amsterdam: Elsevier, 2013: 319–44.
[9]   American Water Works Association Research Foundation. Water treatment membrane processes. New York: McGraw-Hill, 1996.
[10] Subramani A, Jacangelo JG. Emerging desalination technologies for water treatment: A critical review. Water Res. 2015;75:164–87.

[11]   Membranes Market by Type (Polymeric, Ceramics), Technology (RO, UF, MF, NF, Pervaporation, Gas Separation), Application (Water & Wastewater Treatment, Food & Beverage, Medical & Pharmaceutical, Industrial Gas Processing), Region - Global Forecast to 2021. http://market-sandmarkets.com. Accessed 2017 at: http://www.marketsandmarkets.com/Market-Reports/membranes-market-1176.html.

[12]   Koltuniewicz A, Drioli E. Membrane in clean technology. Theory and practice, vols 1–2. Weinheim: Wiley-VCH Verlag GmbH & Co. KGaA, 2008.

[13]   Tchobanoglous G, FL B, Stensel HD. Wastewater engineering: treatment and reuse. New York: Metcalf & Eddy, Metcalf, Inc., 2003.

[14]   Bitton G. Wastewater microbiology. New York: John Wiley & Sons, 2005.

[15]   Casey TJ. Unit treatment processes in water and waste engineering. New York: John Wiley & Sons, 1996.

[16]   Wang LK, Pereira NC, Hung Y-T. Biological treatment processes. New York: Humana Press, 2009.

[17]   Gunder B. The membrane-coupled activated sludge process in municipal wastewater treatment. Lancaster: Technomic Publishing Co., Inc., 2001.

[18]   Parsons S. Advanced oxidation processes for water and wastewater treatment. London, England: IWA Publishing, 2004.

[19]   Litter MI, Candal RJ, Meichtry JM. Advanced oxidation technologies: sustainable solutions for environmental treatments. London, UK: CRC Press, Taylor and Francis, 2014.

[20]   Gogate PR, Pandit AB. A review of imperative technologies for wastewater treatment I: oxidation technologies at ambient conditions. Adv Environ Res. 2004;8:501–51.

[21]   Koprivanac N, Kusic H. Hazardous organic pollutants in colored wastewaters. New York: Nova Science Publishers, Inc., 2009.

[22]   Beltran FJ. Ozone-UV radiation-hydrogen peroxide oxidation technologies. In: Tarr MA, editor. Chemical degradation methods for wastes and pollutants, environmental and industrial applications. New York, USA: Marcel Dekker, Inc., 2003: 1–77.

[23]   Mantzavinos D, Psillakis E. Enhancement of biodegradability of industrial wastewaters by chemical oxidation pre-treatment. J Chem Technol Biotechnol. 2004;79:431–54.

[24]   Andreozzi R, Vaprio V, Insola A, Marotta R. Advanced oxidation processes (AOP) for water purification and recovery. Catal Today. 1999;53:51–59.

[25]   Fenton HJH. On a new reaction of tartaric acid. Chem News. 1876;33:190.

[26]   Fenton HJH. Oxidation of tartaric acid in presence of iron. J Chem Soc, Trans. 1894;65: 899–910.

[27]   Tarr MA. Fenton and modified Fenton methods for pollutant degradation. In: Tarr MA, editor. Chemical degradation methods for wastes and pollutants, environmental and industrial applications. New York, USA: Marcel Dekker, Inc., 2003: 165–200.

[28]   Kušić H, Koprivanac N, Lončarić Božić A, Selanec I. Photo-assisted Fenton type processes for the degradation of phenol: a kinetic study. J Hazard Mater. 2006;B136:632–44.

[29]   Kavitha V, Palanivelu K. The role of ferrous ion in Fenton and photo-Fenton processes for the degradation of phenol. Chemosphere. 2004;55:1235–43.

[30]   Juretic Perisic D, Gilja V, Novak Stankov M, Katancic Z, Kusic H, Lavrencic Stangar U, et al. Removal of diclofenac from water by zeolite-assisted advanced oxidation processes. J Photochem Photobiol. 2016;321:238–47.

[31]   Kusic H, Peternel I, Ukic S, Koprivanac N, Bolanca T, Papic S, et al. Modeling of iron activated persulfate oxidation treating reactive azo dye in water matrix. Chem Eng J. 2011;172:109–21.

[32]   Legube B, Karpel Vel Leitner N. Catalytic ozonation: a promising advanced oxidation technology for water treatment. Catal Today. 1999;53:61–72.

[33]   Vilhunen S, Sillanpaa M. Recent developments in photochemical and chemical AOPs in water treatment: a mini review. Rev Environ Sci Biotechnol. 2010;9:323–30.

[34] Pichat P. Photocatalysis and water purification: from fundamentals to recent applications. Germany: Wiley, Weinheim, 2013.

[35] Chong MN, Jin B, Chow CWK, Saint C. Recent developments in photocatalytic water treatment technology: a review. Water Res. 2010;44(10):2997–3027.

[36] Pelaez M, Nolan NT, Pillai SC, Seery MK, Falaras P, Kontos AG, et al. A review on the visible light active titanium dioxide photocatalysts for environmental applications. Appl Catal B: Environ. 2012;125:331–49.

[37] Schneider J, Matsuoka M, Takeuchi M, Zhang J, Horiuchi Y, Anpo M, et al. Understanding TiO2 photocatalysis: mechanisms and materials. Chem Rev. 2014;114:9919–86.

[38] Legrini O, Oliveros E, Braun AM. Photochemical process for water treatment. Chem Rev. 1993;93(2):671–98.

[39] Rodriguez M. Fenton and UV-VIS based advanced oxidation processes in wastewater treatment: Degradation, mineralization and biodegradability enhancement, PhD Thesis, University of Barcelona, 2003.

[40] Lukeš P. Water treatment by pulsed streamer corona discharge, PhD Thesis, Published by Institute of Chemical Technology, Prague and Institute of Plasma Physics AS CR, 2001.

[41] Buxton GV, Greenstock CL, Helman WP, Ross AB. Critical review of rate constants for reactions of hydrated electrons, hydrogen atoms and hydroxyl radicals (.OH/.O-) in aqueous solution. J Phys Chem Ref Data. 1988;17:513–886.

[42] NIST Standard Reference Database 40, NDRL/NIST Solution Kinetics Database; National Institute of Standards and Technology, USA. Accessed at March 8, 2018: http://kinetics.nist.gov/solution/Detail?id=1982FIE/RAG2443-2449:7

Discussion, Questions and Exercises

1. Explain the difference between the criteria and standards for water quality.
2. Which parameters in the characterization of water are required to determine before re-using water or its discharge into the environment?
3. What is the difference between the biological, physical and chemical characterization of the water?
4. What analytical methods are used for the analysis of inorganic and organic traces contaminants and which requirements of that method must be satisfied and why?
5. Concentration of chloride ions in feed, concentrate and permeate streams are 1000, 3800, and 200 mg L^{-1}, respectively. Designed recovery rate for the RO unit is 75 %. According to the concentrations of chloride ions calculate actual recovery rate. (*Solution:* $Y = 78\%$)
6. Concentration of solute in feed water is 500 mg L^{-1} and is treated with nanofiltration membrane. Rejection of this solute with NF membrane is 90 %. Calculate concentration of solute in permeate. (*Solution:* $y_p = 50$ mg L^{-1})
7. RO unit contains 3 membrane modules. Feed flow rate is 1000 m^3 d^{-1}. Calculate permeate and retentate flow rate for each membrane module if recovery for one module is 15 %. Also calculate overall recovery of this process. (*Solution:* $Q_{p,1} = 150.0$ m^3 d^{-1}; $Q_{r,1} = 850.0$ m^3 d^{-1}; $Q_{p,2} = 127.5$ m^3 d^{-1}; $Q_{r,2} = 722.5$ m^3 d^{-1}; $Q_{p,3} = 108.4$ m^3 d^{-1}; $Q_{r,3} = 614.1$ m^3 d^{-1}, and $Y = 38.6\%$)

8. Two-step process is presented in the diagram below together with feed flux and concentrations of DOC and Ca. In 1st and 2nd stage 7 and 3 vessels were used, respectively. In each vessel were 7 NF modules. A recovery for modules is given in Table 8.4 in Example 8.2.

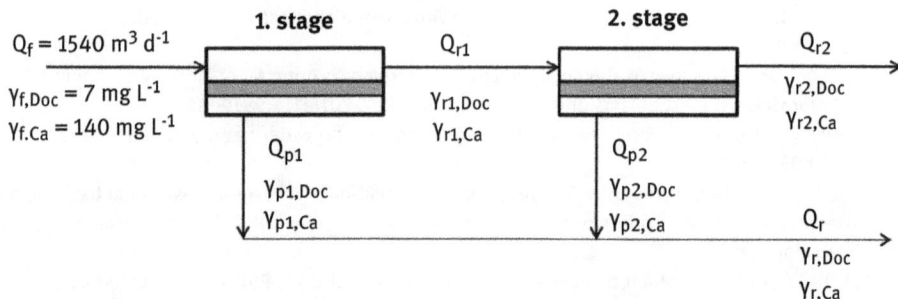

Solution:

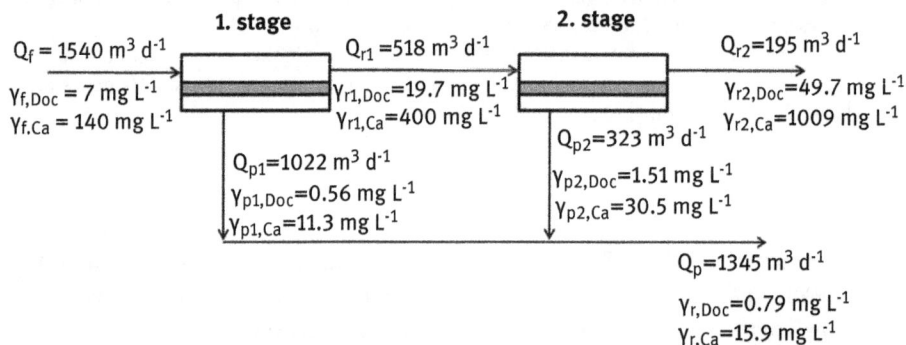

9. What is the relationship between the BOD and COD values in wastewater?
10. What are the advantages of membrane bioreactor over conventional processes for wastewater treatment?
11. What is activated sludge?
12. How to calculate the parameters μ_{max} and K_s?
13. Describe the main advantages and constrains of AOPs in water treatment.
14. What is the main oxidative specie in all AOPs and how it may be generated?
15. What is the operating pH range in Fenton process? What would happen with ferrous and ferric iron at basic conditions?
16. Describe the underlying phenomena of photocatalytic AOPs using semiconductors.

Ivica Kisić, Željka Zgorelec and Aleksandra Percin

9 Soil treatment engineering

Abstract: Soil is loose skin of the Earth, located between the lithosphere and atmosphere, which originated from parent material under the influence of pedogenetic processes. As a conditionally renewable natural resource, soil has a decisive influence on sustainable development of global economy, especially on sustainable agriculture and environmental protection. In recent decades, a growing interest prevails for non-production soil functions, primarily those relating to environmental protection. It especially refers to protection of natural resources whose quality depends directly on soil and soil management. Soil contamination is one of the most dangerous forms of soil degradation with the consequences that are reflected in virtually the entire biosphere, primarily at heterotrophic organisms, and also at mankind as a food consumer. Contamination is correlated with the degree of industrialization and intensity of agrochemical usage. It is typically caused by industrial activity, agricultural chemicals or improper disposal of waste. The negative effects caused by pollution are undeniable: reduced agricultural productivity, polluted water sources and raw materials for food are only a few of the effects of soil degradation, while almost all human diseases (excluding AIDS) may be partly related to the transport of contaminants, in the food chain or the air, to the final recipients – people, plants and animals. The remediation of contaminated soil is a relatively new scientific field which is strongly developing in the last 30 years and becoming a more important subject. In order to achieve quality remediation of contaminated soil it is very important to conduct an inventory as accurately as possible, that is, to determine the current state of soil contamination.

Keywords: soil science, soil contaminants, remediation technologies

9.1 Introduction

While practical knowledge and empirically obtained insights about the soil have their origin in the most ancient history, soil science has developed as an independent science in the mid-nineteenth century. Soil was mentioned in the oldest scientific book *Papyrus Ebers* from the sixteenth century BC, and also Greek and Roman writers have been working on the issue of soil fertility and classification. Ancient History and the Middle Ages were centuries of empiricism. The entire knowledge on soil from that time remains

This article has previously been published in the journal Physical Sciences Reviews. Please cite as: Kisic, I., Zgorelec, Z., Percin, A. Soil Treatment Engineering. *Physical Sciences Reviews* [Online] **2017**, *2* (11). DOI: 10.1515/psr-2016-0124

https://doi.org/10.1515/9783110468038-009

based on the experience, and that knowledge was collected mainly for agricultural purposes. The progress of chemistry has accelerated the development of knowledge on the soil. In the second half of the eighteenth century, when the chemistry was so advanced, knowledge of the major chemical principles and analytical procedures has enabled the determination of the composition of living and dead inorganic and organic substances. Biochemistry has developed and the first studies of the soil begun.

At first, the knowledge of the soil was obtained as a part of the geologic system – as geochemistry or agrogeology, chemistry or general crop science, and later as a collection of different sciences that can be applied in agriculture and forestry. The first scientific work about the soil, *Bodenkunde – oder die Lehre vom Boden* by German agrochemist Carl Sprengel (1787–1859), was published in 1837. Soil science had soil fertility in its focus for a long time, as an expression of indisputably the most important productive role of soil in agriculture and forestry. Only recently, more attention is given to the other roles of soil. Based on all the above reasons, all nations classify soil as the most important national natural resource, that is, as a member of the so-called environmental triad: AIR – SOIL – WATER.

9.2 Basics of soil science

When speaking about the soil it is most important to point out that we obtain more than 95% of food and raw materials for food from the soil. At the same time, a favorable fertile soil for this purpose is a very limited resource (Figure 9.1).

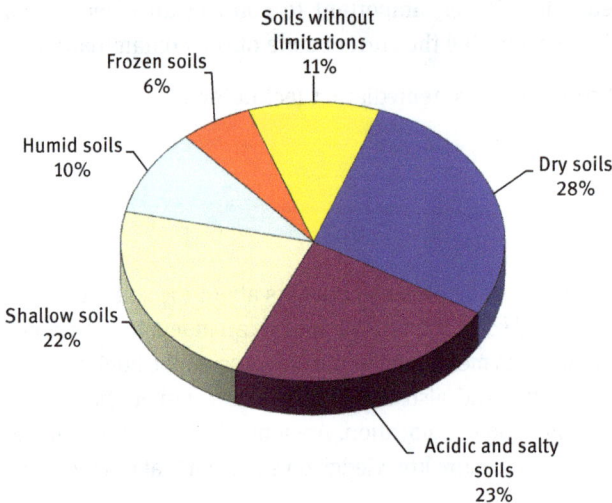

Figure 9.1: The percentage share of the soil surface with different restrictions for agricultural use of the world's total soil area [1]. (Adopted from FAO)

There are several definitions of the soil. The simplest definition is: soil is the surface layer of the Earth's crust altered by the combined influence of climate, air, water, plants and animals. From the ecological point of view, soil is the layer of the Earth's crust that provides optimum conditions for plant growth. Based on the above-mentioned definitions, we can say that soil is loose skin of the Earth, located between the lithosphere and atmosphere, which arose from parent material under the influence of factors of pedogenesis and pedogenesis process (Figure 9.2). Soil is a natural product created by complex processes, first by fragmentation and decomposition of rocks, i.e. primary minerals of parent material, and by synthesis – the formation of new, secondary minerals. The resulting mass can hold water, which allows colonization, primarily by bacteria, fungi, algae and lichen, and finally by higher plants. The names of soil in different languages are: počv (Bulgarian), tlo (Bosnian and Croatian), púda (Czech), Έδαφος (Greek), jord (Danish, Norwegian and Finnish), Boden (German), soil (English), grundo (Esperanto), suelo (Spanish), sol (French), talaj (Hungarian), suolo (Italian), počva (Russian and Macedonian), bodem (Dutch), gleba (Polish and Belarusian), solo (Portuguese), sol (Romanian), poda (Slovak), prst (Slovenian), zemljište (Serbian), toprak (Turkish) and grunt (Ukrainian).

Figure 9.2: Soil [1].
(Photo by: Z. Zgorelec, 2007).

The biggest difference between soil and rock is in the content of organic matter. Agricultural soil contains 0.5–5% of organic matter, and peat soil up to 50% of organic matter. Although its content in agricultural soils is relatively small, the organic matter brings and maintains life in the soil. Without organic matter, soil would no longer be soil but a lively rock from which the soil was created. In this case, soil would lose its primary role, which is the supply of plants with water, air and nutrients [2]. It is necessary to clearly distinguish between the two terms that are frequently replaced and cause confusion – soil and land. The difference is obvious. Soil was already described as a natural product created by processes of soil formation (pedogenesis), and the term land refers to the land area (Figure 9.3). Land is a broader concept – it's a land (terrestrial) surface and bio-productive system that includes soil, vegetation, other biota, and the ecological and hydrological processes that occur in the entire system. We have already mentioned the names of soil in the major European languages, and the names of the land are in French – terre, English – land (landscape), Russian – zemlja, Italian – terreno, Dutch – landen, Polish – kraina or parcela, and German – Erde. And finally, a few words on the concept of pedology. Although it is in many languages a deep-rooted word (pedology), it is the concept that outsiders often replace with the concept of *pedology* as a science of physical and mental development of children. Therefore, the term **soil science** is increasingly used in recent years.

Figure 9.3: Land [1].
(Photo by: I. Kisic, 2011).

9.3 Soil formation

9.3.1 Factors and processes of soil formation

There are four main factors of pedogenesis: lithosphere (parent material), atmosphere, hydrosphere and biosphere. Lithosphere has a passive role, and the other three spheres actively participate in development processes of pedosphere. The atmosphere acts in particular via precipitation, heat, wind and gas; hydrosphere acts via movement and strength of water masses; the biosphere acts via most diverse organisms, primarily with flora, fauna and human activity. The formation of soil is a long process which lasts for thousands of years, and in some substrates such as limestone it lasts for hundreds of thousands, or even a million years. To form one centimeter layer of red soil (*terra rossa*) as the most valuable type of soil on the limestone of the Mediterranean, it requires a flow of 8,000–10,000 years, therefore for 1 m depth it takes 800,000–1,000,000 years. A man can destroy it by improper management in several dozen years or permanently convert it in a few days. We distinguish five basic groups of pedogenetic processes: depletion of lithosphere; formation of organic matter; degradation (*decomposition*) of organic matter; displacement of components (translocation) and the new formation (*neogenesis*).

Weathering of parent materials is changing the physical and chemical properties of the lithosphere.

The compact masses of rock disintegrate and are then made from the same mineral compounds and partly of new mineral compounds. Surface mass significantly increases, and their atoms and atomic groups are more mobile than they were in compact rocks of the lithosphere. Processes of formation of organic matter enrich the mineral erosion with the living organic matter that becomes an important factor in the dynamics and development of soil. Processes of decomposition (rotting, decay) of organic matter result in many chemically and physiologically active compounds that enter in reactions with disintegrated minerals. Processes of transfer can change the position of the mass, or soil particles, on the surface of pedosphere or in its interior. Surface displacement is the most important factor in the development of the external morphology of pedosphere. The most diverse morphological forms are developed according to the forces participating in the surface displacement of pedosphere mass (rain, floods, river, sea, glaciers, wind, etc.). Processes of the new formation of the rock weathering products and the products of degradation of organic matter created new inorganic and organic complexes, of which the most important and most active are the colloidal components of soil – the secondary minerals or clay minerals and humus. These components significantly alter the features of erosion.

9.3.2 Basic soil characteristics

Soil formation was briefly summarized in the previous section. The basic physical, chemical and biological properties of soil and fertility, as a result of these qualities, are presented in this section. According to its physical characteristics, soil is a three-part system that has a solid, liquid and gaseous component. The ratio of individual components is a dynamic value (especially the ratio of liquid and gaseous components), dependent on the mechanical composition of soil, climate, current weather conditions, and the season, as well as on all other external factors. The solid component consists of mineral and organic parts. Mineral part originates from parent material, and the organic is more or less a humified organic matter. The liquid component of soil is water, more specifically, an aqueous solution of soil.

The most important <u>physical parameter</u> of the soil is the granulometric composition of the soil. The solid component of the soil is by nature a polydisperse system that consists of particles of different dimensions, from the ions, molecules and colloids that are invisible to the naked eye, to the particles of stone. Each particle, that can't continue to break into pieces by weak mechanical forces (or peptization), is called a mechanical or granulometric element. Mechanical components in the soil are rarely found separately and they are connected to the larger particles – structural aggregates. The term mechanical composition or texture means the content of individual fractions in the soil in weight percent (Table 9.1). As the lower and upper limits for the classification of mechanical fractions according to the particle size are conventional values, different classification is applied in the world. Textural classes are determined by, among other things, the Atterberg triangle that has data on the percentage of silt, sand and clay graphically applied on every side of it, and the intersection can show a texture mark, as shown in Figure 9.4. According to their characteristics, all classes can be divided into three basic groups, or types of soil: sand, loam and clay soils. It is widely accepted that the clay soils are labeled as heavy soils, and sandy soils as light soils. The increased soil fragmentation (dispersion) in the same conditions increases the holding power of water, adsorption capacity, the amount of related nutrients, swelling, stickiness and plasticity, and decreases the permeability of water – internal natural drainage. **Shortly, the chemical properties of the soil are improved and the physical properties deteriorate.**

Table 9.1: International classification of granulometric composition [3].

Soil with a diameter greater than 2 mm		Soil with a diameter less than 2 mm	
Rock	More than 20 mm	Coarse sand	2.0–0.2 mm
Gravel	20–2.0 mm	Fine sand	0.2–0.02 mm
		Dust/powder	0.02–0.002 mm
		Clay	<0.002 mm
Stony soils		Stony soils	
I. Absolutely stony soils: more than 90% of the stone		I. Very stony soils: 30–50% of the stone	
II. Very stony soils: 70–90% of the stone		II. Stony soils: 10–30% of the stone	
III. Stony soils: 50–70% of the stone		III. Poorly stony soils: less than 10% of the stone	

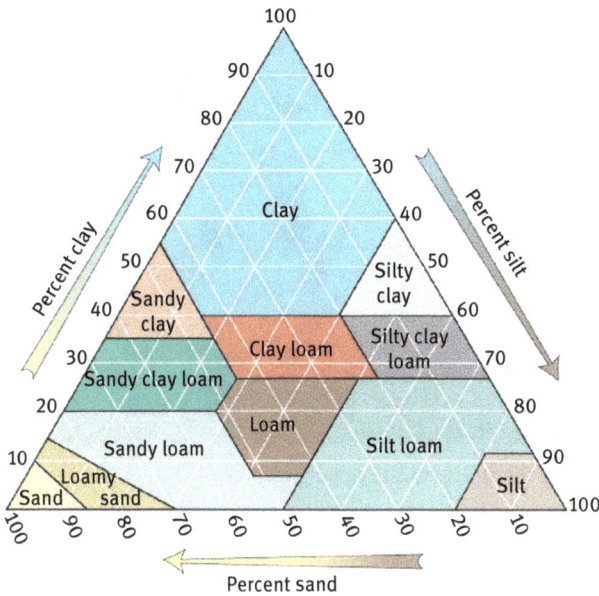

Figure 9.4: Atterberg's triangle for determining the mechanical composition [4].

9.3.2.1 Soil structure

Mechanical elements are not free or loose in the soil, separated from one another, but are connected to larger clusters – structural aggregate. The term structure of the soil includes the size, form and manner of distribution of structural aggregates in the soil. Classification of structure can be carried out on the basis of different criteria – the shape and size of pores, the microstructure and the like. According to the shape, they can be *cubical*, the ones with all three axes equally developed, *prismatic* or *columnar*, the vertical axis is more developed so they are elongated, and *flat*, the horizontal axis is longer than the vertical (Figure 9.5).

Structure is one of the most important factors of soil fertility. It can significantly correct the unfair qualities of the soil caused by mechanical composition. Soils with a favorable structure contain more air and are suitable for processing, so that the root system can develop in them optimally. Structural soil is porous, with a favorable ratio of micro-pores and macro-pores, and that relationship is stable even in conditions of increased humidity. Soils with a stable structure have a good water regime and are permeable to water. They receive and retain water well and they can create water reserves in the soil because they are not subject to drought. The excess of water from such soils is drained through a system of macro-pores. Such soils are, therefore, well drained. A stable structure is an important indicator of the air regime of soil, it provides aeration, and thanks to that the favorable thermal regime. Soils with a

(a)   (b)

Figure 9.5: Various shapes and sizes of the structural soil aggregates.
(Photo by: I. Kisic, 2005 & 2010).

favorable and stable structure are less subject to the wind and water erosion. Soils with unfavorable structures have exactly the reverse features, bad air-water regime and poor drainage, and, at times, reduction processes occur in them due to prolonged water retention.

Soil density is a number that shows how many times the weight of the soil is denser than the weight of an equal volume of water. There is a difference between **soil particle density (SPD)** and **bulk (dry) density (BD)**. SPD indicates the number of times that the mass of soil without pores, therefore only the mass of the rigid soil component is denser than the mass of an equal volume of water. This value does not depend on the mechanical composition, structure and compaction of the soil, it is a constant value associated with the content of organic matter. Soils with a higher content of organic matter have a lower SPD than soils with low organic matter. In arable soils SPD values are in the range from 2.2 g/cm^3 to 2.9 g/cm^3. The term BD stands for how many times the mass of a volume of natural soil with pores is denser than the mass of an equal volume of water. It is determined in the laboratory so that the mass of a known volume of natural soil is divided with that volume. The value of BD is not constant; it changes with the soil tillage (loosening) and compaction and depends on the SPD and the porosity of the soil. The average value of BD in arable soils is from 1.4 g/cm^3 to 1.6 g/cm^3 and, generally, is decreasing with depth. The total porosity of the soil and the various evaluations of some soil parameters are determined based on the SPD. **Based on the data on the volume density of the soil, the weight of arable layer and the known quantity of a particular parameter, the total amount of the specified parameter per hectare is easily determined. Therefore, the volume density of the soil is a fundamental tool in balancing the investigated parameters in the soil**. The free spaces between structural aggregates are called pores or voids of the soil, and their total content in volume percentages is called **porosity of the soil (P)**. It is determined by the formula:

$$P\ \% = (SPD - BD) : SPD \times 100$$

The pores can be of different sizes, from large cracks and burrows of fauna, to capillary and micro-capillary pores. Favorable air-water conditions in the soil arise only when there's optimal ratio of macro-pores and micro-pores. It is considered to be the ratio of 1:1, or 3:2 in favor of capillary pores. The porosity of the soil depends on many factors, primarily on the mechanical composition, structure and content of organic matter.

Air of the soil consists of different gases that come from the atmosphere into the soil or are formed in the soil. Generally, the most important difference in relation to the air from the atmosphere is that it is fully saturated with water vapor for the most part of the year, richer in carbon dioxide and poorer in oxygen. The composition of the air from the atmosphere and the air from the soil is presented in Table 9.2. This table shows that the composition of air from the soil is very variable and many factors influence on it. Although the air from soil is derived mainly from atmosphere, a part of the gases are also created in the soil as a result of microbiological-biochemical processes. The sum of the content of carbon dioxide (CO_2) and oxygen (O_2) in the air of the soil and the atmosphere is approximately equal. Thus, if the soil has more oxygen there will be less carbon dioxide, and vice versa. Therefore, the ratio of CO_2 and O_2 in the soil depends on the soil aeration. CO_2 is released into the soil by respiration of the plant roots and microbial decomposition of organic matter. That is why CO_2 content is higher in summer than in winter. CO_2 content is higher in humic

Table 9.2: The composition of air of the atmosphere and pedosphere [3].

Gas	Volume ratio in the atmosphere %	Volume ratio in the pedosphere %
Nitrogen (N_2)	78.09	78–80
Oxygen (O_2)	20.95	0.1–20
Argon (Ar)*	0.93	
Carbon dioxide (CO_2)	0.03	0.1–15
Neon (Ne)*	0.018	
Helium (He)*	0.0052	
Methane (CH_4)	0.0015	
Krypton (Kr)*	0.001	
Nitrogen (I) oxide (N_2O)	0.005	
Hydrogen (H_2)	0.0005	
Ozone (O_3)	0.0004	
Xenon (Xe)*	0.000008	

soil because it has a high biological activity, than in mineral soil that has a low biological activity. It's higher in the wet than in the dry soils, in heavy soils than in lighter soils, in structureless than in well-ventilated, structural soils. The high content of carbon dioxide itself has no adverse effect on plant growth and the damage to plants comes indirectly from lack of oxygen.

In soils with low aeration in conditions of reduction, in small amounts, the following compounds are formed – ammonia (NH_3), methane (CH_4), hydrogen sulfide (H_2S), hydrogen (H_2) and other gases. Their appearance is an indicator of unfavorable conditions in the soil, of disturbed air-water conditions in the soil, because they can occur only under conditions of lack of oxygen. Favorable air-water regime of soil displays shades of brown, yellow and black, while shades of gray, blue, green and yellow indicate on the unfavorable air-water regime of soil (Figure 9.6).

(a)   (b)

Figure 9.6: Favorable (left – a) and unfavorable air-water regime of soil (right – b) [1]. (Photo by: F. Basic, 1984 and 1986).

Basic chemical properties of soil, e.g. organic matter (humus) in the soil, absorptive ability of soil and soil reaction, will be shown in continuation. In a broader sense, the term humus involves all dead organic matter of soil, and in the narrow sense it is a special organic matter of dark color, formed by humification processes. These are very complex degradation processes of the original organic matter and the synthesis of new, more complex compounds under the influence of soil microorganisms. The composition of humus is complex and cannot be expressed by a single chemical formula, although the adequate procedures can excrete chemical compounds known as components of humus. Thereafter remains a certain part of which chemical nature is not fully known nor understood. With regard to the content of individual fractions, there are two major forms of humus in the soil, although in nature there are many transitional forms. These are mild or mature, and acidic or raw humus. Mild or mature humus, composed of humic (gray) acids, saturated with bases, and salts of humic acids, or humates, is resistant to degradation. It accumulates in the soil and gives it the best physical, chemical and biological characteristics. It is created in the presence of high-quality organic matter that is rich in bases, such as the remains of grass steppe vegetation and deciduous trees, and soils in which, of the microorganisms, bacteria dominate. Acidic or raw humus occurs in conditions of humid climate on leached rocks that are poor in bases, especially under coniferous vegetation. It is dominated by fulvic acids, which are mostly free due to the lack of base, and they give soil unfavorable features. The process of humification in such conditions is a result of microbial activity of fungi. Raw humus is not a favorable form of humus. Between these two extreme forms, there is a series of transitional forms, whose characteristics and impact on the soil depend on the soil characteristics, how rich they are in bases, the type of vegetation, the number and type of microorganisms, climate, altitude and other conditions. Another important feature, which is also a criterion for evaluating the quality of humus, is the ratio of carbon and nitrogen, or the C:N ratio. The optimal ratio of C:N in the soil is 10:1. The value of C:N ratio in the soil higher than 15 is an indicator that plants have a limited content of nitrogen in the soil, while the value of C:N ratio lower than 10 is an indicator of a limited decomposition of organic matter in soil.

Although content of humus in relation to the mineral part is significantly smaller, and the depth in which it is produced is relatively small (up to approximately 20 cm deep), due to its characteristics, especially colloidal nature, humus is the most active component of soil, other than clay. In terms of compost, the ideal value of C:N ratio is 30 and it indicates a sufficient amount of food available for micro-organisms. Table 9.3 shows that, for example, compost consisting of pine needles, straw, coffee grounds and others, have the ideal value of C:N ratio.

Humus has a multiple positive effect on the soil:
- It is the main factor of creating and maintaining a stable crumbly structure, high porosity, optimal ratio of micro-pores and macro-pores, and thus the most favorable air-water settings and good drainage – permeability of soil for water;
- It has a large capacity for water and high adsorption capacity;
- It has a beneficial effect on warming and thermal properties of soil because of its dark color;

- It contains nutrients that released during the process of mineralization, increase the nutritional potential of soil;
- It has a beneficial effect on soil microbial activity.

Table 9.3: Features of some of the materials for making compost [1, 5].

Material	C/N ratio	The ability of degradation	Humidity in the original form
The leaves of alder	25–30	Good	Medium
The leaves of ash	25–30	Good	Medium
The leaves of linden, beech and oak	40–60	Good	Medium
Pine needles	30	Medium	Low
The leaves and roots of potato	25–30	Good	Good after wilting
Smaller branches from the garden and orchard	30–60	Good	Medium to excessive
Grass from the mown lawn	12–25	Good	Good after wilting
Straw	20–30	Good	Very low
The bark (of trees)	100–130	Medium	Low
Sawdust of fir	200–230	Medium	Low
The remains of pruning in the orchard	100–150	Bad	Very low
The remains from the garden	13	Good	Excessive
The straw of wheat	150	Good	Very low
The straw of barley	100	Good	Very low
The straw of oat	50	Good	Very low
The straw of rye	65	Good	Very low
Poultry manure	13–18	Good	Very low
Cattle manure	20	Good	Medium
Horse manure	25	Good	Good
Kitchen waste	12–20	Good	Generally excessive
Coffee grounds	25–30	Good	Good
Cardboard	200–500	Good	Low
Grape husks	40–60	Good	Low
The remains of the pressed fruit trees	30–80	Medium to good	Low

The term **sorption or sorption ability** implies the ability of soil to hold in a variety of substances – ions and molecules in the soil solution, colloids suspended in the water, particles of larger dimensions and microorganisms. Considering the forces involved in the process of sorption, sorption is divided into several groups. Mechanical sorption occurs when soil with its system of pores acts as a natural filter. The particles, which have larger dimensions than the dimensions of the pores, are retained in the soil. According to the system of mechanical sorption, a part of clay particles leached from the upper horizons is retained in soil. Physical sorption is a result of retention of molecules of the compounds dissolved in the water, or gases on the surface of soil particles, under the influence of the forces of the surface attraction. These forces arise at the border between the rigid component and the soil solution. Chemical sorption occurs so that the compounds chemically transform from the readily soluble to the less soluble form, or the reaction occurs with the cations that are bound to the adsorption complex of soil. Biological sorption happens as a result of the influence of higher plants and microorganisms. The plant receives nutrients and other substances in soluble form from the soil solution. By building its underground and above-ground organs, the plant accumulates these substances, primarily biogenic elements in insoluble, organic form. Biological sorption prevents leaching of nutrients and other substances from the soil, and thus increases their accumulation in the humus horizon of the soil. Physicochemical sorption, also called sorption of cations or adsorption, is the most important way of sorption in soil. Adsorption is the ability of colloidal particles of negative charge (acidoid) to bind cations from the soil solution on its surface, by using the physical and chemical forces. The plant receives cations through the root so they cannot be washed out of the soil, and they can be replaced by equivalent amounts of cations from the soil solution. All organic and mineral colloids, which have the ability of sorption of cations, are called adsorption or cation-exchange capacity of the soil (CEC). The composition of the adsorption complex includes clay minerals – secondary alumosilicates, humus, organic and mineral colloids, amorphous colloidal particles (allophane, hydroxides of iron, aluminum, silicon, etc.) and smaller fragments of primary minerals of colloidal dimensions.

Soil reaction shows the relation of the concentration of hydrogen ions and hydroxyl ions. If the concentration of H+ ions is greater, the result is a lower pH and the reaction is more acidic. If the concentration of OH- ions is greater, the result is a higher pH and the reaction is more alkaline (Table 9.4).

Table 9.4: Schematic representation of the reaction of some media [1].

Acidic							Neutral	Alkaline					
0	1	2	3	4	5	6	7	8	9	10	11	12	13 14
Car battery acid	Lemon juice		Wine		Normal rain	Distilled water		Soda water	Soft soap		Ammonia		Alkali

The reaction of soil solution is quantitatively expressed in pH units which represent the negative logarithm of the hydrogen ion concentration in moles per liter of solution (M, or mol/L or mol/dm^3).

It can be noticed from Table 9.5 that neutral soils can be in the range of pH from 5.9 to 6.3 when distilled water was used as the extracting agent; 5.5 to 6.2 if a solution of 0.01 M $CaCl_2$ is used for extraction, and 6.5 to 7.2 in the soils where the reaction of soil is determined in suspension of 1 M KCl. Most cultivated plants correspond to a neutral, slightly acidic or slightly alkaline reaction (Table 9.6 and Figures 9.7–9.10).

Table 9.5: Interpretation of the of soil reaction considering the various extracting agents [6, 7].

Media	Value	Evaluation	Media	Value	Evaluation	Media	Value	Evaluation
pH_{H2O}	<5.2	very acidic soils	pH_{KCl}	<4.5	very acidic soils	pH_{CaCl2}	<4.3	extremely acidic soils
pH_{H2O}	5.2–5.5	acidic soils	pH_{KCl}	4.5–5.5	acidic soils	pH_{CaCl2}	4.3–4.5	very strong acidic soils
pH_{H2O}	5.5–5.9	slightly acidic soils	pH_{KCl}	5.5–6.5	slightly acidic soils	pH_{CaCl2}	4.5–4.8	very acidic soils
pH_{H2O}	5.9–6.3	neutral soils	pH_{KCl}	6.5–7.2	neutral soils	pH_{CaCl2}	4.8–5,0	medium acidic soils
pH_{H2O}	6.3–6.6	slightly alkaline soils	pH_{KCl}	7.2–7.7	slightly alkaline soils	pH_{CaCl2}	5.0–5.5	slightly acidic soils
pH_{H2O}	6.6–7.2	alkaline soils	pH_{KCl}	>7.7	alkaline soils	pH_{CaCl2}	5.5–6.2	neutral soils
pH_{H2O}	>7.2	very alkaline soils				pH_{CaCl2}	6.2–6.7	slightly alkaline soils
						pH_{CaCl2}	6.7–7.0	medium alkaline soils
						pH_{CaCl2}	7.0–7.3	very alkaline soils
						pH_{CaCl2}	>7.3	very strong alkaline soils

Table 9.6: Optimum and tolerant reaction of the soil for the cultivation of certain crops [8].

Plant	Optimum pH	Tolerant pH	Plant	Optimum pH	Tolerant pH
Arable crops					
Alfalfa	6.5–7.5	6.0–8.0	Rice	5.0–6.5	4.5–8.0
Castor	6.0–7.5		Millet	5.5–7.0	5.0–8.5
Sugarcane	6.0–7.0	4.5–8.0	Sunflower	6.0–7.0	5.5–7.5
White clover	5.0–5.5	5.5–8.0	Wheat		5.3–7.0
Potato		4.5–7.5	Yellow lupine		4.8–6.2
Barley		5.3–7.0	Rye		5.3–7.0
Soy		5.3–7.0	Tobacco		5.3–7.0
Flax		5.6–7.3	Hop		5.3–7.0
Melon		5.6–7.3	Rapeseed		5.6–7.3
Flowers					
Anthurium	5.5–6.5	5.0–7.5	Hibiscus	6.0–7.0	5.0–8.0
Blueberry	4.0–5.5		Rhododendron	4.0–5.5	

(continued)

Table 9.6: (*continued*)

Plant	Optimum pH	Tolerant pH	Plant	Optimum pH	Tolerant pH
Azalea	4.5–5.0		Ixora	6.0–7.5	5.0–8.0
Begonia	5.5–7.0		Bougainvillea	5.5–7.0	5.0–8.0
Camellia	4.5–5.5		Magnolia	5.5–6.5	5.0–7.0
Oxeye	6.0–7.5	5.0–8.0	Oleander	6.0–7.5	
Orchid	4.0–5.0		Geranium	6.0–7.0	5.5–8.0
Rose	5.5–7.0		Pomegranate	6.0–7.5	
Vegetables					
Asparagus	5.5–7.0	5.0–8.0	Rape	6.0–7.0	5.5–7.5
White pepper	5.5–6.5	5.0–7.0	Broccoli	6.0–7.5	5.5–7.5
Cabbage	6.0–7.0	5.5–7.0	Red onion	6.0–6.5	5.5–7.0
Carrot	5.5–6.5	5.0–7.0	Bean	5.5–6.5	5.0–7.0
Cauliflower	6.0–7.0	5.5–7.5	Celery	5.8–7.0	5.0–7.5
Cucumber	5.5–6.5	5.0–7.0	Sweet maize	5.5–7.0	5.0–8.0
Sweet Potato	5.0–6.0	5.0–7.0	Tomato	5.5–7.0	5.0–7.5
Salad	6.0–7.0	5.5–7.5	Watermelon	5.0–6.5	5.0–7.0

Figure 9.7: Wheat at different values of soil reaction. (The pH measured in $CaCl_2$ from left to right is 5.72; 10.61; 9.14; 8.39 and 8.09.) [1]

Figure 9.8: Barley at different values of soil reaction. (The pH measured in $CaCl_2$ from left to right is 5.65; 10.24; 8.92; 8.41 and 7.96.) [1]

Figure 9.9: Tomato at different values of soil reaction. (The pH measured in $CaCl_2$ from left to right is 5.65; 9.51; 8.47; 7.94 and 7.95.) [1]

Figure 9.10: Pepper at different values of soil reaction. (The pH measured in CaCl$_2$ from left to right is 5.72; 9.64; 8.51; 7.97 and 8.01.) [1]

9.3.2.2 Soil biology

Living organisms in the soil, or on it, are not considered an integral part of the soil – pedosphere. They belong to the biosphere, although they have a significant influence on all the features of the soil, and they themselves are affected by them and structured, in a certain way, precisely with these features. Animal organisms, fauna or pedofauna strongly affect the important features of the soil. Members of the soil loose the soil, creating in it a number of corridors and thus increasing the permeability of water and air.

Earthworms *(Lumbricus terrestris)* have a particularly important role in soil. Their number varies from a few thousand to a few million, and they weight from a few hundred kilograms to one or maximum two tons per hectare [9]. Soils that are rich in humus, with a neutral to slightly acid reaction, suit them well. They agglutinant the mass of the soil in the digestive tract and create specific aggregates. It is believed that 30 tons of soil passes through the digestive tract of earthworms per year, and it means that all weight of the soil to a depth of 30 cm passes through the digestive tract of earthworms in 60 to 100 years. Earthworms have a very beneficial effect on the soil, and their number and appearance are good indicators of favorable conditions in the soil. There are no earthworms in dry and sandy soils. Various worms, larvae and insects dominate in such soils. The richness of species and their distribution in the depth of the soil can be seen from the view of the number and types of microorganisms, which are located in the mass of only one gram of the soil at different depths, and are shown in Table 9.7.

Table 9.7: Biological diversity of one gram of soil considering the number of microorganisms [10].

The depth of soil, cm	Number of microorganisms/gram of soil				
	Aerobic bacteria	Anaerobic bacteria	Actinomycetes	Fungi	Algae
3–8	7,800,000	1,950,000	2,080,000	119,000	25,000
20–25	1,800,000	379,000	245,000	50,000	5,000
35–40	472,000	98,000	49,000	14,000	500
65–75	10,000	1,000	5,000	6,000	100
135–145	1,000	400	–	3,000	–

The formation and development of soil, as well as its physical, chemical and biological features were presented so far. Soil is a natural creation and a very dynamic system in which life is ruled by certain laws that today, at the beginning of the twenty-first century, are not quite clear nor scientifically determined. **Nothing is accidental in the soil**. Many phenomena in pedosphere are just part of the general system of circulation of matter and energy, in which the soil is not an isolated system. On the contrary, all changes in the lithosphere, biosphere, hydrosphere and atmosphere affect the dynamics of soil and its characteristics. Through examining the physical, chemical and biological properties, we stressed the importance of each of them for the **soil fertility**. Soil fertility is the ability of the soil to satisfy the needs of plants for nutrients, water, air and heat, that is, to ensure suitable conditions for the development of the underground and aboveground parts of the plant. **According to this definition, fertility is a general indicator of soil properties – synthesis of chemical, physical, water, air and thermal properties**. Soil characteristics that are essential for its fertility are: soil reaction (pH), content and form of humus, sorption capability of soil for nutrients, content of physiologically active nutrients, porosity, mechanical composition – in particular the amount and type of clay minerals. All those indicators are measurable, can be quantified and, on this basis, we can assess fertility as collective feature which is affected by all of these features of the soil.

9.3.3 Roles of soil

Soil science has developed as a part of agronomy and forestry profession. It is understandable that its focus was those soil characteristics that made it more or less suitable for growing plants, i.e. the focus was its production role. The problems of environmental damage (especially degradation and soil contamination), particularly of natural resources, in a certain way, have pushed this role. Only in recent decades, there is a growing interest in non-productive roles of the soil, in the first place those relating to environmental protection, in particular the protection of natural resources whose quality depends directly on soil and soil management. Below all the roles of soil are presented, although they are hardly separable from one another [11].

9.3.3.1 Role of soil in the formation of organic matter

The most important, indispensable and primary role of soil is supplying plants with water, air and nutrients, which enables the production of biomass – production of organic matter through photosynthesis. In this role the soil is an indispensable factor for the maintenance of life on Earth, for the plant/crop production in primary economic branches – agriculture and forestry. Production of organic matter in agriculture and forestry settles food and non-food needs of humans. Specifically, in this role soil allows us to supply food (bread, meat, milk, eggs, and mushrooms), drinks (wine, beer, and tea), beverages (fruit juices, juices of various vegetables and other plants), fiber (wool, textile plants – cotton, linen, and hemp), medicinal plants and herbs and energy (firewood, biodiesel, alcohol as fuel). Furthermore, soil provides raw material for wood and food industry (flour, oil, sugar, fibers, caoutchouc).

9.3.3.2 Organic-regulatory role of soil

The soil has a significant place in the biological cycle of matter and energy. Soil is situated between lithosphere and atmosphere, and it has a direct contact with hydrosphere, anthroposphere and biosphere. As such, soil acts as an acceptor of substances that are (intentionally or unintentionally) emitted from these spheres into the environment and are ecologically relevant to all members of the biosphere, whether they have positive or negative impact. These substances can accumulate in soil, so soil acts as their collector due to mechanical, physical and physicochemical sorption of the substances. Soil can modify collected substances, especially the organic ones, with the help of the microbial complex. That way, soil has the role of exchanger (transformer) of these substances.

9.3.3.3 Soil as a filter of water

Soil is an effective universal natural filter for water that penetrates through the soil into underground. For the functioning of terrestrial and aquatic ecosystems, especially for the protection of groundwater from different pollution, this is a particularly important feature of soil. By using colloidal complex, soil binds the different substances that in the process of natural circulation of the substance, or the food chain, arrive in the soil in the form of *dry aerodeposition* as dust, or rain water as *wet deposition* or, even more dangerous, as acid rain. This also applies to environmentally risky substances, therefore, different pollutants. The wider significance of the filtration and buffering capacity of soil can be seen from the fact that 65% of the total population in Europe is supplied with drinking water from groundwater.

9.3.3.4 Climatic-regulatory role of soil

The soil is the central link in the chain of biotransformation of organic carbon; it affects the content and the total amount of CO_2 and other gases that cause the so-called *greenhouse effect*. Globally, the total amount of soil organic carbon (pedosphere) is three times higher than in the above-ground biological mass. It is believed that about 25% of the emitted carbon is derived from the soil – agriculture. Due to the limitations that are provided for carbon dioxide emissions, land management of tomorrow will receive new tasks in regulating the amount of carbon emitted into the atmosphere, and it will be by reducing its emissions.

9.3.3.5 Soil as a source of genetic wealth and protection of biological variety

Soil is the habitat and genetic reservation of many microorganisms and macroorganisms of the soil. The fertile soil must have the appropriate biological activity and show large biodiversity. Biological soil degradation is inextricably linked to the degradation of the physical and chemical properties of soil. Weight of organic carbon in the soil is estimated to be 1,550 Gt (gigatonnes); it is two times higher than in the atmosphere and three times higher than in all living organisms (biosphere) on Earth [12].

(a)   (b)

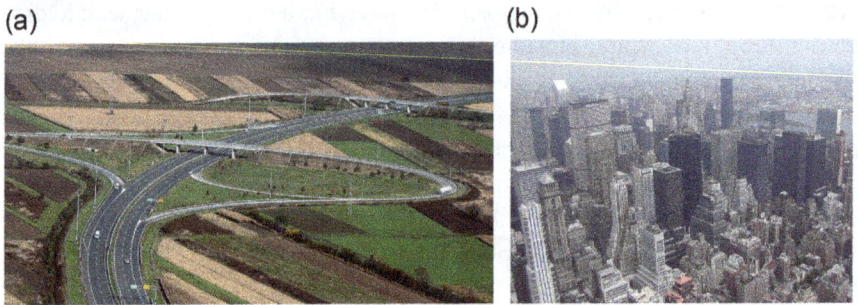

Figure 9.11: Most often are the new roads, as well as the cities in the past, situated on the best soils [1]. (Photos 11 by: I. Kisic, 2000–2010).

(a)   (b)

Figure 9.12: The permanent conversion of soil comes, among other things, as a result of building factories and expanding landfills [1].
(Photos 12 by: I. Kisic, 2000–2010).

9.3.3.6 Spatial role of soil

Soil characteristics play a key role in the use of landscape today, and also in the past (Figures 9.11 and 9.12). Pedosphere provides space for the expansion of urban areas, roads, recreational areas, landfills (for waste disposal) and others. It is believed, for example, that about 2% of the total area of soil in Europe is under buildings. The range is from 0.5% in Ireland, 12% in Hungary, 13% in Italy to 14% in the Netherlands [13].

9.3.3.7 Soil as a source of raw materials

Soil is an important source of raw materials, especially for the building industry. The exploitation of these raw materials is always associated with damage to soil by open pits, or overlapping other fertile soils with these materials. For example, 0.05 to 0.1% of the soil surface in Europe is destroyed by open pits for mining purposes [13]. According to the Soil Atlas 2015, 24 billion tons of fertile soil is lost every year because of misuse, i.e. because of *soil sealing* [14]. *Soil sealing* is one of the most important types of soil degradation today.

9.3.4 Potential soil contaminants

Soil contamination is one of the most dangerous forms of soil degradation with the consequences that are reflected in virtually the entire biosphere, primarily hetero-trophic organisms, and also mankind as a food consumer. The path that any con-taminant takes from the soil to the plate is very short and food safety is crucial and increasingly important requirement which is becoming more difficult and complex to implement. In order to define the contaminated soil, it can be said that it is the soil in which human activity or natural phenomena has increased the content of harmful substances whose concentrations may be harmful to human activity, that is, for the production of plants or animals (Table 9.8). That is why in the last decade the remediation of contaminated soils is becoming an increasingly important subject [15, 16]. In order to achieve satisfaction rate of remediation of contaminated soil, it is very important to determine current condition of soil as accurately as possible. After completing the remediation process, soil sampling should be repeated, that is, the final state should be determined [1]. Based on the determined values of soil contam-ination after completing the process of remediation, recommendations will be pro-vided for future land use.

   Although there are several classifications of soil contamination, in this section potential soil contaminants are divided in next groups: organic contaminants (poly-cyclic aromatic hydrocarbons and persistent organic pollutants); fuels (hydrocar-bons); explosives; inorganic contaminants (metals and metalloids) including radioactive elements.

   Polycyclic aromatic hydrocarbons (PAHs) or polyaromatic hydrocarbons or polynuclear aromatic hydrocarbons are a large group of cyclic hydrocarbons

Table 9.8: The origin of soil pollution [1, 15].

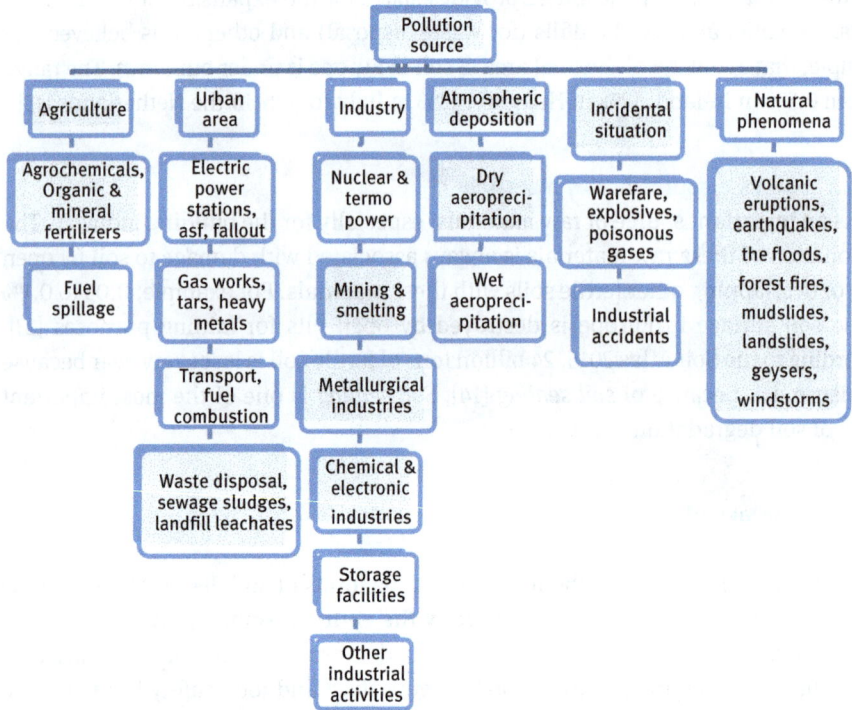

			Pollution source		
Agriculture	Urban area	Industry	Atmospheric deposition	Incidental situation	Natural phenomena
Agrochemicals, Organic & mineral fertilizers	Electric power stations, asf. fallout	Nuclear & termo power	Dry aeropreci-pitation	Warefare, explosives, poisonous gases	Volcanic eruptions, earthquakes, the floods, forest fires, mudslides, landslides, geysers, windstorms
Fuel spillage	Gas works, tars, heavy metals	Mining & smelting	Wet aeropreci-pitation	Industrial accidents	
	Transport, fuel combustion	Metallurgical industries			
	Waste disposal, sewage sludges, landfill leachates	Chemical & electronic industries			
		Storage facilities			
		Other industrial activities			

consisting of 4, 5, 6 or 7 benzene rings fused together. The most common are PAHs with five or six rings. In accordance with their chemical structure they are categorized as persistent organic pollutants [17]. In nature, PAHs are found in small, almost negligible concentrations while the increased content refers only to different anthropogenic activity [18]. Natural activities that can cause increased content of PAHs in the environment include volcanic eruptions, meteorite and comet falls, summer forest fires or the occurrence of other types of large open flames. Increased content of PAHs is usually a result of pyrolysis processes during the combustion, especially of coal and gas in households and other heating facilities, garbage processing, traffic and in some industries (coke, iron and aluminum plants, power plants, galvanizing plants, petroleum and petroleum products processing, production and uses of asphalt and tar). Wherever a lot of thermal energy is needed in the work process an increased content of PAHs can be expected. The highest concentrations of PAHs are found in chimneys in family houses or in large funnels in industrial plants that use coal as raw material. Increased content of PAHs may be measured in major cities as a result of exhaust gases of automobile traffic. Increased content of PAHs can also be determined in dried meat, that is, in well-roasted meat (more accurately, fried), which is prepared on the grill. Some European countries have developed soil quality criteria according to the degree of PAHs contamination and necessary measures of reclamation [19]. The criteria are based on the effect of PAHs on the ecosystem and human health. Criteria for re-cultivation are defined according to the method of soil exploitation in accordance with the established types of soils and their mechanical structure. Values lower than 200 µg/kg of total PAHs are considered to be natural values. The sum of sixteen PAHs between 600 µg/kg and 10,000 µg/kg is considered to be troubling, and values higher than 10,000 µg/kg total PAHs in soil require re-cultivation measures.

Some of **Persistent Organic Pollutants – (POPs)** are toxic synthetic organic aromatic compounds derived from the process of chlorination of biphenyl in the presence of a catalyst [20]. (If their content is increased in soil, due to their high affinity for binding, especially to the organic component of soil, it is very difficult to remove them from soil. At the end of the twentieth century, the United Nations Environment Programme (UNEP) launched an initiative to create the Convention on Persistent Organic Pollutants (http://chm.pops.int/default.aspx). The aim of the Convention is monitoring the production and use of some POPs. According to the Stockholm Convention the following 12 are defined as initial POPs and so-called 'dirty dozen': aldrin; chlordane; dichlorodiphenyltrichloroethan or generally known as DDT; dieldrin; endrin; heptachlor; hexachlorbenzene (HCB); Mirex; toxaphene; polychlorinated biphenyls (PCBs); polychlorinated dibenzo-p-dioxins (PCDDs) and polychlorinated dibenzofurans (PCDFs).

The following pollutants are **hydrocarbons**. Increased content of hydrocarbons in soil may be caused by natural or anthropogenic activity [21]. Unlike other chemicals, especially pesticides, hydrocarbons are not applied to the soil with a specific purpose. Undesirably increased content of hydrocarbons in soil mostly refers to specific incidents (pipeline ruptures, breakdowns during crude oil extraction or breakdowns at refineries), leaks from underground storage tanks for oil and its derivatives, and landfills. Increased content of hydrocarbons in soil can also be caused by processes relating to the extraction of crude oil, natural gas, coal and peat. As noted, the increased content of hydrocarbons is mainly related to human activities, but can sometimes occur by natural processes, for example, decomposition of bituminous shale or other rocks in which the increased content of hydrocarbons was determined.

From the chemical point of view the term **heavy metals** comprises metals with density which is greater than $5\,g/cm^3$ and with an atomic number greater than 20 [22]. From a biological point of view Nieboer and Richardson (1980) have categorized heavy metals to the *essential elements* for at least some organisms (Cr, Mn, Fe, Co, Zn, Mo and V), the *elements necessary for the growth and development of plants* (Mn, Fe, Cu, Zn, Mo and probably Ni) and *phytotoxic elements* (Cd, Hg and Pb) [23]. Therefore, some heavy metals are micronutrients, but at the same time they are phytotoxic. The problem is that there is a fine line between the role of micronutrients and phytotoxicity. There are 65 elements of the periodic table that have metal properties: high thermal conductivity, high density and malleability. Of the ten most abundant elements in the lithosphere, seven are heavy metals.

Among the listed soil degradations, soil contamination with heavy metals is gaining in importance and draws the greatest attention in the public. This term refers to the content of heavy metals in soil in an amount which causes visible or measurable disturbance of some of the already mentioned soil functions. Thereat, all of the soil functions are taken into consideration and in particular the most important characteristics – fertility, suitability for normal growth and development of natural and/or cultivated plants. The plants that are grown in clean soil are suitable for all common ways of use and are completely harmless for every consumer. Otherwise, if the soil is contaminated, the plans have been contaminated through the soil, because the contaminant can be found in soil in chemical form that plants can absorb in an amount that reflects negatively on their growth, development and quality [24].

9.3.5 Remediation technologies for contaminated soils

A widespread and very serious problem of soil contamination by organic and inorganic substances with natural and anthropogenic activities has already been discussed in the previous section of this chapter. Unlike organic contaminants, inorganic contaminants cannot be degraded and they accumulate in/on the adsorption complex or soil organic matter. If their emission occurs from the adsorption complex, they will be absorbed by a plant or they will percolate to groundwater. If the plant ends up in the food chain, there is a risk that a person will consume it directly or indirectly, if a plant is consumed by animals that would later be consumed by humans. The next probable situation is that the contamination can be drained to deeper soil horizons, which will cause groundwater pollution or the contaminant will reach open water courses through the lateral (sub) surface runoff. Therefore, sites that are contaminated by organic and inorganic contaminants, where the rehabilitation was not conducted, should not be used in any form of agriculture. At the same time they are a potential source of pollution of the hydrosphere and atmosphere.

Compared with the atmosphere and hydrosphere, pedosphere has much less and markedly slower possibilities of recovering from toxic effects. Biogeochemical processes in the soil that affect the fate, behavior and bioavailability of potential contaminants are among the most rewarding scientific researches in recent decades. The negative effects caused by pollution are undeniable: reduced agricultural productivity, polluted water sources and raw materials for food are only a few of the effects of soil degradation, while almost all human diseases (excluding AIDS) may be partly related to the transport of contaminants in the food chain or the air to the final recipients – people, plants and animals. Remediation of contaminated soil can be done in three ways by using various remediation technologies:

- reducing the concentration of total pollution to acceptable levels in accordance with the future land use;
- physical /chemical /biological /mechanical isolation of contamination to prevent further reaction (the spread) of contaminated soil with the environment;
- reducing the bioavailability of organic and inorganic impurities.

The questions that are always asked before the implementation of remediation are: To what levels should the soil be cleaned? When is the soil clean? What does clean soil mean for different purposes? If the soil is clean enough for sport or recreation areas, is it clean enough for residential and agriculture uses? When is the soil clean enough for recreational or green areas? The above indicators are variable paradigms of the soil purity that the wider scientific community, land owners and decision makers are facing in the future use of land/area. The framework of soil remediation limit value is shown in Table 9.9.

Table 9.9: Levels of sensitivity for soil contamination limit values at different ways of using [1].

←

Increasing the sensitivity to soil contamination – lower and stricter limit values for certain contaminants in the soil

| Agriculture | Spaces for living and vacation | Commercial business premises | Industrial areas |

Reducing the dependence on soil contamination – higher and more lenient limit values for certain contaminants in the soil

→

This section of chapter covers four types of soil remediation and explains their basic principles/mechanisms of action. The main criteria for choosing/adopting/selecting a technology that will be used are the type and amount of contamination, the location where remediation would be implemented, but also the type of soil on the location where this technology should be applied. Potentially possible and economically viable technologies are shown in Table 9.10.

9.3.6 Types, forms, techniques and technologies of remediation

Choosing a technology depends mostly on: type and category of contamination, spatial distribution of contamination (surface, volume and location of contamination – the proximity of surface/ground water), soil type (soil pH, soil organic matter, texture – type of clay), period of exposure to potential contamination, future land use and defined legal framework on the required level of remediation in a specific country.

Table 9.10: Remediation technologies of the contaminated soil [1].

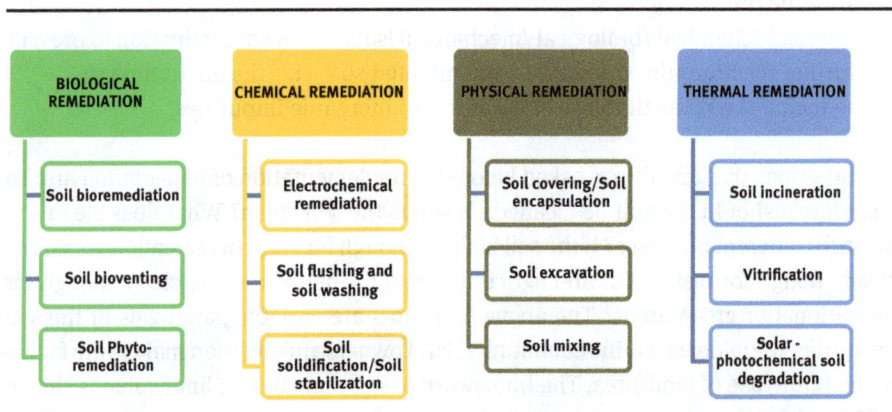

BIOLOGICAL REMEDIATION	CHEMICAL REMEDIATION	PHYSICAL REMEDIATION	THERMAL REMEDIATION
Soil bioremediation	Electrochemical remediation	Soil covering/Soil encapsulation	Soil incineration
Soil bioventing	Soil flushing and soil washing	Soil excavation	Vitrification
Soil Phyto-remediation	Soil solidification/Soil stabilization	Soil mixing	Solar - photochemical soil degradation

Based on the above parameters a decision is made whether to apply *in situ, on situ* or *ex situ* remediation. *In situ* remediation is performed directly at the site without contaminated soil excavation, that is, without disturbance to the soil structure. *On situ* is a form of remediation which is also performed at the site of contamination, but the soil is excavated. In this case, the contaminated soil is more exposed to external factors, and there is a possibility of more rapid spread of pollution, more by wind erosion than water erosion. *Ex situ* remediation is performed in a way that the contaminated soil is excavated and then transported to a special landfill where one of the appropriate forms of remediation is carried out. Due to many reasons (cost of remediation, environmental safety, public health, etc.) methods of remediation that are carried out *in situ* or *on situ* are more rewarding.

Many *in situ* or *on situ* technologies are used in order to stabilize the contaminants in soil by adding various additives, for example, lime, clays, zeolites. Such techniques do not reduce the total amount of contaminants in soil (this primarily refers to metals and metalloids), but they act in a way that they minimize the bioavailability of their fractions. Other *in situ* methods remove contaminants from soil by runoff, or by washing and using surfactants. This process enables collecting the contaminants from the soil by means of various attractants, since the contaminants are now dissolved in water or bound to the solid components. *Ex situ* remediation is based on the stabilization by creating a less polluted soil that can be disposed of in municipal waste landfills. Soil washing is one of the *ex situ* techniques of extraction which results in soil that is cleared of contaminants and which can be returned to its original area or soil that can be disposed of in non-hazardous waste landfills, depending on the remediation efficacy.

Biological methods include any type of remediation that uses microorganisms and plants. There are several processes in which microorganisms and plants can be used for remediation of contaminated soil. Depending on the strategy that is used to purify soil, these processes are blocking or removing contaminants from the soil with the help of plants or microorganisms. Bioremediation technology that uses microorganisms is carried out in two ways: *in situ* or *ex situ*. *In situ* bioremediation uses indigenous microorganisms. It is based on treating the soil in the area where contaminants are identified. The process involves soil excavation or pumping water prior to the bioremediation treatment. The main goal of aerobic *in situ* bioremediation is the supply of oxygen and nutrients necessary for the growth of microorganisms in order to enable the effective degradation, which is achieved by bio-ventilation or by injecting hydrogen peroxide or any other reagent. *Ex situ* bioremediation involves the removal and transport of contaminants which is then processed at a different location. In comparison with the *in situ* bioremediation, it is faster and easier to control and it enables degradation of a wider range of substances. The procedure involves excavation, transport to the landfill, soil treatment before and sometimes after a bioremediation process. The most common procedure is to mix the contaminated soil with an appropriate amount of water or liquid fertilizer in a special bioreactor with the addition of microorganisms.

Restoration with plants – phytoremediation is a technology that uses plants and their rhizosphere microorganisms that live on the roots for the removal, degradation or retention of harmful chemical substances in soil, groundwater and surface waters and the atmosphere [25]. It is believed that the progress of biotechnology has slowed down, as a result of insufficient understanding of the complex relationship between rhizosphere and mechanisms that are based on the ability of plants to absorb and translocate metals from contaminated environments. Some plants that are very effective in the laboratory are not good contaminant accumulators in field conditions. Some plants are good contaminant accumulators in one climate, but have failed to give satisfactory results in another. Everyone involved in this issue points out that this is the method of the future, but also, there are still many unknown facts and even more questions than there are answers for wider application of this method in practice. However, it is the most promising method. For example, winter wheat, winter barley, soybean and tillage practice (ploughing and harrowing) can partially solved the problems of PAHs accumulated in soils due to the contamination by crude oil and drilling fluids [26]. Plants decompose organic contaminants or stabilize them acting as filters. Just like any other method, phytoremediation has its advantages and disadvantages (Table 9.11). The advantage of phytoremediation is the fact that it is the most acceptable method for the environment because it uses plants [28]. It is economically viable because it is one of the cheapest technologies.

Table 9.12 shows a list of plant species that can be used for phytoremediation. The literature states that 400–500 plants have the ability of hyper-accumulation heavy metals (which is about 0.15% of all known species). Hyper-accumulators are plant species that are tolerant to high concentrations of toxic substances in the roots and above-ground plant mass. Most hyper-accumulative plants are present in

Table 9.11: Advantages and disadvantages of phytoremediation [1, 27].

Advantages	Disadvantages
– Cost reduced over traditional methods	– Long remediation time requirement (up to 15 years)
– Low secondary waste volume	– Effective depth limited by plant roots
– Improved aesthetics	– Phytotoxicity limitations
– Habitat creation – biodiversity, green technology	– Fate of contaminants often unclear
– More publicly accepted	– Climate dependent/variable
– Provide erosion control	– Potential transfer of contaminants (i.e. to animals or air)
– Prevent runoff	– Harvesting and disposal of metals in biomass as hazardous waste may be required, although generally not
– Reduce dust emission	– Larger treatment footprint
– Reduce risk of exposure to soil	– Indigenous plants
– Less destructive impact (applied *in-situ*)	– Groundwater contamination possibility

Table 9.12: Plant species for phytoremediation of various soil contaminations [1].

Soil contaminant	Plant species
Cadmium	Brown or Indian mustard, Ethiopian mustard, some species of willow and birch, *Thlaspi arvense*, *Thlaspi caerulescens*
Chrome^{6+}	Brown or Indian mustard
^{137}Cesium	Amaranthus, Brown or Indian mustard, Sunflower, Cabbage, some species of willow
Copper	Brown or Indian mustard, some species of willow, Oilseed rape, Ribwort Plantain
Mercury	Brown or Indian mustard, Sunflower, Hybrid poplar, some species of willow
Nickel	Brown or Indian mustard, Spinach, Cabbage, Peas, Barley, Beans, *Ricinus communis*, *Thlaspi arvense*
Lead	Brown or Indian mustard, Black mustard, Sunflower, Peas, *Thlaspi rotundifolium*), Buckwheat, Hybrid poplar, some types of corn
Arsenic	Hybrid poplar, some types of alder, Aspen, Willow, Ferns, Oilseed rape
Selenium	Canola, Brown mustard, Kenaf
Uranus	Chinese cabbage, Brown mustard, Sunflower, Kale
Zinc	Oats, Brown and Indian mustard, Barley, some species of willow, *Thlaspi arvense*, Oilseed rape, Ribwort Plantain, *Thlaspi caerulescens*
Thallium	*Iberis intermedia* or *Iberis linifolia*
Volatile and semi-volatile contaminants	Alfalfa, Clover, Rye, Sorghum, some species of willow, Poplar, Alder, Blueberry, Spruce, Fescue grass, Elderberry, Mulberry
Agrochemicals	Alfalfa, Hybrid poplar, Willow, Mulberry
Explosive	Rye, Sorghum, Clover, Alfalfa, some species of willow, Poplar, Aspen, Cypress, Sedge

soils that have developed on the serpentinite or soils that have a lot of hemimor-phite minerals in parent materials. These plant species have the ability to rapidly translocate elements through the root to the above ground parts of the plant. A major problem for these species is that they produce relatively little overground plant mass and are characterized by slow growth. Another problem is that most of them selectively bind a particular metal, and the third problem is that they can be used only in their natural habitats. They come from a broad range of families, and most of them are members of cabbage family (*Brassicaceae*), legume family (*Fabaceae*), mint family (*Lamiaceae*), grass family (*Poaceae*) and spurge family (*Euphorbiaceae*).

Some species may be more resistant to several heavy metals, for example, colonial bent (*Agrostis tenuis*) and ribwort plantain (*Plantago lanceolata*) that are resistant to Zn, Cu, Cd and Ni. Rascio and Navari-Izzo (2011) reported that more than 400 plants can be nickel hyper-accumulators, 26 plants can be cobalt hyper-accumulators, 24 copper hyper-accumulators, 18 zinc hyper-accumulators, 8 manganese hyper-accumulators, and five plants can be lead and cadmium hyper-accumulators [29]. In order for a plant to be classified in group of hyper-accumulators, it must have a minimum of 0.001% Hg, 0.01% Cd and Se, 0.1% As, Co, Cu, Cr, Ni, Pb, Sb, Se and Tl, and 1% Zn and Mn in its dry matter of the above ground plant mass [29]. Also, in order for a plant to be classified in the group of hyper-accumulators, it should not show any changes caused by the increased content of heavy metals in the plant. The list of potential plants that are hyper-accumulators for heavy metals from the soil is not definitive. Almost every day, studies on a new plant species that has the ability to accumulate high amount of heavy metals from soil can be found in literature. For example, two investigations were performed in order to determine the influence of combustion residues from thermoelectric power plant (coal ash) on soil contamination and potential biological remediation. Zgorelec et al. (2008) indicate that high amounts of arsenic (592,6 mg/kg) and nickel (111,6 mg/kg) in pure coal ash have significantly influenced on metal content in soil, but also that some cultivar of soybean can accumulate higher amounts of nickel in grain than edible parts of barley cultivars [30]. In terms of wild plants and con-taminated sites by coal ash, Dellantonio et al. (2008) reported remarkable results of hyper-accumulation of boron in leafage of Salix Alba and Salix Caprea [31]. Scientists within International interdisciplinary Project RECOAL («Reintegration of Coal Ash Disposal Sites and Mitigation of Pollution in the West Balkan Area», funded by the 6th Framework Programme of the European Union, European Commission (# 509173 – CORDIS-EU-FP6- Specific Targeted Research)) also investi-gated pilot filter column using clinker and bauxite-bed filters aimed at reducing the pH of the effluent water and a vertical flow soil filter (Figure 9.13), planted with reed (*Phragmites spp.*) and willows (*Salix spp.*), to remove suspended solids and trace elements. The results showed that arsenic was reduced for up to 90%, boron for up to 37 % and the pH was further reduced for 2pH [32].

Figure 9.13: Soil-plant filter for phytoextraction system in field conditions.
RECOAL Project, in-situ research to reduce negative impacts of the coal combustion residues
disposal sites on surface and ground waters. (Photo by: Z. Zgorelec).

Also, energy crops such as *Arundo donax, Miscanthus x giganteus, Panicum virgatum, Pennisetum purpureum, Sida hermaphrodita* and *Sorghum x drummondii* are recommended in phyto-extraction (cadmium, chromium, copper, lead, mercury, nickel and zinc), rhizo-filtration, stabilization and accumulation of heavy metals. Disadvantage of these plants is the fact that they do not translocate significant amount of heavy metals from the rhizosphere into the root system of plants [33]. While some plants have the genetic potential to remove a variety of organic and inorganic contaminants from soil, they show some negative characterists from the viewpoint of biotechnology. Most hyper-accumulative plants have a modest above ground plant mass and even smaller root network or they are species that grow slowly and their habitats are poorly represented in larger areas. Therefore, as considered by some authors, the advocates of genetically modified organisms (GMOs), we need to refocus on genetic engineering in order to solve this problem. They propose the transfer of genes responsible for the phenotype of hyper-accumulation, from the species that are characterized by small underground habitus and slow growth, into the species that have high biomass production, but low capacity of hyper-accumulation of heavy metals and other contaminants.

Proponents of GMO believe the GMO technology can solve the problem of heavy metals and other contaminants in soil as it would include incorporating genes of plants that are resistant to increased concentrations of heavy metals in soil in agricultural and other cultures. Other authors, opponents of GMO, refute this view by the fact that unforeseeable and unforeseen changes would take place in nutrient substances, flavors and forms of agricultural and other crops, which would result in catastrophic consequences for human health and the environment. This procedure of using GMO plants may potentially give satisfactory results in the near future, but in the long run it would lead to incalculable consequences for human nutrition and especially for the environment.

Electro-remediation was originally developed for high performance conversion of certain radioactive organic waste into the environmentally acceptable waste [34]. Electro-kinetic technologies are being applied since 1930s, and initially they were used for the extraction of heavy metals in landfills of sewage sludge from urban waste water. Also, this method is successfully used for the removal of contaminants, especially heavy metals, from the contaminated groundwater. Of all the types of potential soil contamination, the highest quality results in the removal of heavy metals are achieved by this method [35]. Electro-remediation is the process by which heavy metals are extracted from soil using a low-intensity current through a network of cathode and anode in the contaminated soil, in order to create a voltage gradient (Figure 9.14). As a result of the electric field, the mobility of metals and their motion toward the electrode occurs, thus enabling their extraction from soil.

Figure 9.14: Electro-remediation of soil [26].
(Photo by: I. Kisic, 2010).

Since 1980s they have been used for remediation of soils contaminated with inorganic and organic contaminants [36]. Electrodes can be placed in vertical or horizontal position. By applying a direct current to the electrodes, an electric field is created between the anode and cathode. An electric field affects the soil, water and metals, leading to the following processes in the soil: motion of ions toward oppositely charged electrodes (electro-migration); water flow (electro-osmosis) and motion of charged particles toward the electrodes (electrophoresis). As a result of an electric potential the ions begin to move toward the oppositely charged electrodes: cations toward the cathode, and anions toward the anode. The result of the movement of ions depends on the diffusion coefficient and the ion mobility. Therefore, H+ and OH– ions have a higher diffusion coefficient and increased ion mobility in water, in comparison to metals. Diffusion coefficient and mobility of H^+ ions is higher compared to the OH^- ions, allowing H^+ ions to dominate the system. Transport of H^+ and OH^- ions from the anode and the cathode has a significant influence on physicochemical properties of soil mass and the metal associated with them.

The success of electro-remediation depends on the knowledge of physicochemical properties of soil, current humidity, the environment of contaminated soil and the properties and concentration of metals. Mineral composition, mechanical properties of soil and the anomalies on the surface, such as gravel and coarse sand, PVC and asbestos cement pipes, large concentrations of Fe and other materials, can effect on the electroosmotic flow. Soils that are more negatively charged will accumulate contamination at the end of the cathode, while positively charged soils will encourage reverse electroosmotic processes and the accumulation of contaminants on the anode. Contaminants in the soil, especially metals that are tightly bound by cation exchange capacity (CEC), are very difficult to be removed by this process. In order to maximize desorption, different radicals, solvents and chelates are used in the process and are introduced into the soil during the electro-kinetic remediation.

Much better results are achieved by applying this method in soils that have lower cation-exchange capacity, two-layer clay minerals (kaolinite), lighter mechanical composition in relation to soil with higher cation exchange complex, heavier mechanical composition and dominance of three-layer clay minerals [36, 37].

Soil flushing is the process of extraction of soil contaminants using appropriate solutions. Contaminated soil is flooded by water or certain solutions that move the contamination in the area of their concentration in the borehole, where it is removed and processed by further procedures. The process begins by drilling injection and extraction wells in the contaminated soil. The number, location and depth of wells depend on a number of geological factors. Flooding apparatus is draining the solution into the injection wells. When the solution passes through contaminated soil it moves the metals toward the extraction well. The eluates are accumulated in the borehole – the solution with the contaminants that is later extracted from the soil to the surface.

The type of solution to be applied in the process is determined depending on the category and type of contamination. In the process of soil flushing it is common to use only water or water with additives, such as acids (low pH), bases (high pH) or surfactants (e.g. detergents, emulsions). Water is used for contaminants that easily dissolve in water. The acidic solution is a mixture of water with an acid, e.g. hydrochloric and nitric acid, etc., and is used to remove metals and organic components. The alkaline solution is a mixture of water with bases, such as NaOH, and it is applied for the removal of phenols and some metals. The application of these chemicals is the main constraint for a wider application of this technology. Even though the mentioned chemicals are diluted, they drastically change the chemical properties of soil, especially soil reaction, and during and upon the completion of a successful remediation process there is always the possibility that a certain amount of these chemicals is flushed into the groundwater.

Solidification and stabilization are based on solidification or reducing the mobility of contaminants in soil. Of all the soil contaminants the best and highest quality results were obtained from solidification of heavy metals. It consists of *in situ* and *ex situ* technologies that inhibit or slow down the release of metals from contaminated soil. Solidification immobilizes metals in contaminated soil by binding them to binding agents, while stabilization converts soluble, mobile and toxic metals into less mobile and inert forms [21, 38]. These two methods are a combination of physical (solidification) and chemical processes (stabilization) and are most often used together in order to reduce the impact of metals on the environment. Methods of solidification and stabilization do not reduce the concentration of metals in soil, but they reduce the risk of processes that are harmful to organisms by reducing the potential bioavailability of metals. The process includes immobilization with chemicals that react with metals and introducing reagents (impurities) into the contaminated soil. Immobilization prevents the mobility of metals and reduces their impact on the environment. Preliminary research is necessary to determine which of the metals are found in contaminated soil, and then to determine the appropriate type of stabilization material that will reduce the solubility of toxic contaminants. The procedure can be performed *in situ* by introducing impurities in contaminated soil or *ex situ* by excavation and mixing the soil with admixtures using machines and disposing solidified soil in previously prepared landfills or compaction in place, depending on the outcome of the procedure.

From the physical methods for remediation of contaminated soil, this section of chapter will address covering or encapsulation of contaminated soil, i.e. mixing the soil. Unlike other remediation technologies in physical remediation the changes of chemical and biological soil properties are very small. The system of covering the contaminated soil is one of the most common forms of remediation technologies on smaller areas. This technique involves covering the contaminated area by a multi-layered cover in order to provide physical and chemical protection for the soil from wind and water erosion and to avoid direct contact of such soil with flora and fauna.

Designing a system for covering the land is specific and depends on its future use. The most critical components of the system are covering layers that create a barrier and drainage layers. At the top there is a layer of soil on which the grass or other plants are seeded, while underneath there is a layer of gravel and pipes to promote drainage. This technology is similar (almost the same) to that applied for the closure of municipal landfills. In practice, it is most commonly carried out along with phyto-stabilization and/or phyto-volatilization. Covering of an area of land can consist of a single layer of fertile soil and a complex multi-layered system of soil and geosynthetics. In areas of dry climate, simpler systems are applied and in wet areas more complex systems are applied, while their complexity depends on the type and degree of contamination. The materials in the construction of coverage system include low-pass and high-pass geosynthetic materials. Low-pass materials prevent the water from reaching the contaminated soil. High-pass materials collect the water that filters into a capsule, and it can be temporary or permanent. Temporary capsules are installed before the final coverage in order to reduce the creation of eluate until a more effective remediation is performed. These capsules are usually necessary to reduce the infiltration. For the final cover layer a more stable substance is used (usually the soil) and in this way subsequent maintenance costs are reduced. It is not advisable to use this remediation method for depositing highly contaminated soil, especially for liquid contaminants. Inorganic or liquid pesticides containing about 5% organic material should be first solidified or stabilized, and only then covered using this technology.

Soil mixing reduces the concentration of contaminants by dilution to a level which is not harmful to the environment [39–41]. The application of such technology most clearly reflects the famous sentence: *The solution for pollution is dilution.* This can be achieved by bringing in pure soil and mixing it with contaminated soil or by deployment of clean material that is already in the area. Another method of dilution is based on agro-technical operations – plowing, disc harrowing, harrowing, milling or subsoiling, during which the vertical mixing of contaminated surface layer and less contaminated deeper soil layers occurs. This method also reduces surface contamination, which is no surprise considering the technology. As this is a very simple and inexpensive method of remediation of contaminated sites, it is often applied in many contaminated sites. This method is particularly preferred by investors who have caused pollution since it can be carried out by their own human and technical resources. Problems arise when trying to define the sites where this method can be applied. If a contaminated site is located near the surface or groundwater, this method of remediation of contaminated sites should not be applied. By using soil mixing technology very good results were achieved in remediation of soils contaminated with volatile and semi-volatile contaminants and pollution from fuel [26, 42].

Thermal remediation is the use of technology that is performed at very high temperatures, above 1,000 °C. In this way, the contaminated soil is *burned* and contaminants are converted into less harmful forms or they evaporate from the soil. It is very difficult for the agronomists to accept such a method because it is relatively successful in solving the problem of potential contamination in soil. At the same time, the high temperature in soil destroys the soil edaphon. In this way, the soil loses its essential role – a substrate for growing plants.

Incineration of contaminated soil has the advantage of reducing the volume to the minimum before depositing. To avoid evaporation, the incineration temperature must not be higher than the boiling point of certain pollutants. In most cases, the contaminated soil is burned in cement kilns of recent production, and sometimes in thermal power plants. Extremely alkaline conditions in cement kilns are ideal for degradation of chlorinated organic waste [43]. Due to the complexity of implementation of this method, satisfactory results were achieved only with soil that was markedly contaminated by polychlorinated biphenyls (concentrations greater than 50 mg/kg soil). When all this is done properly, the destruction of chlorinated compounds in cement kilns can be more than 99% successful with no adverse effects on the quality of the discharged gases from cement factories.

The burning process can, for example, reduce the volume of hyper-accumulating plants. It is customary to burn heavy metal hyper-accumulating plants in cement kilns or in steel furnaces. The ashes of the burned biomass containing 20–40% of metals are considered a valuable ore, while a similar ash with only 2–4% of metals is not commercially viable and should be disposed of [44].

The most effective, least painful and least expensive soil remediation technology is prevention. Unfortunately, there are many papers and lectures on this subject, but no one is respectful of this technology.

References

[1]   Kisic I. Remediation of polluted soil. Zagreb, Croatia: Faculty of Agriculture University of Zagreb. University book (in Croatian language) 2012:276 References are renumbered as per style. Please check and confirm.

[2]   Basic F. The soils of Croatia. Netherlands: Springer. World Soils Book Series 2013.

[3]   Resulovic H, Custovic H. Pedologija. Sarajevo: Univerzitet u Sarajevu. BiH 2005:318.

[4]   Philipmarshall.net. Teaching/Architectural Conservation II (HP 382)/Masonry/Soil texture/ Figure 5. Feb 2016 Standard soil-texture triangle. Accessed: 3 Feb 2016 Available at: http:// philipmarshall.net/Teaching/rwuhp382/masonry/soil_texture.htm.

[5]   Del Fabro A. Small bible of biological garden 2004. Leo commerce (translation of: Orto biologico)64.

[6]   Reimann, C., Siewers U., Tarvainen T., Bityukova L., Eriksson J., Giucis, A. et al. Agricultural soil in Northern Europe: A geochemical atlas. Stuttgart, Germany: 2003:279.

[7]   Škoric A. Composition and characteristics of the soil. Zagreb: Faculty of Agricultural Sciences. (In Croatian language) 1991:136.

[8]   Bašić F, Herceg N. Principles of agronomy. Zagreb: Synopsis. University book (in Croatian language) 2010:454.

[9]   Birkas M. Environmentally–sound adaptable tillage. Budapest: Akademiai Kiado, 2005:355.

[10]   Alexander M. Introduction to soil microbiology. New York: John Wiley & Sons, 1977:467.

[11] Blum WE. Functions of soil for society and the environment. Rev Environ Sci Bio Technol. 2005;4:75–79.

[12] Lal R. Soil carbon sequestration impacts on global climate change and food security. Science. 2004;304:1623–1627.

[13] Van–Camp L, Bujarrabal B, Gentile AR, Reports of the technical working groups established under the thematic strategy for soil protection, vol. IV: contamination and land management. EUR 21319 EN/4. Luxembourg: Office for Official Publications of the European Communities, 2004.

[14] Soil atlas. Facts and figures about earth, land and fields. Berlin, Germany: Heinrich Böll Foundation. and the Institute for Advanced Sustainability Studies, Potsdam, Germany 2015.

[15] Mirsal IA. Soil pollution – origin, monitoring & remediation. Berlin, Heidelberg: Springer, 2008.

[16] Bardos PR, Bakker MM, Slenders HL, Nathanail PC. Sustainability and remediation. In: Swartjes FA, editor(s). Dealing with contaminated sites–from theory towards practical application. Dordrecht, Heidelberg, London, New York: Springer, 2011:889–948.

[17] Wilcke W. Polycyclic aromatic hydrocarbons (PAHs) in soil – a review. J Plant Nutr Soil Sci. 2000;163:229–248.

[18] Samanta SK, Singh OV, Jain RK. Polycyclic aromatic hydrocarbons: Environmental pollution and bioremediation. Trends Biotechnol. 2002;20(6):243–248.

[19] Vegter JJ, Lowe J, Kasamas H. CLARINET – contaminated land rehabilitation network for environmental technologies. Mar 2002. Accessed: 20 Mar 2014 Available at: http://www.eugris.info/displayproject.asp?Projectid=4420.

[20] Kraus M, Wilcke W. Persistent organic pollutants in soil density fractions: distribution and sorption strength. Chemosphere. 2005;59:1507–1515.

[21] Riser-Roberts E. Remediation of petroleum contaminated soils. Boston: Lewis Publishers. Biological, Physical and Chemical Processes 1998.

[22] Nieboer E, Richardson DH. The replacement of the nondescript term „heavy metals" by a biologically and chemically significant classification of metal ions. Environ Pollut. 1980;1:3–26.

[23] Adriano DC. Trace elements in terrestrial environments: biogeochemistry, bioavailability and risks of metals. New York: Springer–Verlag, 2001:861.

[24] Kabata-Pendias A, Mukherjee AB. Trace elements from soil to human. Berlin, Heidelberg, New York: Springer, 2007.

[25] Prasad MN, Freitas HM. Metal hyperaccumulation in plants – Biodiversity prospecting for phytoremediation technology. Electron J Biotechnol. 2003 April 17;6(3) Accessed: 17 April 2014. Available at: http://www.ejbiotechnology.info/content/vol6/issue3/index.html.

[26] Kisic I, Mesic S, Basic F, The effect of drilling fluids and crude oil on some chemical characteristics of soil and crops. Geoderma. 2009;149:209–216.

[27] Green C, Hoffnagle A. Phytoremediation field studies database for chlorinated solvents, pesticides, explosives and metals. U.S. Environmental Protection Agency, Office of Superfund Remediation and Technology, Innovation Washington, DC, Mar 2004. Accessed: 18 Mar 2014 Available at: http://www.cluin.org/download/studentpapers/hoffnagle–phytoremediation.pdf.

[28] Zgorelec Ž. Phytoaccumulation of Metals and Metalloids from Soil Polluted by Coal Ash. Zagreb: University of Zagreb Faculty of Agriculture, 2009. Doctoral thesis105.

[29] Rascio N, Navari–Izzo F. Heavy metal hyperaccumulating plants: How and why do they it? And what makes them so interesting? Plant Sci. 2011;180(2):169–181.

[30] Zgorelec Z, Basic F, Kisic I, Wenzel WW, Custovic H. Arsenic and nickel enrichment coeficients for crops growing on coal ash. Cereal Res Commun. 2008;36(Part 2 Suppl. 5):1219–1222.

[31] Dellantonio A, Fitz WJ, Custovic H, Environmental risks of farmed and barren alkaline coal ash landfills in Tuzla, Bosnia and Herzegovina. Environ Pollut. 2008;153:677–686.

[32] Wenzel WW, Fitz WJ, Dellantonio A, Handbook on treatment of coal ash disposal sites. Vienna, Austria: cordis.europa.eu. (Handbook – findings from Deliverable HEIS, D6, D7, D10 and D11.) 2008.

[33] Prelac M, Bilandžija N, Zgorelec Ž. Potencijal fitoremedijacije teških metala iz tla pomoću Poaceae kultura za proizvodnju energije: Pregledni rad. J central Eur agriculture. 2016;17(3):901–916.

[34] Balazs B, Chiba Z, Hsu P, Lewis P, Murguia L, Adamson M. Destruction of hazardous and mixed wastes using mediated electrochemical oxidation in a Ag (II)HNO3 bench scale system. April 1997 SciTech Connect. Accessed: 9 April 2014 Available at: http://www.osti.gov/scitech/biblio/611759.

[35] Virkutyte J. Heavy Metal Bonding and Remediation Conditions in Electrokinetically Treated Waste Medias. Doctoral dissertation 2005. Available at:. http://epublications.uef.fi/pub/urn_isbn_951-27-0441-2/urn_isbn_951-27-0441-2.pdf.

[36] Page MM, Page CL. Electroremediation of Contaminated Soils. J Environ Eng. 2002; 128(3):208–219.

[37] Bolan NS, Adriano DC, Mani AP, Duraisamy A. Immobilization and phytoavailability of cadmium in variable charge soils. II. Effect of lime compost. Plant Soil. 2003;251:187–198.

[38] Durn G, Gaurina–Medjimurec N, Fröschl H, Veil JA, Veronek B, Mesic S. Improvements in treatment of waste from petroleum industry in Croatia. J Energy Resour Technol. 2008;130(2): 1–10.

[39] Kisic I, Jurisic A, Durn G, Mesic H, Mesic S. Effects of hydrocarbons on temporal changes in soil and crops. Afr J Agric Res. 2010;5(14):1821–1829.

[40] Rhykerd RL, Crews B, McInnes KJ, Weaver RW. Impact of bulking agents, forced aeration, and tillage on remediation of oil–contaminated soil. Bioresour Technol. 1999;67(3):279–285.

[41] Serrano A, Gallego M, González JL. Assessment of natural attenuation of volatile aromatic hydrocarbons in agricultural soil contaminated with diesel fuel. Environ Pollut. 2006;144(1):203–209.

[42] Kisic I, Mesic S, Brkic V, Mesic H, Bertovic L. Growing crops on polluted soil. Cereal Res Commun. 2008;36:1215–1218.

[43] Lodolo A, Gonzales–Valencia E, Miertus S. Overview of remediation technologies for persistent toxic substances. Arch Ind Hyg Toxicol. 2001;52(2):253–280.

[44] Chaney RL, Malik M, Li YM, Brown SL, Angle JS, Baker AJ. Phytoremediation of soil metals. Curr Opinions Biotechnol. 1997;8:279–284.

[45] ISO 13877. Soil quality – determination of polynuclear aromatic hydrocarbons – method using high-performance liquid chromatography 1998.

[46] ISO 18287. Soil quality – determination of polycyclic aromatic hydrocarbons (PAH) – gas chromatography method with mass spectrometric detection (GC-MS) 2006.

[47] ISO 11466. Soil quality – extraction of trace elements soluble in aqua regia 1995.

[48] ISO 11407. Soil quality – determination of cadmium, chromium, cobalt, copper, lead, manganese, nickel and zinc – Flame and electrothermal atomic absorption spectrometric methods 1998.

[49] ISO/DIS 22036. Soil quality – determination of trace elements in extracts of soil by inductively coupled plasma – atomic emission spectrometry (ICP – AES) 2006.

Practical tasks

Task No 1:

According to data presented in table calculate the CN ratio in compost mixture. And answer to the two questions:

a.   Does compost mixture have optimal CN ratio?

b. If not what amount (kg) of old fruits branches should be added in compost mixture in order to achieve optimal CN ratio?

Material	Carbon (%)	Nitrogen (%)	Moisture (%)	Mass (kg)
Horse manure	40	1.6	75	56
Food/kitchen waste	14	0.92	63	249
Old fruits branches	53	1.0	15	70
Cardboard	43	0.1	9	100

Task No 2:

Using the soil textural triangle (Atterberg's triangle) determine the textural classes of soil if soil consists:

a. 60% of clay, 20% of silt and 80% of sand _____

b. 10% of clay, 30% of silt and 70% of sand _____

Depending on the textural classes in which soil will (a) or (b) inorganic contaminants be more accumulated and why?

Interesting unknowns about soil

–   Swedish scientist Jöns Jacob Berzelius (1779–1848) defined the soil as an enormous chemical laboratory in which countless processes of degradation and synthesis take place simultaneously and continuously.

–   The Egyptian Queen Cleopatra declared earthworms, the inhabitants of soil, saints. Aristotle called them the intestines of the earth, and Charles Darwin felt that they belong to an important position in the development of the world.

–   As already mentioned, pedosphere has much less and markedly slower possibilities of recovering from toxic effects and the fact that soils are an irreplaceable factors of sustaining life on earth, the Food and Agriculture Organization (FAO) of the United Nations declared 2015 the International Year of Soils. The goal was to increase awareness and understanding of the importance of soil for food safety/security and essential ecosystem functions.

–   For all these reasons, in order to continue the efforts made during the International Year of the Soils, the International Union of Soil Sciences (IUSS) proclaimed The International Decade of Soils (2015–2024). International Decade of soils is result of the 'Vienna Soil Declaration' of Dec. 7, 2015. In that Declaration the IUSS has identified the key roles of soil in addressing the major resource, environmental, health and social problems which humanity is currently facing.

Study questions

- In terms of the adverse effects on humans and animals, which group of pollutants in soil (organic or inorganic) represents the greater threat to health? Explain why.
- Enumerate three ways to perform remediation of contaminated soils.
- List the four major technologies of soil remediation and explain the criteria for selection of individual technology.
- What is the main difference between *On situ* and *Ex situ* remediation technologies?
- Explain why the biological restoration with plants is the most promising method and enumerate several plants that can successfully accumulate heavy metals from soils.
- Electro-remediation as a method for conversion of heavy metals is used in (circle the correct statement):
  (a) Sewage sludge remediation
  (b) Remediation of groundwater
  (c) Soil remediation
- An electric field between the cathode and anode leads to some processes in soil. These processes are:
  (a) _____
  (b) _____
  (c) _____
- Explain the main role of adding different radicals, solvents and chelates in processes of electro-remediation.
- List some of disadvantages of using the chemicals in the flushing remediation methods.
- Although the methods of solidification and stabilization do not directly reduce the concentration of metals in soil, they indirectly perform a very important role. Explain advantage of these methods.
- Enumerate the methods which are integral part of physical methods.
- In terms of the primary role of soil (production), specify the main negative impact of the implementation the thermal remediation. This method is most effective in removal of _____ (organic soil contaminant).

A few words about analytical problems

Determination of organic contaminants (for example PAHs) or inorganic contaminants (for example lead content) in soils in first step includes extraction or digestion procedures. Second step includes an analysis of extracts or solutions of digested samples. There are various analytical methods that are applicable for these purposes but standards methods for analysis of listed contaminants are described by several ISO norms. For instance some PAHs in soil samples can be extracted with acetone or toluene in a Soxhlet apparatus and determine using high-performance liquid chromatography [45]. Also, PAHs can be determined by means of gas chromatography method with mass spectrometric detection (GC-MS) in combination of two extracts (acetone and petroleum ether) according to [46]. As previously are mentioned, cadmium, mercury and lead are consisting group of phototoxic elements and their determination in soil first include digestion procedure. For instance, lead (Pb) can be extracted in aqua regia [47] from soils which are containing less than 20 % (m/m) organic carbon and determined by flame and electrothermal atomic absorption spectrometric method [48] or using inductively coupled plasma – atomic emission spectrometry (ICP-AES) according [49].

Anita Šalić, Ana Jurinjak Tušek and Bruno Zelić

10 Modeling of environmental processes

Abstract: What's the weather tomorrow? How will the population grow? When will the economy situation get better? Can this factory work better? These are just some everyday questions that models can give us answers to. Basically, we get in touch with model outcomes (answers) every day, and most of us are not even aware of it. When talking about environmental processes, models, especially mathematical models, play a significant role in process design, optimization, development, future predictions, etc. In this chapter, a short introduction into models (types of the models and their application) with accent on mathematical modeling in environmental engineering will be given. Some basic advices on how to approach mathematical modeling, how to use different types of models and how to apply different methods for solving of mathematical models will be discussed on the examples describing the environmental processes.

Keywords: mathematical modeling, environmental processes, numerical methods

> The scientists do not try to explain, they hardly even try to interpret, they mainly make models. By a model is meant a mathematical construct which, with the addition of certain verb al interpretations, describes observed phenomena. The justification of such a mathematical construct is solely and precisely that it is expected to work. (John von Neumann)

10.1 Models

Models and modeling are two simple words that usually cause discomfort, fear, respect, excitement, joy and a whole lot of different emotions in engineering field. Negative that emerges from facing the unknown and positive that shines after that first, usually wrong, solution of the problem. In order to understand some basic principles that hide behind those two words, first we have to answer several important questions:
- What are the models?
- Why do we need the models and modeling?
- How to approach modeling?

The first question is What are the models? [1] which will say *"really nothing more than an imitation or reality."* They can be verbal, schematic, physical, computer and mathematical models. Other authors will say that dividing them into only three

https://doi.org/10.1515/9783110468038-010

groups – verbal, physical and mathematical – is enough. Basically, many different authors/teachers/researches will classify them into different groups. But when talking about types of models, one always have to have in mind model characteristics that have a great impact on the solution technique as well as potential area of application. This is the reason why in different research area some models are more present than in the others and why there are different points of view on the classification.

For example, from *the time point of view*, models can be divided into:

- *Static models* – where there is no change in state variable in time (steady state) (i. e. no change in concentration, temperature ...) .
- *Dynamic models* – where changes in state are time dependent.

If the models are observed from *the point of independent variables number*, they can be divided into:

- *One-dimensional (1D) models* – where state variable depends on only one independent variable. For their formulation, usually differential equations are used.
- *Multidimensional models* – where state variable depends on multiple independent variables. For their formulation, usually partial differential equations are used.

Taking into account *number of the unknowns*, models can be divided into:

- *Underdeterminated* – fewer equations than unknowns.
- *Determinated* – problem-solving example.
- *Overdeterminated* – more equations than unknowns.

In Figure 10.1, a more general approach of dividing models into categories is demonstrated.

Physical models are 3D models of real systems (i. e. chemical molecules, small-scale laboratory devices, etc.). They are *static* and usually start as *analogue models* that transfer to *physical models* with *dynamic* feature which then translates to final product, *prototype*.

Abstract models cannot be presented in reality like physical models. They are divided into two groups: *mental* and *symbolic*. *Mental models* are formed in human consciousness and represent subjective interpretation of real system. The most common *mental model* formed in everyday life is dream. *Symbolic models* present a large group of models. They can be divided into *mathematical* and *nonmathematical*. *Mathematical models* are the most common and most important from process point of view. They can again be divided into different categories and used depending on the area of application. In this chapter, the main focus will be given on *mathematical models*.

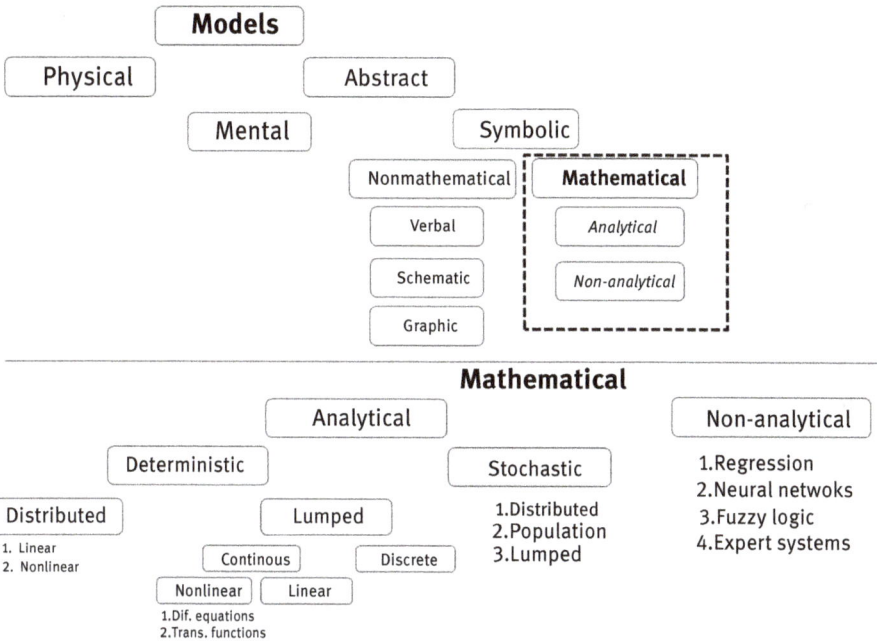

Figure 10.1: Types of models (inspired by [2]).

10.2 Mathematical models and modeling

As mentioned, mathematical models are mostly used when describing processes; therefore, some modifications of previously listed three important questions are necessary. Modified key questions are:

- What are mathematical models and mathematical modeling?
- Why do we need mathematical models and mathematical modeling?
- How to approach mathematical modeling?

Before answering the baove questions, few additional words about mathematical models and types of the mathematical models are necessary. Although they can be divided as presented in Figure 10.1, there are many different approaches (as for models in general).

Some of classifications of the mathematical models according to Hangos and Cameron [1] are:

1. *Deterministic* – models based on cause–effect analysis of process and have "fix data";
2. *Stochastic* – models based on elements that are probabilistic in nature and use random variables and uncertain data.

When based on dependent variables and their dependence on spatial position (Figure 10.2) models could be divided into:

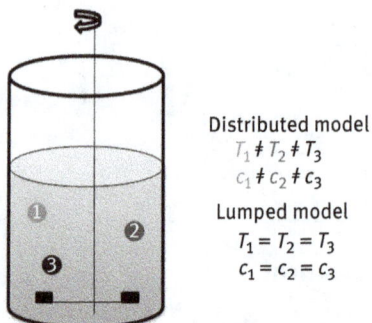

Distributed model
$$T_1 \neq T_2 \neq T_3$$
$$c_1 \neq c_2 \neq c_3$$

Lumped model
$$T_1 = T_2 = T_3$$
$$c_1 = c_2 = c_3$$

Figure 10.2: Difference between distributed and lumped model.

1. *Lumped models* – dependent variables are not function of spatial position (i. e. processes with ideal mixing);
2. *Distributed models* – dependent variables are function of spatial position.

Classification of mathematical models could be based on mathematical description of the process:
1. *Continuous* – dependent variables are defined over continuous space and time;
2. *Discrete* – only defined for discrete values of time and space and can be used for the modeling existence of the system in general.

Additional classification could be based on mathematical structure:
1. *Linear* – there is linear dependence between variables;
2. *Nonlinear* – with nonlinear dependence of process variables.

On the other hand, when talking about environmental processes, they are usually more complex than already mentioned models (due to large number of equations, their mutual connection, etc.). Mathematical models of environmental processes often consist of multiple mathematical objects like equations, graphs and rules. Because of that, when talking about process models, mathematical models are usually divided into three different categories (Figure 10.3): *mechanistic, empirical* and *combined*.

White box models (mechanistic) are being used for description of systems when all necessary information about the system is available and known. The name itself indicates that all is transparent and understandable and that white box models usually do not require experimental data. From the level of complexity, white box models are most demanding since they contain a lot of complex expressions to

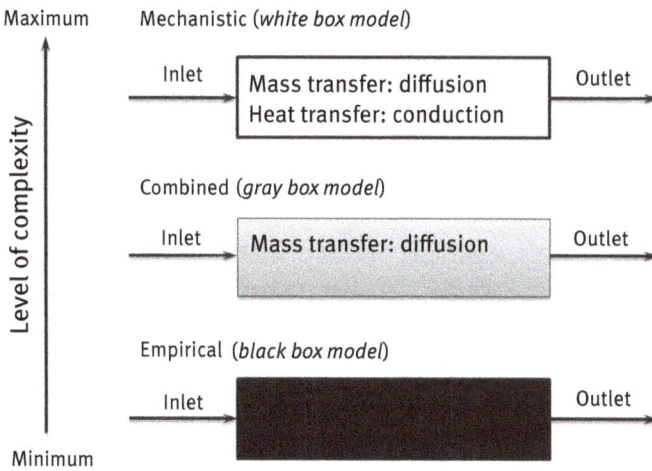

Figure 10.3: Types of process models.

present all what is happening in the system. On the other hand, *black box models (empirical)* are used for the systems where knowledge about the process is based on experimental data and not from understanding the mechanism inside the process. In practice, most of systems are described somewhere between the black box and white box models and they are named *gray box models*. In gray box models, based on previous knowledge about the system, model structure is defined together with some process parameters (i. e. defining kinetic and transport mechanisms). Then, experimental data are used to finish building up the model.

Therefore, if models are "*really nothing more than an imitation or reality,*" What is *mathematical modeling*?

According to some definitions, mathematical modelling is:

> "*an activity, a cognitive activity in which we think about and make models to describe how devices or objects of interest behave.*" (C.L. Dym)

> "*translation of beliefs about how the world functions into the language of mathematics*" (G. Marion)

> "*not based on the real existence of a physical model system, but describes the behavior of the original system by mathematical equations. This is the highest level of abstraction.*"(unknown source)

> "*mathematical (quantitative and qualitative) expression (mathematical equations, statistical relations, computer programs) that is used to determine environment interactions (inlet variables), system status (state variables) and purpose of control (outlet variables)*" (B. Zelić)

In general, combining definitions, mathematical modeling is an *active process* used to describe real systems by translating them to mathematical expressions. Nowadays, various relevant textbooks are dedicated to modeling and modeling applications, like

Himmelblau and Bishop [3], Davis [4], Riggs [5], Rice and Do [6], Aris [7] and Denn [8]. If you take a look at the year that those books were published, most of them got printed before 1990. Although many years have passed, basic principles of mathematical modeling have remained the same. What changes is the speed of solving them (Figure 10.4).

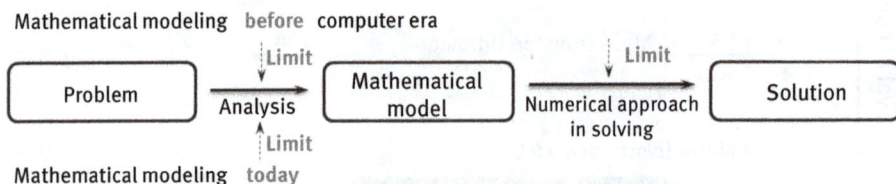

Mathematical modeling before computer era

| Problem | → Analysis | Mathematical model | → Numerical approach in solving | Solution |

Limit

Limit

Limit

Mathematical modeling today

Figure 10.4: Limits in mathematical modeling before computer era and today.

The next big question is *Why do we need mathematical modeling?*

As mentioned, mathematical modeling aims to describe the different aspects of the real world, their interaction and their dynamics through mathematics. Nowadays, mathematical modeling has a key role in different fields such as the environmental engineering and industry, while its potential contribution in many other areas (business, sociology, psychology, medicine, physics, etc.) is becoming more and more evident [9–11]. According to Hangos and Cameron [1], the modeling enterprise links together a purpose P with a subject or physical system S and the system of equations M which represent the model. A series of experiments E can be applied to equations M in order to answer questions about system S.

From the environment and industry point of view, mathematical models can be used in:
- *Process design* – technical, economic and environmental studies of the process; analysis of process parameters impact on the process; optimization of the process; dynamic analysis of the process, etc.
- *Process control* – process control and automatization; optimization of batch reactor control; optimization of multiprocess control, etc.
- *Finding flaws in the process* – reviling the causes for poor quality of the product/ process; reviling the irregularities in the process, etc.
- *Exploration of new solutions*
- *Analyzing the safety of the process* – reviling the danger hot spots in the process; creating the scenarios that can lead to dangerous situations; protocol development for avoiding dangerous situations, etc.
- *Training and teaching of the operator* – simulation of safety protocols for stopping the process in the case of emergency; simulation of different situations that can happen in the process (change of temperature, pressure, inlet concentration ...) and how to react on them in the best way according to the good practice

– *Environment protection* – calculation of the emission; prediction how some hazardous component will spread; evaluation of economic, environmental and sociological impact of pollutions, etc.

Mathematical models can be used in wide range of areas because they are broad and flexible and can be easily modified and extended [10].

Although the first recognizable models were numbers, counting and "writing" numbers (e. g. as marks on bones) is documented since about 30,000 BC the real expansion is linked to the incredible increase of computer calculation speed and progress of scientific computation that allows the translation of a mathematical model – which can be explicitly solved only occasionally – into algorithms that can be treated and solved by ever more powerful computers [9]. But always keep in mind: *If you don't know how to solve something, not even computer can help you.* (A.D. Noel)

And finally, *How to approach mathematical modeling?* is the general question that will be addressed in this subchapter.

10.3 Basic rules in mathematical modeling

According to C.A.R. Hoar, "*there are two approaches in developing the mathematical model of the process: first, to develop a simple model that has no significant errors and second, to develop mathematical model that is so complex that it has no errors. First approach is much harder.*" The basic idea in development is that a mathematical model must be able to address universal concepts, such as the conservation of mass or the momentum of a fluid or the moment of inertia of a structure; moreover, in order to obtain a successful numerical simulation, it is necessary to define what level of details must be introduced in the different parts of a model and what simplifications must be carried out to facilitate its integration into different models [9].

The first step, before starting to set up a process model, the problem definition should be clearly stated. Problem definition defines the process, the modeling goal and the validation criteria [1]. While developing the mathematical model, one should always have in mind that modeling is complex, demanding and usually with many "two steps in front one step behind." Returning back to previous steps and trying to find the answers for unusual solutions that were obtained from the model is everyday practice. To get the right model at first attempt is very rare and very unusual phenomena in mathematical modeling.

Like in any other engineering approaches, there are some basic rules that can be followed to make modeling easier and to reduce the number of mistakes. Those basic rules are usually referred as seven-step modeling procedure (define, collect, evaluate, develop, solve, interpret/verify and validate, Figure 10.5).

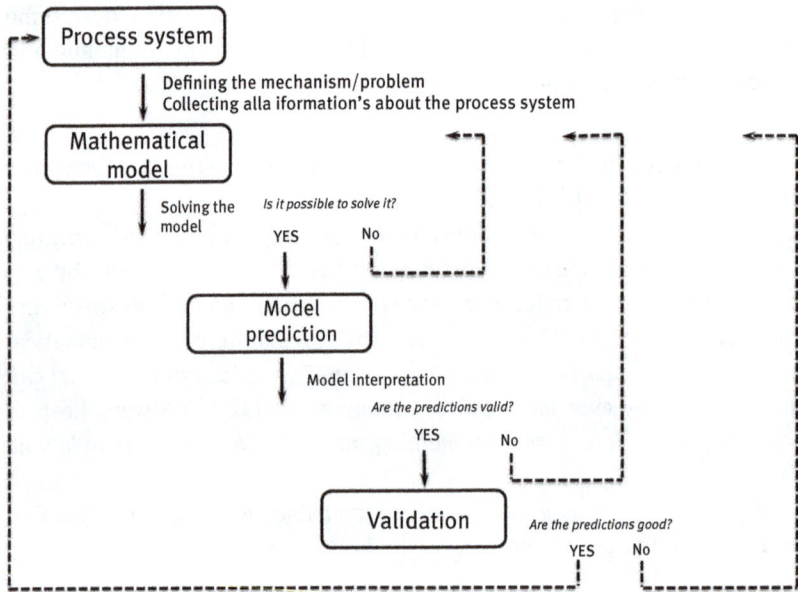

Figure 10.5: Basic scheme of mathematical model development (inspired by [2, 11]).

In Examples 10.1 and 10.2, basic steps (steps are based on books: Hangos and Cameron [1], and Plazl and Lakner [2]), presented in Figure 10.5, will be demonstrated.

Example 10.1.
Pollutant A is being removed from the wastewater in batch reactor that has volume of $V = 250$ L. Degradation of the pollutant A can be described by first-order kinetics. The reaction rate constant is temperature dependent according to Arhenius equation: $k = A \cdot e^{-\frac{Ea}{R \cdot T}}$ where frequency factor is $A = 2.5 \cdot 10^8$ 1/h, and activation energy $Ea = 50$ kJ/mol. Process temperature is kept constant at $T = 25$ °C, and initial concentration of the pollutant A is $c_{A,0} = 5$ mol/L.

a. Compare an analytical and numerical solution for the concentration change of the pollutant A during the time at specific time point listed in table with experimental data.

t/h	0	5	10	20	50	80	100
c_A (analytical)/mol/L							
c_A (numerical)/mol/L							
c_A (experimental)/mol/L	5	1	0.2	0.01	0	0	0

b. Define the half-life time of the pollutant A
c. Define the time necessary to achieve 95% degradation of the pollutant A if process if preformed at 3 °C lower temperature

---

As a first step in approaching to the mathematical modeling (Figure 10.5), *process system* (a part of real world with defined physical boundaries) *and purpose/goal of modeling have to be defined.*

Process system:

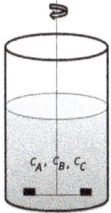

- batch reactor
- first-order reaction, A⟷B
- constant volume, V = const.
- constant density, $\Delta_0 = \Delta$
- constant temperature, T = const.

Figure (a): Batch reactor.
Purpose/goal of modelling: To predict the dynamic change of concentration of reactant A

The second step is *defining the mechanisms* and *information about the process system.* It means all physical, chemical and biochemical processes that are taking place in process system have to be defined. Depending on the purpose of the modeling, they can be defined on molecular (usually used for confirming different theories and are not used in process engineering), microscopic and macroscopic (global process description) level (Figure 10.6).

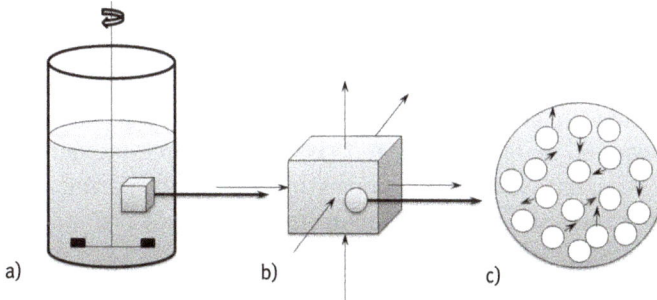

a)   b)   c)

Figure 10.6: Levels of developing mathematical models: (a) macroscopic, (b) microscopic and (c) molecular level.

The most common mechanisms in modeling of the process are:
- reaction kinetics
- heat transfer
- heat transfer by covection
- radiation
- diffusion of mass

- mass and heat transfer through boundary layer
- mass transfer by convection
- change of physical state
- growth and cell death
- metabolic pathways etc.

and those mechanisms present just partial models of entire process model.

As for the collecting information about the process system, all the parameter values and inlet variables, which can be measured, have to be defined. Additionally, there is also a possibility to estimate some of parameters using the advantage of correlation equations (i. e. to calculate reaction rate if the temperature and frequency factor are known). On the other hand, sometimes there is no sufficient experimental/literature data to define parameters or to estimate them in order to model work. In that case, simplifications of previous steps are necessary.

Defining the mechanisms:
Chemical reaction (A⟷B), first order ($r_A = k \cdot c_A$)
Information about the process system:
$c_{A,0} = 5$ mol/L
$V = 250$ L
$T = 20\,°C$

The next step is to define *mathematical model, a set of model equations*. This is actually the step where real systems are translated to mathematical expressions – mathematical modeling. Equations are the universal tool for modeling and they are obtained from the balances of mass, energy or impulse that entered and exited the observed system in definite time and volume. Generation and consumption within the observed system (Figure 10.7) is also taken into account in case of (bio)chemical reactions. Model equations can be either *differential* or *algebraic*. More details about the balance equations, types and solving, will be addressed in Section 5.

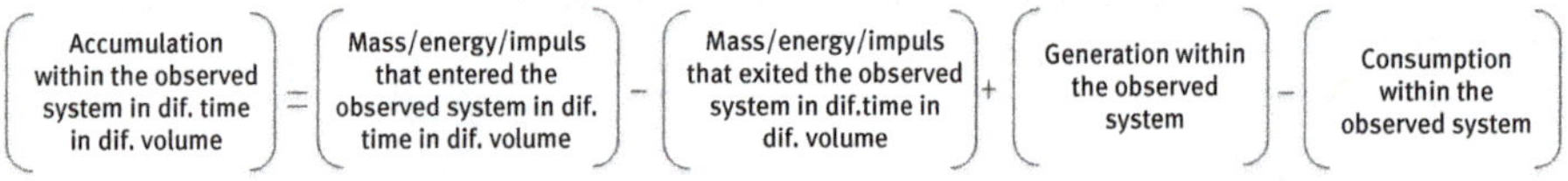

$$\begin{pmatrix} \text{Accumulation} \\ \text{within the observed} \\ \text{system in dif. time} \\ \text{in dif. volume} \end{pmatrix} = \begin{pmatrix} \text{Mass/energy/impuls} \\ \text{that entered the} \\ \text{observed system in dif.} \\ \text{time in dif. volume} \end{pmatrix} - \begin{pmatrix} \text{Mass/energy/impuls} \\ \text{that exited the observed} \\ \text{system in dif.time in} \\ \text{dif. volume} \end{pmatrix} + \begin{pmatrix} \text{Generation within} \\ \text{the observed} \\ \text{system} \end{pmatrix} - \begin{pmatrix} \text{Consumption} \\ \text{within the} \\ \text{observed system} \end{pmatrix}$$

Figure 10.7: General mass balance equation.

For successful definition of mathematical model, it is necessary to clearly describe what are inlet and outlet variables (inputs/outputs; i. e. mass, time, temperature, pressure ...) and parameters (reaction rate constant, activation energy, etc.), the type of space dependence and time dependence (static or dynamic model). Basically, five steps have to be followed:

1. To define model assumptions (i. e. ideal mixing, constant physical and chemical properties ...). This step is crucial because it is connected with model equations and initial conditions that have strong impact on results of model simulation. The most common assumptions are:
   a. Time (stationary or non-stationary process)
   b. Space (1D, 2D or 3D problem)
   c. Types of flow (laminar, plug flow ...)
   d. Mechanism (i. e. reaction rate is negligible in comparison to diffusion rate)
   e. Properties of the matter (i. e. temperature (in)dependence)
   f. Accuracy of estimated parameters, outputs ...
   g. System geometry
2. To determine the system, subsystem and space elements
3. To define the changeable (independent and dependent) values (i.e. *dependent*: reactant concentration, temperature ... and *independent*: time, reactor length...)
4. To write the balance (mass, energy, movement)
5. To write the mechanistic equations and constitutive connections.

Model assumptions:
– ideal mixing
– V = const.; T = const.

Dependent variables:
– c_A

Independent variables: -t
– t

Reactant A mass balance:
$$\frac{dc_A}{dt} = -k \cdot c_A$$

Constitutive connections:
$$k = A \cdot e^{-\frac{Ea}{R \cdot T}} = 2.5 \cdot 10^8 \cdot e^{-\frac{50000}{8.314 \cdot 293.15}} = 0.308 \, 1/h$$

In order to solve the mathematical model, there are two approaches: *analytical* and *numerical*. Analytical approach is used only when models are simple because solving complex models with this approach is either impossible or time-consuming. On the other hand, numerical methods (i. e. Taylor, Euler, Runge–Kutta for solving differential equations, or Laplace, finite element method (FEM), finite volume method (FVM) and finite difference method (FDM) for solving partial differential equation) enable solving even the systems with large number of equations (i. e. to predict weather it is necessary to solve 10^5–10^6 partial differential equation simultaneously). As mentioned before, incredible increase of computer calculation speed and progress of scientific computation shortened the time of solving numerous of mathematical model equations on level of minutes and even seconds.

Solving the model:

(a) Analytical solution

$$\frac{dc_A}{c_A} = -k \cdot dt$$

$$\int_{c_{AO}}^{c_A} \frac{dc_A}{c_A} = -k \cdot \int_0^t dt$$

$$\ln c_A - \ln c_{A,\,0} = -k \cdot t$$
$$\ln c_A = -k \cdot t + \ln c_{A,\,0}$$
$$c_A = c_{A,\,0} \cdot e^{-k \cdot t}$$

(b) Numerical solution (Wolfram Mathematica 10.1 code)

```
A = 2.5·10^8;
Ea = 50·10^3;
R = 8.314;
T = (20 + 273.15);
c_A,0 = 5;
t_f = 100;
k = A·Exp[−Ea/(RT)];
model = NDSolve[{c_A'[t] = = −k·c_A[t],c_A[0] = = 5},c_A,{t,0,t_f}]
kc_A[t_]: = c_A[t]/.model
Plot[kc_A[t],{t,0,t_f},PlotRange>All,AxesLabel->{"t/h","c_A/mol/L"}]
```

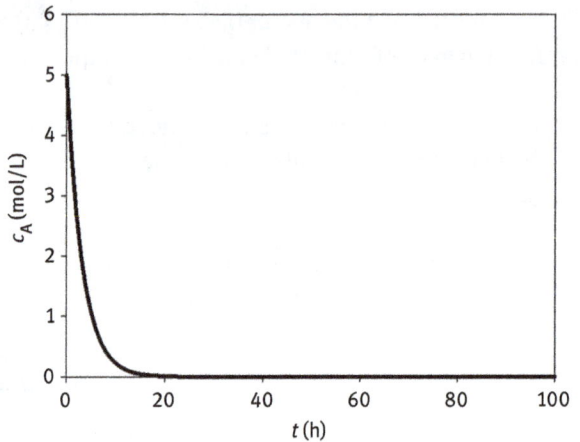

Figure (b): Graphical solution of mathematical model

Table 10.1: Comparison between analytical and numerical solution of mathematical model for pollutant A concentration profile.

t/h	0	5	10	20	50	80	100
c_A (analytical)/mol/L	5	1.072	0.229	0.011	$1.025 \cdot 10^{-6}$	$9.953 \cdot 10^{-11}$	$2.102 \cdot 10^{-13}$
c_A (numerical)/mol/L	5	1.072	0.230	0.011	$1.029 \cdot 10^{-6}$	$9.725 \cdot 10^{-11}$	$2.939 \cdot 10^{-13}$

The next step is model *interpretation* or *verification of model solution*. It is necessary to evaluate if the model behaves correctly and if the obtained answer is expected. If the answer on one of these two questions is no, a stepback is necessary in order to perform some corrections of the model. If the answers are yes, the final step of mathematical modeling is *validation*. There are various ways to validate a model and they depend on process system, purpose of modelling and ability to collect independent data about the system. Model validation can be performed by:

- Verifying experimentally the assumptions for the simplification,
- Comparing the model behavior with the process behavior,

- Developing analytical models for simplified cases and comparing their behavior,
- Comparing with other models using a common problem and
- Comparing the model directly with process data.

A poor approach is to validate a model based on the data that were used for model development and model parameter estimation.

Additionally, there are some model properties that enable its validation:
- Accuracy – model is accurate if the model response is satisfactory close to the response of real system,
- Realism in system description – model is based on assumptions about mechanisms that are close to those in real system,
- Precision – a unique solution (i. e. a result of model simulation is only one x–y dependence not a bundle of curves),
- Robust – model is not sensitive to "noise" that is present in input data and,
- Generality – possibility to apply model on wide range of problems.

Table 10.2: Comparison of analytical and numerical solution of mathematical model with experimental data for pollutant A concentration profile.

t/h	0	5	10	20	50	80	100
c_A (analytical)/mol/L	5	1.072	0.229	0.011	$1.025 \cdot 10^{-6}$	$9.953 \cdot 10^{-11}$	$2.102 \cdot 10^{-13}$
c_A (numerical)/mol/	5	1.072	0.230	0.011	$1.029 \cdot 10^{-6}$	$9.725 \cdot 10^{-11}$	$2.939 \cdot 10^{-13}$
c_A (experimental)/mol/L	5	1	0.2	0.01	0	0	0

```
figure1 = Plot[kcA[t],{t,0,tf},PlotRange->All,AxesLabel->{"t/h","cA/mol/L"}];
data = {{0,5},{5,1},{10,0.2},{20,0.01},{50,0},{80,0},{100,0}};
figure2 = ListPlot[data];
Show[{figure1,figure2}]
```

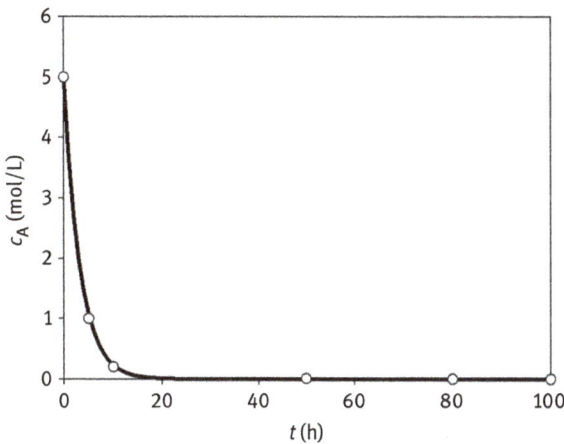

Figure (c): Graphical comparison of solution of mathematical model and experimental data.

Once the model is developed and validated, it can be used for many different purposes: optimization, process control, process design, finding flaws in the process, etc. By applying mathematical models, analysis and measurements that are time-consuming or even impossible to perform in reality become possible and solvable in a short period of time. The number of possibilities is endless.

Model application:

(b) Component A half-life

$c_A = c_{A,0} \cdot e^{-k \cdot t}$

$0.5 \cdot c_{A,0} = c_{A,0} \cdot e^{-0.308 \cdot t}$

$\ln 0.5 = -0.308 \cdot t$

$t = 2.250$ h

(c) Time to achieve 95% degradation at T = 17 °C

$k = A \cdot e^{-\frac{E_a}{R \cdot T}} = 2.5 \cdot 10^8 \cdot e^{-\frac{50000}{8.314 \cdot 290.15}} = 0.249 \ 1/h$

$c_A = c_{A,0} \cdot e^{-k \cdot t}$

$0.05 \cdot c_{A,0} = c_{A,0} \cdot e^{-0.249 \cdot t}$

$\ln 0.05 = -0.249 \cdot t$

$t = 12.031$ h

Example 10.2.

Pollutant A is being removed from the wastewater of pharmaceutical industry in batch reactor that has volume of $V = 200$ L. Degradation of the pollutant A can be described with second-order kinetics with reaction rate constant $k = 0.005$ mol/L·h. Initial concentration of the pollutant A is $c_{A,0} = 5$ mol/L.

a. Compare an analytical and numerical solution for the concentration change of the pollutant A during the time period of 100 h at specific time point listed in table.

b. Define the time necessary to achieve 95% degradation of the pollutant A.

c. Define the time necessary to achieve 95% degradation of the pollutant A if process would be performed at continuous stirred tank reactor that has volume of $V = 100$ L at constant flow rate of $q = 30$ L/h.

d. Wastewater treatment system operates under constant temperature conditions ($T = 20$ °C). In case of temperature control failure, reverse reaction mechanics occurs ($2A \rightleftarrows B$). Calculate the concentration of the pollutant A if the temperature regulation failure occurs at the tenth hour of the degradation and it lasts for 3 h. In the moment of the failure, the concentration of the pollutant A in reactor is $c_A = 4$ mol/L and the concentration of the component B is $c_B = 1$ mol/L. Reaction rate constant of reverse reaction is $k_2 = 0.001$ mol/L·h.

(a) Change of concentration of reactant A in time in batch reactor

Process system:

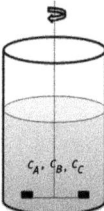

– batch reactor
– constant volume, V = const.
– constant density, $\Delta_0 = \Delta$
– constant temperature, T = const.

Figure (a): Batch reactor.

Purpose/goal of modeling: To predict change of concentration of reactant A in time

Defining the mechanisms: Chemical reaction (A $\leftrightarrow$ B), second order ($r_A = k \cdot c_A^2$)

Information's about the process system:
$c_{A,0} = 5$ mol/L
$V = 200$ L
$T = 20$ °C
Model assumptions:
– ideal mixing
– $V =$ const.; $T =$ const.
Masass balance:
$\frac{dc_A}{dt} = -k \cdot c_A^2$

Solving the model:

(a) Analytical solution

$\frac{dc_A}{dt} = -k \cdot c_A^2$

$\frac{dc_A}{c_A^2} = -k \cdot dt$

$\int_{c_{A,0}}^{c_A} \frac{dc_A}{c_A^2} = -k \cdot \int_0^t dt$

$-\left(\frac{1}{c_A} - \frac{1}{c_{A,0}}\right) = -k \cdot t$

$\frac{1}{c_A} = \frac{1}{c_{A,0}} + k \cdot t$

(b) Numerical solution

$k = 0.005$;

$c_{AO} = 5$;

$t_f = 100$;

model = NDSolve[{c_A'[t] = = -k×c_A^2[t],c_A[0] = = 5},c_A,{t,0,t_f}]

kc_A[t_]: = c_A[t]/.model

Plot[kc_A[t],{t,0,t_f},PlotRange->All,AxesLabel->{"t/h","c_A/mol/L"}]

Table 10.3: Comparison between analytical and numerical solution of mathematical model for pollutant A concentration profile.

t/h	0	5	10	20	50	80	100
c_A (analytical)/mol/L	5	4.444	4.000	3.333	2.222	1.667	1.428
c_A (numerical)/mol/L	5	4.444	4.000	3.333	2.222	1.667	1.428

Application:
(b) Time necessary to achieve 95% degradation of the pollutant A

$\frac{1}{c_A} = \frac{1}{c_{A,0}} + k \cdot t$

$\frac{1}{0.1 \cdot c_{A,0}} = \frac{1}{c_{A,0}} + 0.005 \cdot t$

(c) Time necessary to achieve 95% degradation of the pollutant A in continuous stirred tank reactor

$\frac{d}{dt}(V \cdot c_A) = q \cdot c_{A,0} - q \cdot c_A - V \cdot r_A$

$V \cdot \frac{dc_A}{dt} = q \cdot (c_{A,0} - c_A) - V \cdot k \cdot c_A^2$

$\frac{dc_A}{dt} = \frac{q}{V} \cdot (c_{A,0} - c_A) - V \cdot k \cdot c_A^2$

$\frac{dc_A}{dt} = \frac{q}{V} \cdot (c_{A,0} - c_A) - k \cdot c_A^2$

$$\int_{c_{A,0}}^{c_A} \frac{dc_A}{\frac{q}{V} \cdot (c_{A,0} - c_A) - k \cdot c_A^2} = \int_0^t dt$$

$t = 0.192$ h

(d) Change of concentration of reactant A in batch reactor after reaction mechanism change

$\frac{dc_A}{dt} = -2 \cdot k \cdot c_A^2 + k_2 \cdot c_B$

$\frac{dc_B}{dt} = 2 \cdot k \cdot c_A^2 - k_2 \cdot c_B$

$\frac{1}{\alpha_A} \frac{dc_A}{dt} = \frac{1}{\alpha_B} \frac{dc_B}{dt} \leftrightarrow c_B = 0.5 \cdot (c_{A,0} - c_A) + c_{B,0}$

$\frac{dc_A}{dt} = -2 \cdot k \cdot c_A^2 + k_2 \cdot 0.5 \cdot (c_{A,0} - c_A) + c_{B,0}$

$c_A = 3.574$ mol/L

Example 10.3.
Exothermic reaction of the burning the waste ($\rho = 1200$ kg/m^3) containing toluene is taking place in tubular reactor with active volume of $V = 1500$ L. Reaction can be described by first-order kinetics. Reactor has diameter of $R = 0.4$ m, flow rate through reactor is $q = 200$ L/min and concentration of the toluene at the inflow is $c_i = 10$ mol/L. Temperature at the inflow is $T_i = 40$ °C, and the temperature in heat exchanger $T_o = 130$ °C. Heat transfer coefficient is $k_Q = 5.2$ W/m$^2 \cdot$ K and specific heat capacity $c_P = 4.5$ kJ/kg$\cdot$K. Reaction enthalpy is $(-\Delta H_A) = 80$ kJ/mol and parameters of the Arrhenius model are $A = 3.6 \cdot 10^3$ 1/s, $Ea = 42$ kJ/mol.
a.   Set the balances for the mass and heat change over the time.
b.   Set the balances for the mass and heat change in dimensionless form and calculate the specific dimensionless numbers.
c.   Define the distribution of conversion and temperature.
d.   Calculate the conversion, concentration of the toluene and the temperature at the reactor outflow.

(a) Balances for the mass and heat change over the time in a tubular reactor

mass inflow:	$F_1 = S \cdot v \cdot c(t,x)$	
mass outflow:	$F_2 = S \cdot v \cdot c(t, x + \Delta x)$	
reaction:	$F_R = \Delta x \cdot S \cdot r(c)$	
mass accumulation:	$m = S \cdot \Delta x \cdot c(t, x)$	
mass change over time:	$\frac{\Delta m}{\Delta t} = F_1 - F_2 - F_R$	

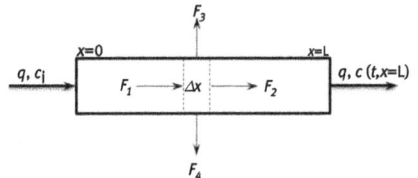

Figure (d): Tubular reactor.

$\frac{\Delta}{\Delta t}(S \cdot \Delta x \cdot c(t,x)) = S \cdot v \cdot c(t,x) - S \cdot v \cdot c(t, x + \Delta x) - \Delta x \cdot S \cdot r(c)$

$\frac{\Delta}{\Delta t} c(t,x) = -v \cdot \frac{c(t, x + \Delta x) - c(t,x)}{\Delta x} - r(c)$

$\lim_{\Delta t \to 0} \left(\frac{\Delta}{\Delta t} c(t,x) \right) = -v \cdot \lim_{\Delta x \to 0} \left(\frac{c(t, x + \Delta x) - c(t,x)}{\Delta x} \right) - r(c)$

$\frac{\partial}{\partial t} c(t,x) = -v \cdot \frac{\partial c(t,x)}{\partial x} - r(c)$

Initial conditions:	$c(t = 0, x) = c_0(x)$
Boundary conditions:	$c(t, x = 0) = c_i(t)$
heat inflow:	$F_1 = S \cdot v \cdot \rho \cdot c_p \cdot T(t,x)$
heat outflow:	$F_2 = S \cdot v \cdot \rho \cdot c_p \cdot T(t, x + \Delta x)$

heat transfer through
heat exchanger wall: $\quad F_3 + F_4 = 2 \cdot R \cdot \pi \cdot \Delta x \cdot k_Q \cdot (T(t,x) - T_p)$

reaction heat: $\quad Q_R = (-\Delta H_A) \cdot S \cdot \Delta x \cdot r(c)$

heat accumulation: $\quad Q = S \cdot \Delta x \cdot \rho \cdot c_p \cdot T(t,x)$

heat change over time: $\quad \frac{\Delta Q}{\Delta t} = F_1 - F_2 + Q_R - (F_3 + F_4)$

$$\frac{\Delta}{\Delta t}(S \cdot \Delta x \cdot \rho \cdot c_p \cdot T(t,x)) = S \cdot v \cdot \rho \cdot c_p \cdot T(t,x) - S \cdot v \cdot \rho \cdot c_p \cdot T(t, x + \Delta x) + (-\Delta H_A) \cdot S \cdot \Delta x \cdot r(c)$$
$$- 2 \cdot R \cdot \pi \cdot \Delta x \cdot k_Q \cdot (T(t,x) - T_p)$$

$$\frac{\Delta}{\Delta t} T(t,x) = -v \cdot \frac{T(t, x + \Delta x) - T(t,x)}{\Delta x} + \frac{(-\Delta H_A) \cdot r(c)}{\rho \cdot c_p} - \frac{2 \cdot k_Q \cdot (T(t,x) - T_p)}{\rho \cdot c_p \cdot R}$$

$$\lim_{\Delta t \to 0} \left(\frac{\Delta}{\Delta t} T(t,x) \right) = -v \cdot \lim_{\Delta x \to 0} \left(\frac{T(t, x + \Delta x) - T(t,x)}{\Delta x} \right) + \frac{(-\Delta H_A) \cdot r(c)}{\rho \cdot c_p} - \frac{2 \cdot k_Q \cdot (T(t,x) - T_p)}{\rho \cdot c_p \cdot R}$$

$$\frac{\partial}{\partial t} T(t,x) = -v \cdot \frac{\partial T(t,x)}{\partial x} + \frac{(-\Delta H_A) \cdot r(c)}{\rho \cdot c_p} - \frac{2 \cdot k_Q \cdot (T(t,x) - T_p)}{\rho \cdot c_p \cdot R}$$

Initial conditions: $\qquad T(t = 0, x) = T_0(x)$

Boundary conditions: $\qquad T(t, x = 0) = T_i(t)$

Steady-state balances: $\qquad -v \cdot \frac{\partial c}{\partial x} - A \cdot e^{-\frac{E_a}{R \cdot T}} \cdot c = 0$

$$-v \cdot \frac{\partial T}{\partial x} + \frac{(-\Delta H_A)}{\rho \cdot c_p} \cdot A \cdot e^{-\frac{E_a}{R \cdot T}} \cdot c - \frac{2 \cdot k_Q \cdot (T - T_p)}{\rho \cdot c_p \cdot R} = 0$$

(a) Balances for the mass and heat change over the time in a tubular reactor in dimensionless form

mass change steady-state concentration: $-v \cdot \frac{\partial c}{\partial x} - A \cdot e^{-\frac{E_a}{R \cdot T}} \cdot c = 0$

relative concentration: $\quad y = \frac{c(t,x)}{c_i}$ $\qquad\qquad$ relative time: $\qquad \tau = \frac{t}{L/v} = \frac{t}{\tau_0}$

relative distance: $\qquad \xi = \frac{x}{L}$ $\qquad\qquad\quad$ relative temperature: $\quad \theta = \frac{T(t,x)}{T_R}$

$$-v \cdot \frac{c_i \cdot \partial \left(\frac{c}{c_i} \right)}{L \cdot \partial \left(\frac{x}{L} \right)} - A \cdot e^{-\frac{E_a}{R \cdot T}} \cdot c_i \cdot e^{-\frac{E_a}{R \cdot T_R} \left(\frac{1}{\theta} - 1 \right)} \cdot \frac{c}{c_i} = 0$$

$$-\frac{\partial y}{\partial \xi} - Da \cdot e^{-\varepsilon \left(\frac{1}{\theta} - 1 \right)} \cdot y = 0$$

Damkohler number: $\qquad\qquad\qquad\qquad\qquad$ Arrhenius number:

$$Da = \tau \cdot A \cdot e^{-\frac{E_a}{R \cdot T_0}} \qquad\qquad\qquad\qquad\qquad \varepsilon = \frac{E_a}{R \cdot T_0}$$

$$-v \cdot \frac{T_R}{L} \cdot \frac{T_i \cdot \partial \left(\frac{T}{T_R} \right)}{\partial \left(\frac{x}{L} \right)} + \frac{(-\Delta H_A) \cdot c_i}{\rho \cdot c_p} A \cdot e^{-\frac{E_a}{R \cdot T}} \cdot e^{-\frac{E_a}{R \cdot T_R} \left(\frac{1}{\theta} - 1 \right)} \cdot \frac{c}{c_i} - \frac{2 \cdot k_Q \cdot T_R}{R \cdot \rho \cdot c_p} \cdot \left(\frac{T}{T_R} - \frac{T_p}{T_R} \right) = 0$$

$$-\frac{\partial \theta}{\partial \xi} + \beta \cdot Da \cdot e^{-\varepsilon \left(\frac{1}{\theta} - 1 \right)} \cdot y - St \cdot (\theta - \theta_P) = 0$$

Dimensionless reaction temperature: $\qquad\qquad\qquad$ Stanton number:

$$\beta = \frac{(-\Delta H_A) \cdot c_i}{\rho \cdot c_p \cdot T_R} \qquad\qquad\qquad\qquad\qquad St = \frac{2 \cdot k_Q \cdot L}{R \cdot \rho \cdot c_p \cdot v}$$

Balances expressed with the conversion:

$$\frac{\partial x}{\partial \tau} = -\frac{\partial x}{\partial \xi} + Da \cdot e^{-\varepsilon \left(\frac{1}{\theta} - 1 \right)} \cdot (1 - x) = 0$$

$$\frac{\partial \theta}{\partial \tau} = -\frac{\partial \theta}{\partial \xi} + \beta \cdot Da \cdot e^{-\varepsilon \left(\frac{1}{\theta} - 1 \right)} \cdot (1 - x) - St \cdot (\theta - 1) = 0$$

Initial conditions: $x(\tau = 0, \xi) = x_0(\xi)$ $\qquad\qquad$ Boundary conditions: $x(\tau, \xi = 0) = 0$

$\qquad\qquad\qquad\quad \theta(\tau = 0, \xi) = \theta_0(\xi)$ $\qquad\qquad\qquad\qquad\qquad\qquad\quad \theta(\tau, \xi = 0) = \theta_i(\tau)$

Calculation of dimensionless numbers using Wolfram Mathematica 10.0.

$V = 1500 \times 10^{-3}$; (*reactor length*)
$q = 200 \times 10^{-3}/60$; $L = N[V/(r^2 \times P_i)]$;
$r = 0.4$; (*average residence time*)
$c_i = 10 \times 10^3$; $\tau = V/q$;
$R = 8.314$; (*Damkohler number*)
$T_i = 40 + 273.15$; $Da = \tau \times k_0 \times Exp[-(Ea/(R \times T_0))]$;
$T_0 = 130 + 273.15$; (*dimensionless reaction temperature*)
$kQ = 5.2$; $\beta = (\Delta H \times ci)/(\rho \times c_p \times T_0)$;
$p = 1200$; (*Arrhenius number*)
$c_p = 4.5 \times 10^3$; $\epsilon = Ea/(R \times T_0)$;
$k_0 = 3.6 \times 10^3$; (*Stantons number*)
$Ea = 45 \times 10^3$; $St = (kQ \times 2 \times r \times Pi \times L)/(q \times c_p \times \rho)$;
$\Delta H = 80 \times 10^3$;

Print["reactor length = ",L]
Print["average residence time = ",τ]
Print["Damkohler number = ",Da]
Print["dimensionless reaction temperature = ",β]
Print["Arrhenius number = ",ϵ]
Print["Stantions number = ",St]

reactor length = 2.98416
average residence time = 450
Damkohler number = 2.39234
dimensionless reaction temperature = 0.367476
Arrhenius number = 13.4257
Stantions number = 0.00216667

(c) Distribution of conversion and temperature over the time and reactor length
model = NDSolve[{$x'[\xi] == Da \times Exp[-\epsilon \times (1/\theta[\xi]-1)] \times (1-x[\xi])$,
$\quad \theta'[\xi] == -St \times (\theta[\xi]-1)+\beta \times Da \times Exp[-\epsilon \times (1/\theta[\xi]-1)] \times (1-x[\xi])$,
$\quad x[0] == 0$,
$\quad \theta[0] == T_i/T_0\}, \{x,\theta\},\{\xi,0,1\}]$

$kx[\xi\_] := x[\xi]/.$model
$k\theta[\xi\_] := \theta[\xi]/.$model
Plot[{$kx[\xi],k\theta[\xi]$},{$\xi,0,1$},PlotRange->
All,PlotStyle->
{RGBColor[1,0,0],RGBColor[0,0,1]}]

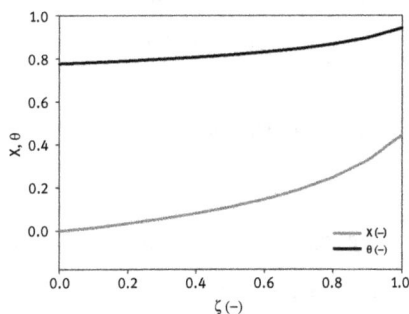

Figure (e): Distribution of conversion and temperature

(d) Calculation of the conversion, concentration of the toluene and the temperature at the reactor outflow.

xoutflow = $kx[1]$;

coutflow = $ci\times(1-kx[1])/1000$;

Toutflow = $k\theta[1]\times To$;

Print["conversion at the reactor outflow = ",xoutflow]

Print["concentration at the reactor outflow = ",coutflow]

Print["temperature at the reactor outflow = ",Toutflow]

conversion at the reactor outflow = {0.443806}

concentration at the reactor outflow = {5.56194}

temperature at the reactor outflow = {379.048}

10.4 Balance equations – Solving mathematical models

As mentioned, universal tool for modeling are equations which are obtained by the balances for mass, energy or impulse that entered and exited the observed system in differential time and volume and generation and consumption within the observed system (Figure 10.8). They can be either *differential* or *algebraic* equations or in matrix-vector form. When talking about mathematical modeling of processes, differential equations are mostly used because algebraic equations are just not sufficient enough to describe all connections between inlet and outlet variable and all that is going on in the system. What type of equation will be used depends on (Figure 10.8):

- dependent variables and their dependence on spatial position (lumped models or distributed models),
- time dependence (stationary or dynamic),
- purpose of modeling and
- on how complex the process is.

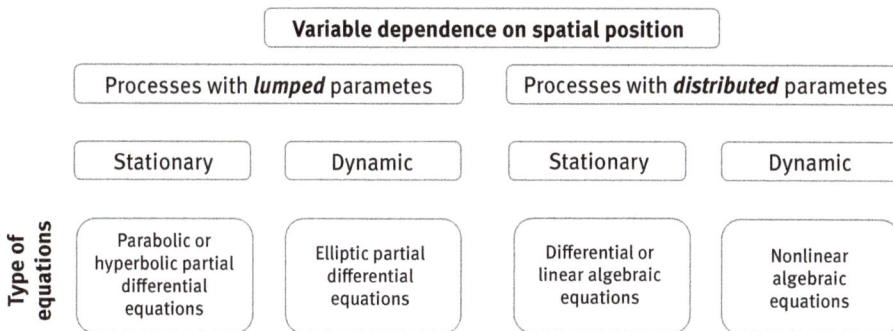

Variable dependence on spatial position			
Processes with *lumped* parametes		Processes with *distributed* parametes	
Stationary	Dynamic	Stationary	Dynamic
Parabolic or hyperbolic partial differential equations	Elliptic partial differential equations	Differential or linear algebraic equations	Nonlinear algebraic equations

Type of equations

Figure 10.8: Types of equations used for describing mathematical models of the process.

In order to solve different types of equations, there are again different mathematical/numerical approaches. In this chapter, only list of methods for solving them with short explanation together with some examples will be given due to many examples available.

To solve linear equations, methods can be divided into:

- *direct* – that provide exact solution
  - Gauss elimination (solving principle is demonstrated in Example 10.4)
    - Equations are written in matrix form and the method is based on a sequence of operations performed on the corresponding matrix of coefficients.

    General form:

$$\begin{pmatrix} a_{11} & a_{12} & a_{13} & \cdots & a_{1n} \\ a_{21} & a_{22} & a_{23} & \cdots & a_{2n} \\ a_{31} & a_{32} & a_{33} & \cdots & a_{3n} \\ \cdots & \cdots & \cdots & \cdots & \cdots \\ a_{n1} & a_{n2} & a_{n3} & \cdots & a_{nn} \end{pmatrix} \cdot \begin{pmatrix} x_1 \\ x_2 \\ x_3 \\ \cdots \\ x_n \end{pmatrix} = \begin{pmatrix} b_1 \\ b_2 \\ b_3 \\ \cdots \\ b_n \end{pmatrix}$$

  - Gauss–Jordan method
    - Elements of transformations in Gauss elimination are expend until a matrix is convert into reduced row echelon form
- *iterative/indirect* – that start with assumption of solution and then compute iteratively a sequence of solutions that get closer and closer to the final solution
  - Gauss–Seidle method or Liebmann method or the method of successive displacement
    - In each iteration, a possible solution to the system with a particular error is obtained. Method is then repeated again until the solution is more accurate and error smaller. If the system does not converge and with every next step error is growing, the system cannot be solved by this method
  - Jacobi method
    - is an algorithm for determining the solutions of a diagonally dominant system of linear equations. Each diagonal element is solved for, and an approximate value is plugged in. The process is then iterated until it converges.

Example 10.4.

Degradation of chemical contamination of component A to component C is three steps reaction performed in batch reactor and it follows the mechanism:

A ⇌ B   (1)
B → C   (2)
C → B   (3)

All reactions follow first-order kinetics ($r = k \cdot c$) where reaction rates are: $k_{A \leftrightarrow B} = 5$ mol/L, $k_{B \leftrightarrow A} = 7$ mol/L, $k_{B \leftrightarrow C} = 3$ mol/L and $k_{C \leftrightarrow B} = 2$ mol/L. If, after 5 h, accumulations of components A, B and C are:

$\frac{dc_A}{dt} = 18 \text{ mol/L} \cdot h$

$\frac{dc_B}{dt} = -28 \text{ mol/L} \cdot h$

$\frac{dc_C}{dt} = 10 \text{ mol/L} \cdot h$

define mathematical model of the process and calculate numerical concentration off all components in that moment using Gauss elimination method.

Process system:

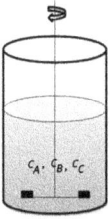

– Batch reactor
– first-order reaction

$A \rightleftarrows B$ (1)
$B \rightarrow C$ (2)
$C \rightarrow B$ (3)

Figure (a): Batch reactor.
Purpose/goal of modeling:
To calculate numerical concentration of components A, B and C 5 h after the reaction started.
Defining the mechanisms:
Chemical reactions (A $\rightleftarrows$ B, B$\leftrightarrow$C and C$\leftrightarrow$B), first order kinetics ($r = k \cdot c$)
Information's about the process system:
$k_{A\leftrightarrow B} = 5 \text{ mol/L}$, $k_{B\leftrightarrow A} = 7 \text{ mol/L}$, $k_{B\leftrightarrow C} = 3 \text{ mol/L}$ and $k_{C\leftrightarrow B} = 2 \text{ mol/L}$ $t = 5$ h

$\frac{dc_A}{dt} = 18 \text{ mol/L} \cdot h$

$\frac{dc_B}{dt} = -28 \text{ mol/L} \cdot h$

$\frac{dc_C}{dt} = 10 \text{ mol/L} \cdot h$

Model assumptions:
– ideal mixing
– V = const.; T = const.; Δ = const.

Mass balances:

$\frac{dc_A}{dt} = -k_{A \rightarrow B} \cdot c_A + k_{B \rightarrow A} \cdot c_B$

$\frac{dc_B}{dt} = k_{A \rightarrow B} \cdot c_A - k_{B \rightarrow A} \cdot c_B - k_{B \rightarrow C} \cdot c_B + k_{C \rightarrow B} \cdot c_C$

$\frac{dc_C}{dt} = -k_{C \rightarrow B} \cdot c_C$

Gauss elimination method:

$\frac{dc_A}{dt} = -k_{A \rightarrow B} \cdot c_A + k_{B \rightarrow A} \cdot c_B = -5 \cdot c_A + 7 \cdot c_B$

$\frac{dc_B}{dt} = k_{A \rightarrow B} \cdot c_A - k_{B \rightarrow A} \cdot c_B - k_{B \rightarrow C} \cdot c_B + k_{C \rightarrow B} \cdot c_C$

$\quad\quad = 5 \cdot c_A - 7 \cdot c_B - 3 \cdot c_B + 2 \cdot c_C = 5 \cdot c_A - 10 \cdot c_B + 2 \cdot c_C$

$\frac{dc_C}{dt} = k_{B \rightarrow C} \cdot c_B - k_{C \rightarrow B} \cdot c_C = 3 \cdot c_B - 2 \cdot c_C$

Equation 1: $-5 \cdot c_A + 7 \cdot c_B + 0 \cdot c_C = 18$

Equation 2: $5 \cdot c_A - 10 \cdot c_B + 2 \cdot c_C = -28$

Equation 3: $0 \cdot c_A + 3 \cdot c_B - 2 \cdot c_C = 10$

$$\begin{pmatrix} -5 & 7 & 0 \\ 5 & -10 & 2 \\ 0 & 3 & -2 \end{pmatrix} \begin{pmatrix} c_A \\ c_B \\ c_C \end{pmatrix} = \begin{pmatrix} 18 \\ -28 \\ 10 \end{pmatrix} \Rightarrow \begin{pmatrix} -5 & 7 & 0 & 18 \\ 5 & -10 & 2 & -28 \\ 0 & 3 & -2 & 10 \end{pmatrix}$$

Elements of the second row reduce by the values of the first.

$$\begin{pmatrix} -5 & 7 & 0 & 18 \\ 0 & -3 & 2 & -10 \\ 0 & 3 & -2 & 10 \end{pmatrix}$$

Elements of the third row reduce by the values of the second.

$$\begin{pmatrix} -5 & 7 & 0 & 18 \\ 0 & -3 & 2 & -10 \\ 0 & 0 & 0 & 0 \end{pmatrix}$$

Elements of the first row divide by −5 and the second row by −3.

$$\begin{pmatrix} 1 & -1.4 & 0 & -3.6 \\ 0 & 1 & -0.67 & 3.33 \\ 0 & 0 & 0 & 0 \end{pmatrix} \Rightarrow \begin{pmatrix} 1 & -1.4 & 0 \\ 0 & 1 & -0.67 \\ 0 & 0 & 0 \end{pmatrix} \begin{pmatrix} c_A \\ c_B \\ c_C \end{pmatrix} = \begin{pmatrix} 3.6 \\ 3.33 \\ 0 \end{pmatrix}$$

Written in algebraic form:

$c_C = 0$ mol/L

$c_B = 3.33$ mol/L

$c_A = -3.6 + 1.4 \cdot c_B = -3.6 + 4.66 = 1.06$ mol/L

For solving non-linear equations mostly used methods (all iterative) are:
- *iterative/indirect*
  - Jacobi non-linear iteration
  - Wegstein method
  - Newton–Raphson method or Newton's method
  - Regula–Falsi method or false position Method

For solving differential equations mostly used methods are:
- *Analytical methods*
  - n-order differential equations that have n constants and n boundary conditions are solved simply by solving mathematical equations. Analytical solution is the exact solution of differential equation and it satisfies differential equation and boundary conditions.
- *Numerical methods*
  - Taylor meth
  - od is based on the Taylor series that is a representation of a function as an infinite sum of terms that are calculated form the values of the function's derivatives at a single point.

    Simple differential equation $\frac{dy(t)}{dt} = f(t, y(t))$, $y(0) = y_0$ can be approximated with Taylor series as

    $$y(t + \Delta t) = y(t) + \Delta t \cdot y'(t) + \tfrac{1}{2} \cdot \Delta t^2 \cdot y''(t) + \dots + \tfrac{1}{n!} \cdot \Delta t^n \cdot y^{(n)}(\tau)$$

    - Euler method is based on the approximation of the function with the first component of the Taylor series (solving principle is demonstrated in Example 10.5)
    - Runge method (solving principle is demonstrated in Example 10.5)

– Runge–Kutta method also based on the Taylor series but more suitable for the solving of the differential equations than Euler method (solving principle is demonstrated in Example 10.5)

Example 10.5.

Degradation of contaminant A (A⟷P) follows second-order kinetics ($r = k \cdot c^2$) and it is performed in a batch rector. If the reaction rate constant of component A degradation is $k = 2$ 1/h, and initial component concentrations is $c_{A,0} = 1$ mol/L and $c_{P,0} = 0$ mol/L calculate change in concentration of both components using:

a. Euler method

b. Runge method

c. Runge–Kutta method

Develop dynamic mathematical model for both components. Calculate the first three steps of each method; define method step and all the assumptions necessary for solving this problem.

Process system:

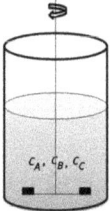

– Batch reactor
– second-order kinetics
A → P

c_A, c_B, c_C

Figure (a): Batch reactor.

Purpose/goal of modeling:

To calculate concentration of components A and C

Defining the mechanisms:

Chemical reactions (A⟷P), second-order kinetics ($r = k \cdot c^2$)

Information's about the process system:

$k = 2$ 1/h, $c_{A,0} = 1$ mol/L, $c_{P,0} = 0$ mol/L

Model assumptions:

– ideal mixing

– V = const.; T = const.; Δ = const.

Mass balances: $\frac{dc_A}{dt} = -k \cdot c_A \Rightarrow y'(c_{A,0}) = -k \cdot c_A$; $\quad \frac{dc_P}{dt} = k \cdot c_A \Rightarrow y'(c_{P,0}) = k \cdot c_A$

Euler method: $h = 0.1$

$t_0 = 0$

$c_{A,0} = 1$ mol/L

$c_{P,0} = 0$ mol/L

$t_1 = t_0 + h = 0 + 0.1 = 0.1$ h

$c_{A,1} = c_{A,0} + h \cdot y'(c_{A,0}) = 1 + 0.1 \cdot (-2 \cdot 1) = 0.8$ mol/L

$c_{P,1} = c_{P,0} + h \cdot y'(c_{P,0}) = 0 + 0.1 \cdot (2 \cdot 1) = 0.2$ mol/L

$t_2 = t_1 + h = 0.1 + 0.1 = 0.2$ h

$c_{A,2} = c_{A,1} + h \cdot y'(c_{A,1}) = 0.8 + 0.1 \cdot (-2 \cdot 1) = 0.64$ mol/L

$c_{P,2} = c_{P,1} + h \cdot y'(c_{P,1}) = 0.2 + 0.1 \cdot (2 \cdot 1) = 0.36$ mol/L

Runge method: h = 0.1

$t_0 = 0$

$c_{A,0} = 1$ mol/L

$c_{P,0} = 0$ mol/L

$t_1 = t_0 + h = 0 + 0.1 = 0.1$ h

for $c_{A,1}$:

$g_{1,1}^{c_A} = y(t_0, c_{A,0}) = -k \cdot c_{A,0}$

$\quad = -2 \cdot 1 = -2$

$y_{1-1/2}^{c_A} = c_{A,0} + \frac{h}{2} \cdot g_{1,1}^{c_A} = 1 + \frac{0.1}{2} \cdot (-2) = 0.9$

$g_{2,1}^{c_A} = y(t_0 + \frac{h}{2}, y_{1-1/2}^{c_A}) = -k \cdot y_{1-1/2}^{c_A}$

$\quad = -2 \cdot 0.9 = -1.8$

$c_{A,1} = c_{A,0} + h \cdot g_{2,1}^{c_A}$

$\quad = 1 + 0.1 \cdot (-1.8) = \mathbf{0.82}$ **mol/L**

$t_2 = t_1 + h = 0 + 0.1 = 0.2$ h

for $c_{A,2}$:

$g_{1,2}^{c_A} = y(t_1, c_{A,1}) = -k \cdot c_{A,1}$

$\quad = -2 \cdot 0.82 = -1.64$

$y_{2-1/2}^{c_A} = c_{A,1} + h2 \cdot g_{1,2}^{c_A}$

$\quad = 0.82 + \frac{0.1}{2} \cdot (-1.64) = 0.738$

$g_{2,2}^{c_A} = y\left(t_1 + \frac{h}{2}, y_{2-1/2}^{c_A}\right) = -k \cdot y_{2-1/2}^{c_A}$

$\quad = -2 \cdot 0.738 = -1.476$

$c_{A,2} = c_{A,1} + h \cdot g_{2,2}^{c_A}$

$\quad = 0.82 + 0.1 \cdot (-1.476) = \mathbf{0.6724}$ **mol/L**

for $c_{P,1}$:

$g_{1,1}^{c_P} = y(t_0, c_{P,0})$

$\quad = k \cdot c_{A,0} = 2 \cdot 1 = 2$

$y_{1-1/2}^{c_P} = c_{P,0} + \frac{h}{2} \cdot g_{1,1}^{c_P}$

$\quad = 0 + \frac{0.1}{2} \cdot 2 = 0.1$

$g_{2,1}^{c_P} = y(t_0 + \frac{h}{2}, y_{1-1/2}^{c_A})$

$\quad = k \cdot y_{1-1/2}^{c_A} = 2 \cdot 0.9 = 1.8$

$c_{P,1} = c_{P,0} + h \cdot g_{2,1}^{c_P}$

$\quad = 0 + 0.1 \cdot 1.8 = \mathbf{0.18}$ **mol/L**

for $c_{P,2}$:

$g_{1,2}^{c_P} = y(t_1, c_{P,1})$

$\quad = k \cdot c_{A,1} = 2 \cdot 0.82 = 1.64$

$y_{2-1/2}^{c_P} = c_{P,1} + \frac{h}{2} \cdot g_{1,2}^{c_P}$

$\quad = 0.18 + \frac{0.1}{2} \cdot 1.64 = 0.262$

$g_{2,2}^{c_P} = y(t_0 + \frac{h}{2}, y_{2-1/2}^{c_A})$

$\quad = k \cdot y_{1-1/2}^{c_A} = 2 \cdot 0.738 = 1.476$

$c_{P,2} = c_{P,1} + h \cdot g_{2,2}^{c_P}$

$\quad = 0.18 + 0.1 \cdot 1.476 = \mathbf{0.3276}$ **mol/L**

Runge–Kutta method: h = 0.1

$t_0 = 0$

$c_{A,0} = 1$ mol/L

$c_{P,0} = 0$ mol/L

$t_1 = t_0 + h = 0 + 0.1 = 0.1$ h

for $c_{A,1}$:

$g_{1,1}^{c_A} = h \cdot f(t_0, c_{A,0}) = h \cdot (-k \cdot c_{A,0}) = 0.1 \cdot (-2 \cdot 1) = -0.2$

$g_{2,1}^{c_A} = h \cdot f\left(t_0 + \frac{h}{2}, c_{A,0} + \frac{g_{1,1}^{c_A}}{2}\right) = h \cdot (-k \cdot (c_{A,0} + \frac{g_{1,1}^{c_A}}{2})) = 0.1 \cdot (-2 \cdot (1 + \frac{-0.2}{2})) = -0.18$

$g_{3,1}^{c_A} = h \cdot f\left(t_0 + \frac{h}{2}, c_{A,0} + \frac{g_{2,1}^{c_A}}{2}\right) = h \cdot (-k \cdot (c_{A,0} + \frac{g_{2,1}^{c_A}}{2})) = 0.1 \cdot (-2 \cdot (1 + \frac{-0.18}{2})) = -0.182$

$g_{4,1}^{c_A} = h \cdot f\left(t_0 + h, c_{A,0} + g_{3,1}^{c_A}\right) = h \cdot (-k \cdot (c_{A,0} + g_{3,1}^{c_A})) = 0.1 \cdot (-2 \cdot (1 + (-0.182)) = -0.1636$

$c_{A,1} = c_{A,0} + \frac{1}{6} \cdot (g_{1,1}^{c_A} + 2 \cdot g_{2,1}^{c_A} + 2 \cdot g_{3,1}^{c_A} + g_{4,1}^{c_A}) = \mathbf{0.8188}$ **mol/L** *for $c_{P,1}$:*

$g_{1,1}{}^{c_P} = h \cdot f(t_0, c_{A,0}) = h \cdot k \cdot c_{A,0} = 0.1 \cdot 2 \cdot 1 = 0.2$

$g_{2,1}{}^{c_P} = h \cdot f\left(t_0 + \frac{h}{2}, c_{A,0} + \frac{g_{1,1}^{c_A}}{2}\right) = h \cdot k \cdot \left(c_{A,0} + \frac{g_{1,1}^{c_A}}{2}\right) = 0.1 \cdot 2 \cdot \left(1 + \frac{-0.2}{2}\right) = 0.18$

$g_{3,1}{}^{c_P} = h \cdot f\left(t_0 + \frac{h}{2}, c_{A,0} + \frac{g_{2,1}^{c_A}}{2}\right) = h \cdot k \cdot \left(c_{A,0} + \frac{g_{2,1}^{c_A}}{2}\right) = 0.1 \cdot 2 \cdot \left(1 + \frac{-0.182}{2}\right) = 0.182$

$g_{4,1}{}^{c_A} = h \cdot f\left(t_0 + h, c_{A,0} + g_{3,1}^{c_A}\right) = h \cdot k \cdot \left(c_{A,0} + g_{3,1}^{c_A}\right) = 0.1 \cdot 2 \cdot (1 + (-0.182)) = 0.1636$

$$c_{P,1} = c_{P,0} + \tfrac{1}{6} \cdot \left(g_{1,1}{}^{c_P} + 2 \cdot g_{2,1}{}^{c_P} + 2 \cdot g_{3,1}{}^{c_P} + g_{4,1}{}^{c_P}\right) = \textbf{0.1813 mol/L}$$

$t_2 = t_1 + h = 0 + 0.1 = 0.2 \text{ h}$

for $c_{A,2}$:

$g_{1,2}{}^{c_A} = h \cdot f(t_1, c_{A,1}) = h \cdot (-k \cdot c_{A,0}) = 0.1 \cdot (-2 \cdot 0.8188) = -0.1638$

$g_{2,2}{}^{c_A} = h \cdot f\left(t_1 + \frac{h}{2}, c_{A,1} + \frac{g_{1,2}^{c_A}}{2}\right) = h \cdot \left(-k \cdot \left(c_{A,1} + \frac{g_{1,2}^{c_A}}{2}\right)\right)$

$= 0.1 \cdot \left(-2 \cdot \left(0.8188 + \frac{-0.1638}{2}\right)\right) = -0.1474$

$g_{3,2}{}^{c_A} = h \cdot f\left(t_1 + \frac{h}{2}, c_{A,1} + \frac{g_{2,2}^{c_A}}{2}\right) = h \cdot \left(-k \cdot \left(c_{A,1} + \frac{g_{2,2}^{c_A}}{2}\right)\right)$

$= 0.1 \cdot \left(-2 \cdot \left(0.8188 + \frac{-0.1474}{2}\right)\right) = -0.1490$

$g_{4,2}{}^{c_A} = h \cdot f\left(t_1 + h, c_{A,1} + g_{3,2}^{c_A}\right) = h \cdot \left(-k \cdot \left(c_{A,1} + g_{3,2}^{c_A}\right)\right)$

$= 0.1 \cdot (-2 \cdot (0.8188 + (-0.1490))) = -0.1340$

$$c_{A,2} = c_{A,1} + \frac{1}{6} \cdot \left(g_{1,2}{}^{c_A} + 2 \cdot g_{2,2}{}^{c_A} + 2 \cdot g_{3,2}{}^{c_A} + g_{4,2}{}^{c_A}\right) = \textbf{0.6704 mol/L}$$

for $c_{P,2}$:

$g_{1,2}{}^{c_P} = h \cdot f(t_1, c_{A,1}) = h \cdot k \cdot c_{A,1} = 0.1 \cdot 2 \cdot 0.8188 = 0.1638$

$g_{2,2}{}^{c_P} = h \cdot f\left(t_1 + \frac{h}{2}, c_{A,1} + \frac{g_{1,2}^{c_A}}{2}\right) = h \cdot k \cdot \left(c_{A,1} + \frac{g_{1,2}^{c_A}}{2}\right) = 0.1 \cdot 2 \cdot \left(0.8188 + \frac{-0.1638}{2}\right) = 0.1474$

$g_{3,2}{}^{c_P} = h \cdot f\left(t_1 + \frac{h}{2}, c_{A,1} + \frac{g_{2,2}^{c_A}}{2}\right) = h \cdot k \cdot \left(c_{A,1} + \frac{g_{2,2}^{c_A}}{2}\right) = 0.1 \cdot 2 \cdot \left(0.8188 + \frac{-0.1474}{2}\right) = 0.1490$

$g_{4,2}{}^{c_A} = h \cdot f\left(t_1 + h, c_{A,1} + g_{3,3}^{c_A}\right) = h \cdot k \cdot \left(c_{A,1} + g_{3,2}^{c_A}\right) = 0.1 \cdot 2 \cdot (0.8188 + (-0.1490)) = 0.134$

$$c_{P,2} = c_{P,1} + \frac{1}{6} \cdot \left(g_{1,2}{}^{c_P} + 2 \cdot g_{2,2}{}^{c_P} + 2 \cdot g_{3,2}{}^{c_P} + g_{4,2}{}^{c_P}\right) = \textbf{0.3297 mol/L}$$

For solving partial differential equations mostly the used methods are:

- *Numerical methods* – Transport phenomena are mostly described with mathematical models in the form of partial differential equations. The aim of solving partial differential equations is finding the function and boundary conditions that satisfy the relationships between different derivations in specific space and/ or time [12]. It is very complicated to get the solution that satisfies differential equations for analyzed dimension and that is why numerical methods based on the discrimination are used. Most frequently used are: method of lines, FDM, FEM, FVM and method of collocations.

- *Method of lines* (solving principle is demonstrated in Example 10.6) is semi-discretization method. It is based on the discretization in one dimension. After the discretization, the obtained system of differential equation is solved by using one of the methods for the solution of the differential equations.
- *The FDM* (solving principle is demonstrated in Example 10.6) is based on the approximation of the derivates with the series of algebraic equations using the Taylor's polynomial [13]. Precision of the method is enlarged by using more components of the Taylor's polynomial and by reducing the distance between the nods of the discretization. Using the described procedure continuous variables are replaced by discrete variables and the obtained solution is not continuous for the whole analyzed domain but values for the discrete nods (points) are obtained. For nonlinear systems after the discretization, the series of the nonlinear algebraic equations is obtained that can be solved using Newton method. Finite difference discretization can be performed: (i) approximation in three points forward, (ii) approximation in three points backward and (iii) approximation in three central points.
- *FEM* is used for the solution of the problems of the field. It cuts a structure into several elements (pieces of the structure) and then reconnects elements at nodes as if nodes were pins that hold elements together. This procedure results in a set of simultaneous algebraic equations.
- *FVM* is a discretization technique for partial differential equations that are developed from physical conservation laws. This method uses a volume integral formulation of the problem with a finite partitioning set of volumes to discretize the equations. FVM is in common use for discretizing computational fluid dynamics equations [14].

Example 10.6.

In case of the discharge of the pollutants in the river, organic compounds are degraded by aerobic microorganisms presented in the river. The organic compound consumption causes dissolved oxygen concentration reduction in the river. Oxygen consumption takes place parallel with the oxygen transfer from the gas phase to the liquid through interfacial area. With the assumption that the river estuary can be modeled as the ideal tubular reactor analyze the change of the organic compound concentrations and dissolved oxygen concentration over the time and position. The organic compound degradation can be described with the Monod kinetics. Parameters of the Monod kinetics are: maximum specific growth rate $\mu = 0.05$ 1/h and Monod constant $K_S = 0.0001$ g/L. Average river velocity is $v = 7500$ dm/h. Ratio of the organic compound degradation over the oxygen consumption is $Y_{S/O} = 0.2$ and ratio of the biomass growth over the organic compound degradation is $Y_{X/S} = 1$. Initial concentration of the organic compounds is $c_{S0} = 0.04$ g/L, the initial concentration of the biomass is $c_{X,0} = 0.002$ g/L, oxygen saturation concentration is $c_{0,s} = 0.0087$ g/L and oxygen transfer rate is $k_L a = 0.025$ 1/h.

a. Develop the balances for the organic compound change, dissolved oxygen concentration change and biomass concentration change over the time and position.

b. Solve the balances using the method of lines.

(a) Balances for the change of organic compound concentration, dissolved oxygen concentration and biomass concentration over the time and position.

General expression for the balances in ideal the ideal tubular reactor: $\frac{\partial y(t,x)}{\partial t} = -v \cdot \frac{\partial y(t,x)}{\partial x} \pm r(y(t,x))$

Organic compounds concentration change: $\frac{\partial y_S(t,x)}{\partial t} = -v \cdot \frac{\partial y_S(t,x)}{\partial x} - \frac{1}{Y_{X/S}} \mu_{max} \cdot \frac{y_S(t,x)}{K_S + y_S(t,x)} \cdot y_X(t,x)$

Biomass concentration change: $\frac{\partial y_X(t,x)}{\partial t} = -v \cdot \frac{\partial y_X(t,x)}{\partial x} + \mu_{max} \cdot \frac{y_S(t,x)}{K_S + y_S(t,x)} \cdot y_X(t,x)$

Dissolved oxygen concentration change:

$\frac{\partial y_O(t,x)}{\partial t} = -v \cdot \frac{\partial y_O(t,x)}{\partial x} + k_L a \cdot (y_{0S} - y_0) - \frac{1}{Y_{S/O}} \cdot \mu_{max} \cdot \frac{y_S(t,x)}{K_S + y_S(t,x)} \cdot y_X(t,x)$

(b) Solution of the balances using the method of lines.

Discretization of the first derivation: $\frac{\partial y(t,x)}{\partial x} = \frac{y(t,x+h) - y(t,x)}{h}$

Step of the discretization: $h = 0.1$

Discretised form of balances:

$\frac{\partial y_{S,0}}{\partial t} = 0$

$\frac{\partial y_{S,1}}{\partial t} = -v \cdot \frac{y_{S,1} - y_{S,0}}{h} - \frac{1}{Y_{X/S}}$
$\cdot \mu_{max} \cdot \frac{y_{S,1}}{K_S + y_{S,1}} \cdot y_{X,1}$

$\frac{\partial y_{S,2}}{\partial t} = -v \cdot \frac{y_{S,2} - y_{S,1}}{h} - \frac{1}{Y_{X/S}}$
$\cdot \mu_{max} \cdot \frac{y_{S,2}}{K_S + y_{S,2}} \cdot y_{X,2}$

$\vdots$

$\frac{\partial y_{S,10}}{\partial t} = -v \cdot \frac{y_{S,10} - y_{S,9}}{h} - \frac{1}{Y_{X/S}}$
$\cdot \mu_{max} \cdot \frac{y_{S,10}}{K_S + y_{S,10}} \cdot y_{X,10}$

$\frac{\partial y_{X,0}}{\partial t} = 0$

$\frac{\partial y_{X,1}}{\partial t} = -v \cdot \frac{y_{X,1} - y_{X,0}}{h} + \mu_{max} \cdot \frac{y_{S,1}}{K_S + y_{S,1}} \cdot y_{X,1}$

$\frac{\partial y_{X,2}}{\partial t} = -v \cdot \frac{y_{X,2} - y_{X,1}}{h} + \mu_{max} \cdot \frac{y_{S,1}}{K_S + y_{S,1}} \cdot y_{X,2}.$

$\vdots$

$\frac{\partial y_{X,10}}{\partial t} = -v \cdot \frac{y_{X,10} - y_{X,9}}{h} + \mu_{max} \cdot \frac{y_{S,1}}{K_S + y_{S,1}} \cdot y_{X,10}$

$\frac{\partial y_{O0}}{\partial t} = 0$

$\frac{\partial y_{O,1}}{\partial t} = -v \cdot \frac{y_{O,1} - y_{O,0}}{h} + k_L a \cdot \left(y_{0,S} - y_{0,1}\right) - \frac{1}{Y_{S/O}} \cdot \mu_{max} \cdot \frac{y_{S,1}}{K_S + y_{S,1}} \cdot y_{X,1}$

$\frac{\partial y_{O,2}}{\partial t} = -v \cdot \frac{y_{O,2} - y_{O,1}}{h} + k_L a \cdot \left(y_{0,S} - y_{0,2}\right) - \frac{1}{Y_{S/O}} \cdot \mu_{max} \cdot \frac{y_{S,1}}{K_S + y_{S,1}} \cdot y_{X,2}$

$\vdots$

$\frac{\partial y_{O,10}}{\partial t} = -v \cdot \frac{y_{O,10} - y_{O,9}}{h} + k_L a \cdot \left(y_{0,S} - y_{0,9}\right) - \frac{1}{Y_{S/O}} \cdot \mu_{max} \cdot \frac{y_{S,1}}{K_S + y_{S,1}} \cdot y_{X,10}$

Wolfram Mathematica 10.1. Code

```
μmax = 0.05;
Ks = 0.0001;
Yso = 0.2;
Yxs = 1;
kLa = 0.025;1
cOs = 0.0087;
h = 350;
tf = 100;
model = NDSolve[{cS0[t] == 0,
cS1[t] == -v*(cS1[t]-cS0[t])/h-1/Yxs*μmax*cS1[t]/(Ks+cS1[t])*cX1[t],
cS2[t] == -v*(cS2[t]-cS1[t])/h-1/Yxs*μmax*cS2[t]/(Ks+cS2[t])*cX2[t],
cS3[t] == -v*(cS3[t]-cS2[t])/h-1/Yxs*μmax*cS3[t]/(Ks+cS3[t])*cX3[t],
cS4[t] == -v*(cS4[t]-cS3[t])/h-1/Yxs*μmax*cS4[t]/(Ks+cS4[t])*cX4[t],
cS5[t] == -v*(cS5[t]-cS4[t])/h-1/Yxs*μmax*cS5[t]/(Ks+cS5[t])*cX5[t],
cS6[t] == -v*(cS6[t]-cS5[t])/h-1/Yxs*μmax*cS6[t]/(Ks+cS6[t])*cX6[t],
cS7[t] == -v*(cS7[t]-cS6[t])/h-1/Yxs*μmax*cS7[t]/(Ks+cS7[t])*cX7[t],
```

```
cS8[t] = = -v*(cS8[t]-cS7[t])/h-1/Yxs*µmax*cS8[t]/(Ks+cS8[t])*cX8[t],
cS9[t] = = -v*(cS9[t]-cS8[t])/h-1/Yxs*µmax*cS9[t]/(Ks+cS9[t])*cX9[t],
cS10[t] = = -v*(cS10[t]-cS9[t])/h-1/Yxs*µmax*cS10[t]/(Ks+cS10[t])*cX10[t],
cX0[t] = = 0,
cX1[t] = = -v*(cX1[t]-cX0[t])/h+µmax*cS1[t]/(Ks+cS1[t])*cX1[t],
cX2[t] = = -v*(cX2[t]-cX1[t])/h+µmax*cS2[t]/(Ks+cS2[t])*cX2[t],
cX3[t] = = -v*(cX3[t]-cX2[t])/h+µmax*cS3[t]/(Ks+cS3[t])*cX3[t],
cX4[t] = = -v*(cX4[t]-cX3[t])/h+µmax*cS4[t]/(Ks+cS4[t])*cX4[t],
cX5[t] = = -v*(cX5[t]-cX4[t])/h+µmax*cS5[t]/(Ks+cS5[t])*cX5[t],
cX6[t] = = -v*(cX6[t]-cX5[t])/h+µmax*cS6[t]/(Ks+cS6[t])*cX6[t],
cX7[t] = = -v*(cX7[t]-cX6[t])/h+µmax*cS7[t]/(Ks+cS7[t])*cX7[t],
cX8[t] = = -v*(cX8[t]-cX7[t])/h+µmax*cS8[t]/(Ks+cS8[t])*cX8[t],
cX9[t] = = -v*(cX9[t]-cX8[t])/h+µmax*cS9[t]/(Ks+cS9[t])*cX9[t],
cX10[t] = = -v*(cX10[t]-cX9[t])/h+µmax*cS10[t]/(Ks+cS10[t])*cX10[t],
cO0[t] = = 0,
cO1[t] = = -v*(cO1[t]-cO0[t])/h+kLa*(cOs-cO1[t]) 1/Yso*µmax*cS1[t]/(Ks+cS1[t])*cX1[t],
cO2[t] = = -v*(cO2[t]-cO1[t])/h+kLa*(cOs-cO2[t])-1/Yso*µmax*cS2[t]/(Ks+cS2[t])*cX2[t],
cO3[t] = = -v*(cO3[t]-cO2[t])/h+kLa*(cOs-cO3[t])-1/Yso*µmax*cS3[t]/(Ks+cS3[t])*cX3[t],
cO4[t] = = -v*(cO4[t]-cO3[t])/h+kLa*(cOs-cO4[t])-1/Yso*µmax*cS4[t]/(Ks+cS4[t])*cX4[t],
cO5[t] = = -v*(cO5[t]-cO4[t])/h+kLa*(cOs-cO5[t])-1/Yso*µmax*cS5[t]/(Ks+cS5[t])*cX5[t],
cO6[t] = = -v*(cO6[t]-cO5[t])/h+kLa*(cOs-cO6[t])-1/Yso*µmax*cS6[t]/(Ks+cS6[t])*cX6[t],
cO7[t] = = -v*(cO7[t]-cO6[t])/h+kLa*(cOs-cO7[t])-1/Yso*µmax*cS7[t]/(Ks+cS2[t])*cX7[t],
cO8[t] = = -v*(cO8[t]-cO7[t])/h+kLa*(cOs-cO8[t])-1/Yso*µmax*cS8[t]/(Ks+cS8[t])*cX8[t],
cO9[t] = = -v*(cO9[t]-cO8[t])/h+kLa*(cOs-cO9[t])-1/Yso*µmax*cS9[t]/(Ks+cS9[t])*cX9[t],
cO10[t] = = -v*(cO10[t]-cO9[t])/h+kLa*(cOs-cO10[t])-1/Yso*µmax*cS10[t]/(Ks+cS10[t])*cX10[t],
cS0[0] = = 0.040, cS1[0] = = 0,cS2[0] = = 0,cS3[0] = = 0,cS4[0] = = 0,cS5[0] = = 0, cS6[0] = = 0,cS7[0]
= = 0,cS8[0] = = 0,cS9[0] = = 0,cS10[0] = = 0,cX0[0] = = 0.002, cX1[0] = = 0,cX2[0] = = 0,cX3[0] = = 0,
cX4[0] = = 0,cX5[0] = = 0,cX6[0] = = 0, cX7[0] = = 0,cX8[0] = = 0,cX9[0] = = 0,cX10[0] = = 0,cO0[0]
= = 0.0078,cO1[0] = = 0, cO2[0] = = 0,cO3[0] = = 0,cO4[0] = = 0,cO5[0] = = 0,cO6[0] = = 0,cO7[0] = = 0,
cO8[0] = = 0,cO9[0] = = 0,cO10[0] = = 0},{cS0,cS1,cS2,cS3,cS4,cS5,cS6,cS7,cS8,cS9,cS10,cX0,cX1,
cX2,cX3,cX4,cX5,cX6,cX7,cX8,cX9,cX10,cO0,cO1,cO2,cO3,cO4, cO5,cO6,cO7,cO8,cO9,cO10},{t,0,tf}];
kcS0[t_]: = cS0[t]/.model kcS1[t_]: = cS1[t]/.model
kcS2[t_]: = cS2[t]/.model kcS3[t_]: = cS3[t]/.model
kcS4[t_]: = cS4[t]/.model kcS5[t_]: = cS5[t]/.model
kcS6[t_]: = cS6[t]/.model kcS7[t_]: = cS7[t]/.model
kcS8[t_]: = cS8[t]/.model kcS9[t_]: = cS9[t]/.model
kcS10[t_]: = cS10[t]/.model kcX0[t_]: = cX0[t]/.model
kcX1[t_]: = cX1[t]/.model kcX2[t_]: = cX2[t]/.model
kcX3[t_]: = cX3[t]/.model kcX4[t_]: = cX4[t]/.model
kcX5[t_]: = cX5[t]/.model kcX6[t_]: = cX6[t]/.model
kcX7[t_]: = cX7[t]/.model kcX8[t_]: = cX8[t]/.model
kcX9[t_]: = cX9[t]/.model kcX10[t_]: = cX10[t]/.model
kcO0[t_]: = cO0[t]/.model kcO1[t_]: = cO1[t]/.model
kcO2[t_]: = cO2[t]/.model kcO3[t_]: = cO3[t]/.model
kcO4[t_]: = cO4[t]/.model kcO5[t_]: = cO5[t]/.model
kcO6[t_]: = cO6[t]/.model kcO7[t_]: = cO7[t]/.model
kcO8[t_]: = cO8[t]/.model kcO9[t_]: = cO9[t]/.model
kcO10[t_]: = cO10[t]/.model
Plot[{kcS0[t],kcS1[t],kcS2[t],kcS3[t],kcS4[t],kcS5[t],kcS6[t],kcS7[t],kcS8[t],kcS9[t],kcS10[t]},{t,0,tf},
AxesLabel->{"t/h","γS/gL-1"}]
```

```
Plot[{kcX0[t],kcX1[t],kcX2[t],kcX3[t],kcX4[t],kcX5[t],kcX6[t],kcX7[t],kcX8[t],kcX9[t],kcX10[t]},{t,0,tf},
AxesLabel->{"t/h","γX/gL-1"},]
Plot[{kcO0[t],kcO1[t],kcO2[t],kcO3[t],kcO4[t],kcO5[t],kcO6[t],kcO7[t],kcO8[t],kcO9[t],kcO10[t]},{t,0,
tf},AxesLabel->{"t/h","γO/gL-1"}]
Print["supstrate concententration",kcS10[tf]]
Print["biomass concententration",kcX10[tf]]
Print["dissolved oxygen concententration",kcO10[tf]]
```

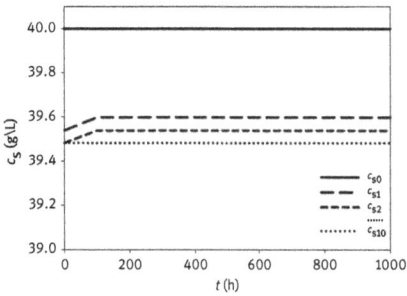

Figure (f): Organic compound concentration change over the time.

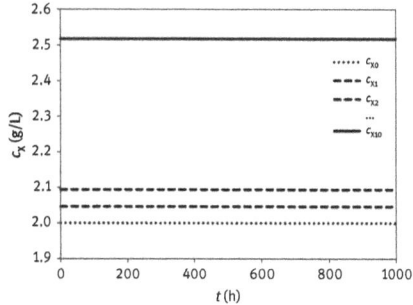

Figure (g): Biomass concentration change over the time.

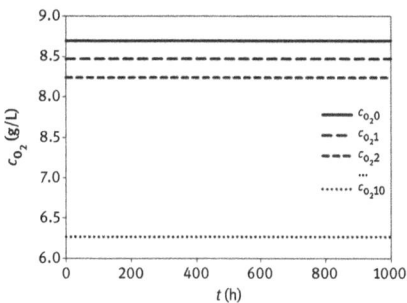

supstrate concententration {0.0399528}
biomass concententration {0.00204715}
dissolved oxygen concententration {0.00663523}

Figure (h): Dissolved oxygen concentration change over the time.

Example 10.7.

After treatment, the surface of the soil with the pesticide the distribution of the pesticide over the time and space was analysed in the segment of the soil to depth of 0.5 m (root zone). It is known that change of the concentration of pesticide depends on the diffusion, convection, microbial degradation, plat uptake, leaching and adsorption. The plat uptake, leaching and adsorption process can be described with first-order kinetics (k_1(plant uptake rate) = 0.084 1/day, k_2(leaching) = 0.008 1/day, k_3(adsorption) = 0.05 1/day), while microbial degradation is described with Monod kinetics. Parameters of the Monod kinetics are: maximum specific growth rate μ_{max} = 0.09 1/day, and Monod constant K_S = 0.000036 g/L. The biomass degradation is also included into the balance with biomass degradation rate of k_d = 0.1881/day. Diffusion coefficient of the used pesticide is D_p = 2.44 · 10^{-4} dm²/day and water velocity is v = 0.001 m/day. Initial concentration of the dissolved pesticide is γ_{So} = 0.000066 g/L and initial biomass concentration is γ_{Xo} = 0.00034 g/L.

a. Set the balances for the dissolved pesticide concentration change and for the biomass concentration change.
b. Solve the balances using the FDM for time period of 20 day and soil segment depth of 0.5 m.

(a) Balances for the dissolved pesticide concentration change and for the biomass concentration change.

$$\frac{\partial \gamma_{pesticide}}{\partial t} = D_p \cdot \frac{\partial^2 \gamma_{pesticide}}{\partial x^2} - v \cdot \frac{\partial \gamma_{pesticide}}{\partial x} - \mu_{max} \cdot \frac{\gamma_{pesticide} \cdot \gamma_{biomass}}{K_S + c_{pesticide}}$$

$$- k_1 \cdot \gamma_{pesticide} - k_2 \cdot \gamma_{pesticide} - k_3 \cdot \gamma_{pesticide}$$

$$\frac{\partial \gamma_{biomass}}{\partial t} = \mu_{max} \cdot \frac{\gamma_{pesticide} \cdot \gamma_{biomass}}{K_S + \gamma_{pesticide}} - k_d \cdot \gamma_{biomass}$$

(b) Finite differences method.

$$\frac{\partial \gamma_{pesticide}}{\partial t} = D_p \cdot \frac{\partial^2 \gamma_{pesticide}}{\partial x^2} - v \cdot \frac{\partial \gamma_{pesticide}}{\partial x} - \mu_{max} \cdot \frac{\gamma_{pesticide} \cdot \gamma_{biomass}}{K_S + \gamma_{pesticide}}$$

$$- k_1 \cdot \gamma_{pesticide} - k_2 \cdot \gamma_{pesticide} - k_3 \cdot \gamma_{pesticide}$$

$$\frac{\gamma S_{i+1,j} - \gamma S_{i-1,j}}{2 \cdot h_t} = D_p \cdot \frac{\gamma S_{i,j+1} - 2 \cdot \gamma S_{i,j} + \gamma S_{i,j-1}}{h_x^2} - v \cdot \frac{\gamma S_{i,j+1} - \gamma S_{i,j-1}}{h_x}$$

$$- \mu_{max} \cdot \frac{\gamma S_{i,j} \cdot \gamma X_{i,j}}{K_S + \gamma S_{i,j}} - k_1 \cdot \gamma S_{i,j} - k_2 \cdot \gamma S_{i,j} - k_3 \cdot \gamma S_{i,j}$$

$$\gamma S_{i+1,j} - \gamma S_{i-1,j} = \frac{Dp \cdot 2 \cdot h_t}{h_x^2} \cdot (\gamma S_{i,j+1} - 2 \cdot \gamma S_{i,j} + \gamma S_{i,j-1})$$

$$- \frac{v \cdot 2 \cdot h_t}{h_x} \cdot (\gamma S_{i,j+1} - \gamma S_{i,j-1}) - (\mu_{max} \cdot 2 \cdot h_t) \cdot \frac{\gamma S_{i,j} \cdot \gamma X_{i,j}}{K_S + \gamma S_{i,j}} -$$

$$- (k_1 \cdot 2 \cdot h_t) \cdot \gamma S_{i,j} - (k_2 \cdot 2 \cdot h_t) \cdot \gamma S_{i,j} - (k_3 \cdot 2 \cdot h_t) \cdot \gamma S_{i,j}$$

$$As = \frac{Dp \cdot 2 \cdot h_t}{h_x^2} \qquad\qquad Cs = (\mu_{max} \cdot 2 \cdot h_t)$$

$$Ds = (k_1 \cdot 2 \cdot h_t)$$

$$Bs = \frac{v \cdot 2 \cdot h_t}{h_x} \qquad\qquad Es = (k_2 \cdot 2 \cdot h_t)$$

$$Fs = (k_3 \cdot 2 \cdot h_t)$$

$$\gamma S_{i+1,j} - \gamma S_{i-1,j} = As \cdot (\gamma S_{i,j+1} - 2 \cdot \gamma S_{i,j} + \gamma S_{i,j-1})$$

$$- Bs \cdot (\gamma S_{i,j+1} - \gamma S_{i,j-1}) - Cs \cdot \frac{\gamma S_{i,j} \cdot \gamma X_{i,j}}{K_S + \gamma S_{i,j}} - Ds \cdot \gamma S_{i,j} - Es \cdot \gamma S_{i,j} - Fs \cdot \gamma S_{i,j}$$

$$\gamma S_{i+1,j} - \gamma S_{i-1,j} = As \cdot (\gamma S_{i,j+1} - 2 \cdot \gamma S_{i,j} + \gamma S_{i,j-1})$$

$$- Bs \cdot (\gamma S_{i,j+1} - \gamma S_{i,j-1})$$

$$- Cs \cdot \frac{\gamma S_{i,j} \cdot \gamma X_{i,j}}{K_S + \gamma S_{i,j}} - Ds \cdot \gamma S_{i,j} - Es \cdot \gamma S_{i,j} - Fs \cdot \gamma S_{i,j}$$

$$\gamma S_{i,j} = \frac{(As - Bs)}{(2 \cdot As + Ds + Es + Fs)} \cdot \gamma S_{i,j+1} + \frac{(As + Bs)}{(2 \cdot As + Ds + Es + Fs)} \cdot \gamma S_{i,j-1}$$

$$- \frac{Cs}{(2 \cdot As + Ds + Es + Fs)} \cdot \frac{\gamma S_{i,j} \cdot \gamma X_{i,j}}{K_S + \gamma S_{i,j}} - - \frac{1}{(2 \cdot As + Ds + Es + Fs)} (\gamma S_{i+1,j} - \gamma S_{i-1,j})$$

$$(2 \cdot As + Ds + Es + Fs) \cdot \gamma S_{i,j} = (As - Bs) \cdot \gamma S_{i,j+1}$$

$$+ (As + Bs) \cdot \gamma S_{i,j-1} - Cs \cdot \frac{\gamma S_{i,j} \cdot \gamma X_{i,j}}{K_S + \gamma S_{i,j}} - (\gamma S_{i+1,j} - \gamma S_{i-1,j})$$

Wolfram Mathematica 10.1.
Code

```
eqnsS = Table[cS_{i,j} == (As-Bs)/
    (2*As+Cs+Es+Fs)*cS_{i,j+1}+(As+Bs)/
dP = 24.4*10^-5;
    (2*As+Cs+Es+Fs)*cS_{i,j-1}Ds/
time = 20;
    (2*As+Cs+Es+Fs)*(cS_{i,j}*cX_{i,j})/
depth = 0.5;
    (Ks+cS_{i,j})-1/(2*As+Cs+Es+Fs)*(cS_{i+1,j}-cS_{i-1,j})/.
    {x->ξi,y>ηj},{i,n_x-1},{j,n_y-1}];
```

```
v = 0.1/10,000;
k1 = 0.084;
mimax = 0.0916;
k3 = 0.008;
k4 = 0.05;
kd = 0.188;
Ks = 0.035574;
cS0 = 0.066;
cX0 = 0.34;
x₀ = 0;
x₁ = time;
y₀ = 0;
y₁ = 0.5;
nₓ = 5;
n_y = 5;
hₓ = N[(x₁-x₀)/nₓ]
h_y = N[(y₁-y₀)/n_y]
Do[ξi = x₀ + i hₓ,{i,0,nₓ}]
Do[ηj = y₀ + j h_y,{j,0,n_y}]
varsS = Table[cS_{i,j},{i,0,nₓ},
{j,0,n_y}];
As = (2*hₓ*dP)/(h_y*h_y)
Bs = (2*hₓ*v)/h_y
Cs = k1*2*hₓ
Ds = mimax*2*hₓ
Es = k3*2*hₓ
Fs = k4*2*hₓ
Ax = mimax*2*hₓ;
Bx = kd*2*hₓ;
```

```
eqnsX = Table[cX_{i,j}
= = Ax/Bx*(cS_{i,j}*cX_{i,j})/(cS_{i,j}+Ks)-1/Bx*(cX_{i+1,j}-cX_{i-1,j})/.{x->ξi,y->ηj},{i,nₓ-1},
{j,n_y-1}]
boundS1 = Table[{(cS_{i,0}-cS_{i-1,1})/h_y = = 0/.x->ξi,(cS_{i,n_y} - cS_{i-1,n_y}-1)/h_y = =
0/.x->ξi},{i,nₓ-1}];
boundS2 = Table[{cS_{0,j} = = cS0/.y->ηj,(cS_{nₓ,}
j - cS_{nₓ} - 1, j)/hₓ = = 0/.y->ηj},{j,0,n_y}];
boundX1 = Table[{(cX_{i,0}-cX_{i-1,1})/h_y = = 0/.x->ξi,(cX_{i, n_y} - cX_{i-1, n_y}-1)/h_y = =
0/.x->ξi},{i,nₓ-1}]
boundX2 = Table[{cX_{0,j} = = cX0/.y->ηj,(cX_{nₓ}, j - cX_{nₓ}-1, j)/hₓ = = 0/.y->ηj},
{j,0,n_y}]
uapprS = ListInterpolation[solS,{{0,time},
{0,depth}}]
Plot3D[uapprS[x,y],{x,0,time},{y,0,depth},
ViewPoint->{2.0,-2.4,0.9},
AxesStyle->{Directive[Black,12,Bold],Directive[Black,12,Bold],
Directive[Black,12,Bold]},Ticks->{{0,5,10,15,time},
{0,0.1,0.2,0.3,0.4,0.5},{0,1,2,3,4,5}}]
```

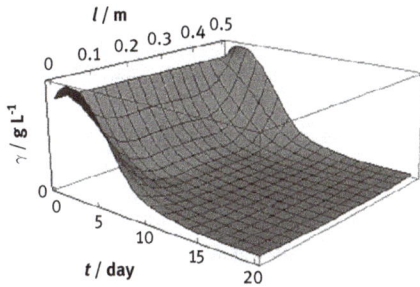

Figure (i): Pesticide concentration change over the time and depth.

10.5 Conclusion

In the end, we can just say that writing this chapter was not easy. Trying to summarize mathematical modelling in just one chapter was something similar with trying to develop a process mathematical model. A lot of inlet information, a lot of approaches, a lot of connections between them and all of that just to get a simple model that only approaches to the real problem. We are aware that this is just a small introduction to mathematical modeling but we hope that we inspired and encouraged you to go and explore all the possibilities that are in front of you. Always have in mind that this is not simple field and it demands a lot of previous knowledge but as Lord Kelvin said: *"I can never satisfy myself until I can make a model of a thing. If I can make a model, I can understand it."* Chase the knowledge.

Important notification

For preparation of this chapter, classroom materials from University of Zagreb, Faculty of Chemical Engineering and Technology, course *Analysis and modelling of the processes* held by prof. B. Zelić were used along with many books considering modeling and mathematical modeling. For additional information, we encourage you to go through books listed in the literature. All the examples (solved and for practice) were created by the authors of this chapter.

Acknowledgement: The authors would like to take this opportunity to express their gratitude to the reviewers' constructive comments and advice. All suggestions are more than welcome, especially when they amplify the quality of the work. Each reviewer comment has been discussed separately followed by the answers. All accepted remarks after first revision have been incorporated directly into the final version of the text and are marked yellow.

References

[1]   Hangos K, Cameron I. Process modelling and model analysis. UK: Academic press, 2001: 1–33.
[2]   Plazl I, Lakner M. Uvod v modeliranje procesov,Taložba FKKT. Slovenia: Univerza v Ljubljani, 2004: 1–48.
[3]   Himmelblau DM, Bishoff KB. Process analysis and simulation. New York: Wiley, 1968.
[4]   Davis ME. Numerical methods and modelling for chemical engineers. New York: Wiley, 1984.
[5]   Riggs JB. An introduction to numerical methods for chemical engineers, 2nd ed. Texas: Texas University Press, 1994.
[6]   Rice RD, Do DD. Applied mathematics and modelling for chemical engineers. New York: Wiley, 1995.
[7]   Aris R. Mathematical modeling techniques. London: Pitman, 1978.
[8]   Denn MM. Process modelling. New York and London: Longman, 1986.
[9]   Mathematical QA. Models in science and engineering. Notices of the AMS. 2009;56(1): 10–19.
[10] Pritsker A Why simulation works, Proceedings on the 1989 Winter Simulation Conference, 1–6.
[11] Dym CL. Principles in mathematical modeling. California: Elsevier, 2004: 1–10.
[12] Payre G. Influence graphs and the generalized finite difference method. Comput Methods Appl Mech Eng. 2007;196:1933–45.
[13] Botte GG, Ritter JA, White RE. Comparison of finite difference and control volume methods for solving differential equations. Comp Chem Eng. 2000;24:2633–54.
[14] Toh KC, Chen XY, Chai JC. Numerical computation of fluid flow and heat transfer in micro-channels. Int J Heat Mass Tran. 2002;45:5133–41.

Examples for practice

Example 1: The liquid phase reaction A↔B is carried out in batch reactor. In a first minute, a reaction initial concentration of reactant A drops from 1.5 mol/L to 1 mol/L.

What is the reaction rate and what is the time necessary to achieve conversion of 95 % if the kinetic is (a) first order and (b) second order with respect to reactant? Graphically describe dynamic change of concentration.

Due to market demands, continuous stirred tank reactor, CSTR, also needs to be tested for the same reaction. What is the volume of that reactor, necessary to achieve the same conversion if the volumetric flow rate of the reactant A is 0.5 L/min and the reaction rates are the same as the one estimated in the previous step? Graphically describe dynamic change from initial to stationary phase and determine hydraulic retention time necessary to achieve mentioned efficiency.

Solution:

Batch:
   a. k_d = 0.405 mol/L·min; t = 7.39 min
   b. k_d = 0.333 mol/L·min; t = 38.03 min

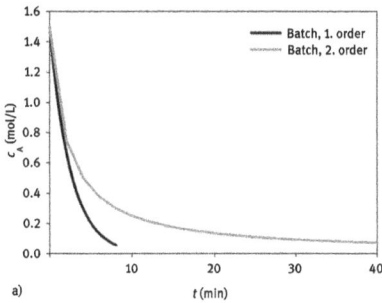

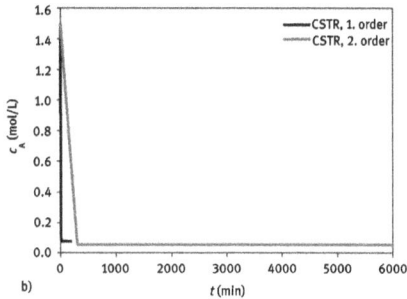

a)   b)

CSTR:
   a. V = 23.45 L; τ = 46.91 min
   b. V = 760.76 L; τ = 1521.7 min

Example 2: Wastewater passes through seven (7) tubular plug flow reactors connected into a series. Length of each reactor is 10 m with cross-section area of 100 dm². Water is polluted by component A (0.010 g/L) that is introduced into the reactor with flow rate of 80 dm²/s. Reaction rate constant is 0.055 L/g s and the process of the pollutant removal is described with second-order kinetics.

How many plug flow reactors have to be connected into a series to achieve 95 % and 99 % pollutant A removal? Graphically describe dynamic change of concentration along length of plug flow reactors.

Solution:
   a. N = 6 reactors for 95 % efficiency
   b. N = 29 reactors for 99 % efficiency (not possible to achieve at present conditions)

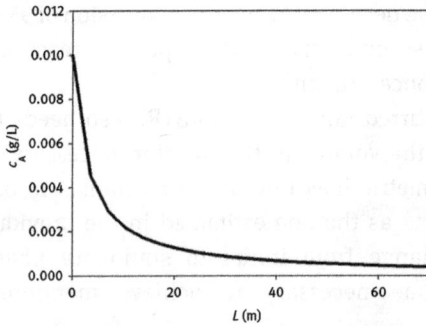

Example 3: Lavender incense stick is burning in a closed room (V = 50,000 L) for 20 minutes. In that time, it releases 50 mg of scent particles in a room. A person which does not like the smell of lavender enters a room and opens the window. If the ventilator is working in the room for 20 min will that be enough to minimize the smell (particles concentration below 0.020 µg/L) so a person can come back to the room? Predict the dynamic change of particle concentration in the room.

Solution:

t = 20 min, $y_{particles}$ = $3 \cdot 10^{-7}$ g/L (limit $y_{particles}$ = 0.00002 g/L)

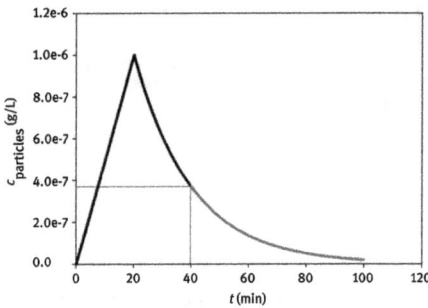

Example 4: A mobile phone indicates that the battery is low (5%) and that charging is necessary. Speed of charging is 2.1%/min. During the first 10 min the phone is just charging but after that the user becomes inpatient and opens several applications. One is using 0.5%/min, second 0.02%/min and third 0.2%/min of battery. How long will it take the battery to fully charge if after 10 min of parallel use application 1 is shut down, after 20 min application 2 and 5 min after the last one is closed? Describe the dynamic change of charging in time.

Solution:

$t = 51$ min

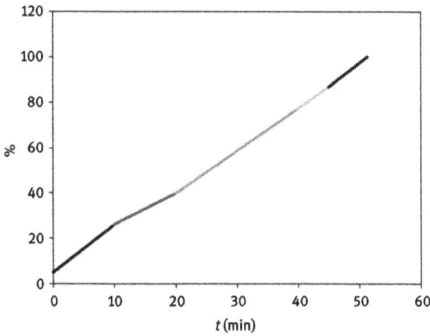

Example 5: During the summer season in some wells level of water can decrease. By decreasing the level of water, concentrations of some compounds harmful for human's, increases. Due to that water becomes harmful for human consumption. Assume a well with inner diameter of 15 dm and the depth of water 40 dm at the beginning of dry season. Concentration of harmful compounds is 0.000001 mmol/L, which is 1000 times lower than the prescribed limit for drinking water. If water evaporates with the rate of 0.70 mol/dm² h, determine the depth of water after which water will become harmful for human consumption.

Solution:

$t = 35.83$ day
$h = 51.75$ dm

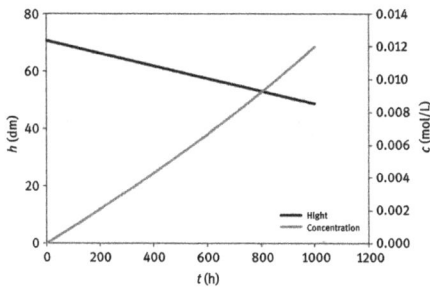

Example 6: After washing the factory for candy production, wastewater analysis showed that this water is not suitable for direct realizing into the nearby river. Due to that, all the water was collected into a tank for microbial wastewater treatment.

Mixed microbial culture (biomass, $y_x = 0.010$ g/L) present in the tank follows Monod kinetics for organic waste (substrate) removal ($y_s = 0.3$ g/L). Define mathematical model that will describe organic waste removal and biomass growth if the estimated kinetic parameters are $\mu_{max} = 5$ d^{-1}, $K_s = 0.045$ g/L, $k_d = 0.002$ h^{-1} and $Y_{x/s} = 0.8$ g/g. Calculate how much time will it take to remove 90 % of initial organic waste. After what time will biomass start dying?

How longer will it take to achieve the same efficiency if after second production cycle concentration of organic waste in the water is 6.7-fold higher? After the first cycle ($t = 2$ h), 75 % of the biomass was removed from the tank and the rest was used for the second cycle.

Solution:

a. t (for 90 % efficiency) = 0.706 ht (biomass dying) = 1.5 h
b. $t = 0.656$ h

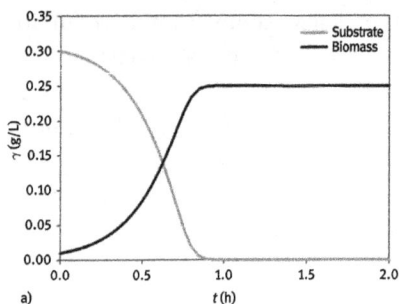

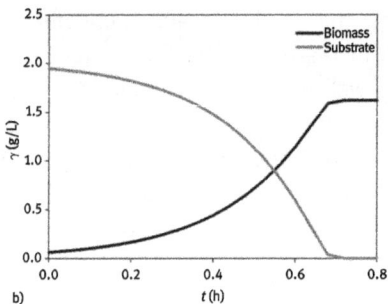

Example 7: Wastewater purification system based on active biomass ($y_x = 0.001$ g/L) is included in building planes of new factory. Expected amount of organic waste (BOD = Biochemical oxygen demand) was calculated to be $y_{BOD} = 0.5$ g/L. Preliminary research demonstrated that present microbial culture follows Monod kinetics with estimated kinetic parameters: $\mu_{max} = 2$ d^{-1}, $K_s = 0.025$ g/L, $k_d = 0.008$ h^{-1} and $Y_{x/s} = 0.5$ g/g. In order to release water in the environment purification efficiency has to be 97 %. Possible solution is one of three reactors:

a. Batch reactor
b. Continuous stirred tank reactor ($V = 1.5 \cdot 10^6$ m^3, $q = 5000$ L/h)
c. Fed-batch reactor ($V = 5 \cdot 10^5$ L) where the substrate is introduced into the reactor with flow rate of 1000 L/h. For all stirred tank reactors only 75 % of total volume is allowed to be used.

What is the best solution?

Solution:

  a. Batch: $t = 2.97$ h

  b. CSTR: $t = 2.96$ h (but $4\tau = 1200$ h)

  c. Fed batch: $t = 3.36$ h (but it takes 375 h to fill the complete reactor volume)

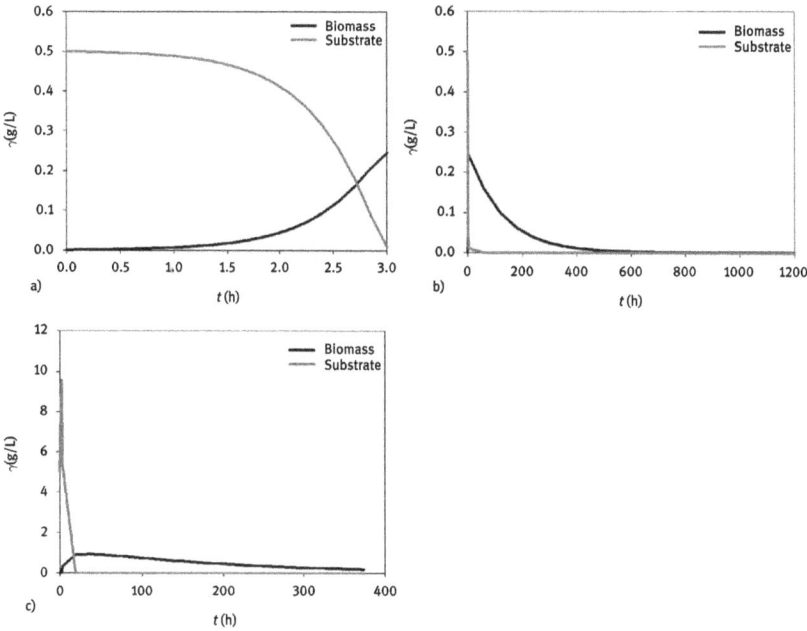

Example 8: A lake ($V = 5 \cdot 10^{13}$ L) is being polluted by two sources. One is the river coming from the north where a pig farm is placed and the second is the river coming from the west where a pesticide factory is releasing process water into the river. Concentration of pollutants that flows into the lake by the first river ($q_v = 1.5 \cdot 10^{12}$ L/year) is 0.150 g/L and by the second river ($q_v = 2.1 \cdot 10^{10}$ L/year) is 0.082 g/L. Some plants that are present in the lake have the possibility to remove some of the organic compounds that are being released into the lake (incorporate pollutants like nitrogen in their metabolic pathway). That process can be described by first order kinetic with $k = 0.02$ year^{-1}. If the pig farm was opened from 1975 to 2000 and pesticide farm from 1981 to 2016, determine the time period when water from the lake is suitable for drinking taking in consideration that the concentration of pollutant in the lake has to be below 0.025 g/L to be safe for human consumption. Assume that the volume of the lake is constant.

Solution:

Time period from 1975 to 1981 and from 2019

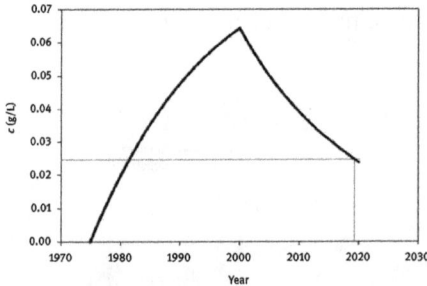

Example 9: After the factory was closed, an artificial lake that was constructed to collect all the wastewater from factory has to be remediated. The concentration of heavy metals (HM) measured in the lake c_{HM} = 300 mmol/L indicated severe pollution. Absorption on zeolite was proposed as a possible solution. During the preliminary measurements, it was determined that the average absorption velocity is 0.009 mmol/L h. Additionally, some heavy metal sedimentation was also noticed and it was determined to be 0.001 mmol/L h. Determine the time necessary to achieve 90% and 95% heavy metal removal from lake with assumption that the lake volume is constant. Demonstrate dynamics of heavy metals concentration change.

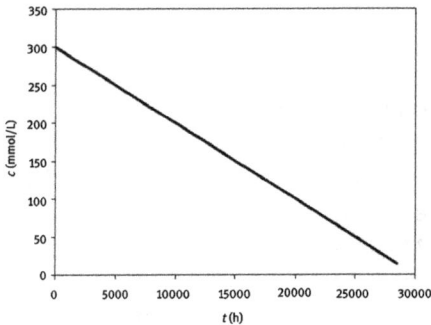

Solution:

   a. t (for achieving 90% removal) = 3.08 g

   b. t (for achieving 95% removal) = 3.25 g

Example 10: In order to burn the waste where benzene is present in concentration of 150 mmol/L tubular incinerator (L = 10 m) was constructed. Reactor works at the temperature of 800 °C. Reaction kinetics is first order (k = 0.05 s^{-1}). Arrhenius number was determined to be $7 \cdot 10^9$ s^{-1}, while the activation energy is 200 kJ/mol. Determine if the constructed incinerator is sufficient to achieve 25 % conversion of benzene if the waste stream is entering with rate of 7.5 m/s and the residence time is 3 s. If it is not sufficient, how long it should be? Demonstrate the dynamic change of benzene concentration along incinerator length.

Solution:

c_b = 26.99 mmol/L (at the exit; X = 82 %)

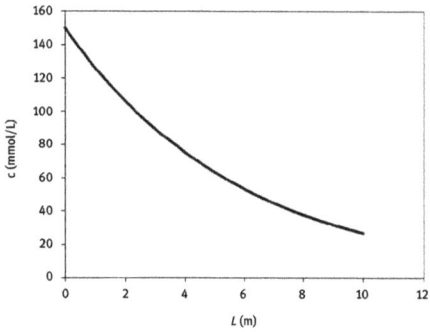

Hrvoje Kušić and Ana Lončarić Božić

11 Risk assessment

Abstract: The chapter introduces risk assessment and examines the basic concepts of risk and hazard, their causes, consequences and probabilities. The methodology of environmental risk assessment including effect assessment, exposure assessment and risk estimation/characterisation is overviewed. Application of risk assessment for chemicals is described with regard to human health and the environment. Some of the commonly used techniques in risk assessments are introduced and illustrated by examples. Environmental risk assessment should provide input to the risk management so well-informed decisions can be made, to protect human health and the environment.

Keywords: risk, hazard, environmental risk assessment, chemical risk, dose-response, risk matrices, logic trees

11.1 Introduction

11.1.1 What is risk?

According to the Merriam-Webster dictionary, word risk is defined as "possibility of loss or injury" and "someone or something that creates or suggests a hazard" [1]. Risk is also defined as "the possibility of incurring misfortune or loss; hazard" [2], "danger, or the possibility of danger, defeat, or loss" [3], "the possibility of suffering harm or loss; danger" and "a factor, thing, element, or course involving uncertain danger; a hazard" [4], "a situation involving exposure to danger" and "the possibility that something unpleasant or unwelcome will happen" [5]. Obviously, terms risk, hazard and possibility are closely related.

In everyday life we are often faced to the "risky" situations which are characterised by uncertainty. Our action or choice can result with undesired consequences. Gambler choses to make a bet and takes a risk of losing the money. When a driver approaching the crossroad accelerates to pass through before the light turns red, he risks an accident or a ticket. Smokers voluntarily increase their health risk.

The term risk is difficult to define precisely because it can be applied to a wide variety of situations and systems [6]. It is suggested that risk cannot be fully defined

This article has previously been published in the journal Physical Sciences Reviews. Please cite as: Lončarić Božić, A., Kušić, H. Risk Assessment. *Physical Sciences Reviews* [Online] **2018**, *3*. DOI: 10.1515/psr-2016-0126

https://doi.org/10.1515/9783110468038-011

until the system itself has been chosen and defined. In such manner risk can be considered as a flexible term, adaptable to the situation in which it is being used. If the loss is something that is well understood, such as death, risk can be simply defined as the probability of that result [7].

In the contest of environmental risk assessment (ERA), risk is broadly defined as the likelihood that a harmful consequence will occur as the result of an action or condition [8, 9].

Generally, risk is a combination of the frequency or probability that the hazard will cause actual harm and the severity of the consequences. Accordingly, risk can be described by following expression:

$$\text{Risk} = f \text{ (Frequency or Probability, Consequences)} \qquad (11.1)$$

whereas both frequency/probability and the consequences refer to the specific hazardous event which is taking place [6].

11.1.2 What is hazard?

Rausand [10] describes hazard as a source of danger, being a property, a situation, or a state. Hazard is not an event but a prerequisite for the occurrence of a hazardous event that may lead to harm. Hazards can include substances, machines, energy forms, or the way certain activities are carried out. For example, a wet floor is a hazard as one could slip and fall. Moving parts of machinery represents a hazard that may cause injuries to the workers, which is often considered within occupational health and safety procedures. A fog is also a hazard as it reduces visibility in the traffic. Many chemicals have properties that make them hazardous such as toxicity, flammability, explosiveness, corrosiveness. However, the use of large amounts of dangerous chemicals is unavoidable in some industry sectors which are vital for a modern industrialised society.

European Seveso III directive [11], aimed at the prevention of major accidents involving dangerous substances and at the mitigation of such events, defines hazard as intrinsic property of a dangerous substance or physical situation, with a potential for creating damage to human health or the environment.

Hazard can be generally defined as the inherent potential for something to cause harm. Hazards are those things that might happen without saying how likely they are to happen. They are possibilities, without probabilities [12]. Hence, the term hazard can be defined as a function of both inherent properties and circumstances that may result in undesired outcome.

The terms risk and hazards are often interchangeably used. A good and simple example of the distinction between the two could be found in the publication of European Environmental Agency [13]. "A large number of chemicals have hazardous properties. Acids may be corrosive or irritant to human beings for example.

The same acid is only a risk to human health if humans are exposed to it. The degree of harm caused by the exposure will depend on the specific exposure scenario. If a human only comes into contact with the acid after it has been heavily diluted, the risk of harm will be minimal but the hazardous property of the chemical will remain unchanged." Hence, the nature of hazard remains the same, but exposure dictates whether harm will actually occur.

11.2 Looking for answers

Risk assessment is a process that results with understanding of risks, their causes, consequences, and their probabilities. Risk is always related to what might happen in the future [10]. Accordingly, it is characterised by some level of uncertainty. The purpose of risk assessment is to provide information so that well-informed decisions can be made.

Basically, risk assessment looks for the answers to three main questions [14]:
– What can happen?
– What is the likelihood of that happening?
– What are the consequences?

Answer to the question "What can happen?" describes what could go wrong. It identifies scenario or hazardous event [10, 14]. The question can be answered in a number of ways and we have to identify all hazardous events that may cause harm. Hazardous event is defined [10] as the first event in a sequence of events that, if not controlled, will lead to the undesired consequences. Figure 11.1 illustrates a sequence of events that led to the spilled milk as an undesired consequence.

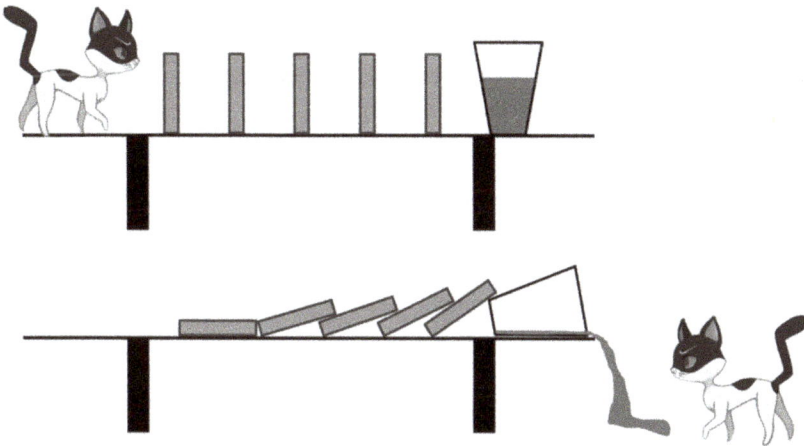

Figure 11.1: Illustration of hazardous event, sequence of events and undesired consequence.

Answer to the second question describes how likely it is that a particular scenario or hazardous event actually happens. When describing the likelihood we can use qualitative statements or probabilities or frequencies [10]. When answering the third question we have to identify the adverse effects or consequences to human health, environment or material goods which can be expressed as some sort of cost [6].

Taking into account that both frequency/probability and the consequences refer to the specific hazardous event, i. e. scenario which is taking place, eq. (11.1) can be modified to eq. (11.2):

$$\text{Risk} = f \,(\text{Scenario, Frequency or Probability, Consequences}) \qquad (11.2)$$

representing the "set of triplets" idea proposed by Kaplan and Garrick [14].

Lindhe [15] gave an example of a "set of triplet" in analysing the risks in drinking water system, where scenario includes a pipe burst with estimated probability of 0.05 that may result with an interruption in the delivery of drinking water to 100 people for 8 hours period as a consequence.

When the consequences of all identified scenarios are considered, the overall risk can be expressed by eq. (11.3) [15]:

$$R = |\{S_i, p_i, X_i\}|_c \qquad (11.3)$$

whereas R stands for risk, S stands for scenario, p stands for probability and X for consequence, index $i = 1,2, \ldots\ldots n$ indicates that there are more than one scenario and index c means *complete* risk.

11.3 Conceptual models

Another question to be answered in risk assessment is: Who (or what) can be affected and how? It depends on our knowledge and belief how the real system in question behaves. One way to formalise this is to develop a conceptual model.

Source–pathway–receptor concept (Figure 11.2) is often used in ERA. It can be easily correlated with the above discussed scenario.

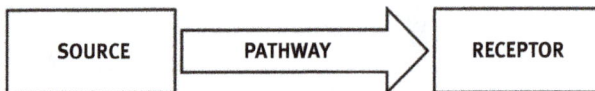

Figure 11.2: Concept source– pathway–receptor used in environmental risk assessment.

In this conceptual model the pathway between a hazard source and a receptor is investigated. The pathway is considered as the linkage by which the receptor could come into contact with the source. If no pathway exists then no risk exists [16].

For a single source affecting the single receptor, the conceptual model might be simple. An example is given in Figure 11.3. It is a simplified representation of a risk of human exposure to the volatile toxic chemical transported from the source through the air.

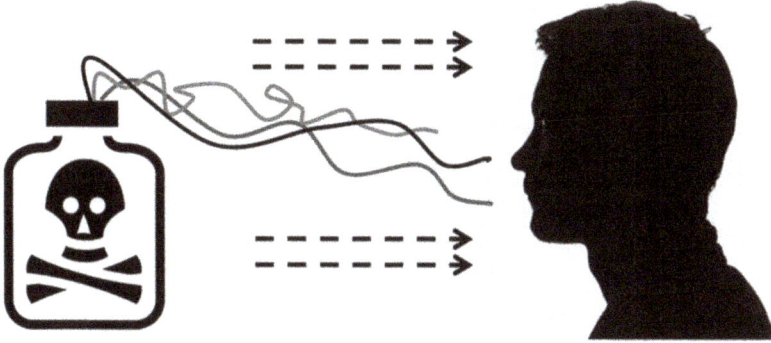

Figure 11.3: Human exposed to volatile toxic chemical through the air path.

However, a number of pathways often need to be considered and conceptual models became more complex. Table 11.1 present examples of some of the possible sources, pathways and receptors [16]:

Table 11.1: Examples of sources, pathways and receptors in environmental risk assessment.

Source	Pathway	Receptor
Industrial emission	Water	People
Leaking pipelines	Air	Ecosystems
Chemical release	Soil	Animals
Contaminated soils	Food chain	Plants Water supply
Contaminated water		

Conceptual model sets boundaries and components of the system and defines the relationships within. It is usually presented in narrative and graphical form. It should include sources and receptors, and exposure pathways. A chain of events must occur to result in exposure. The chain in a collective sense is termed an exposure pathway, the environmental routs which connect source site and receptor [17]. Hence, exposure pathway for a chemical includes chemical release mechanism, transport and transfer mechanism, possible transformation mechanisms, exposure point, and exposure routes.

The example is given in Figure 11.4. The source (1) is a disposed barrel containing hazardous waste chemical. The receptors are residents (2) of the settlement near

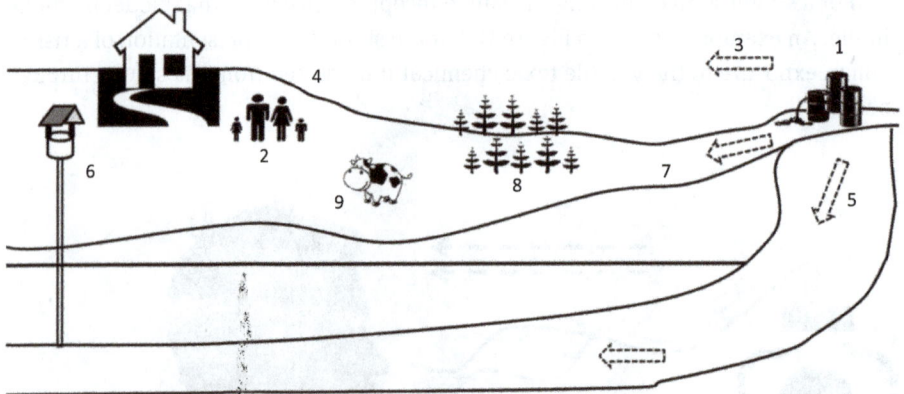

Figure 11.4: Conceptual model of waste disposal location.

disposal location. The chemical can be released by volatilisation and transported through the air (3) to the exposure point, whereas the inhalation by receptor (4) is considered as exposure route. The chemical can be also released by leaching (5) and transported by the ground water flow to the water table. Residential well (6) is considered as an exposure point, whereas the ingestion of contaminated water is exposure route for the residents consuming the potable water from the well. Another considered release mechanism is a spill of chemical (7) which is transported through the soil as a contaminated exposure medium. Spilled chemical can reach the water table through the soil. The chemical can be transferred from contaminated soil to the crops (8), whereas the crop ingestion is an exposure route for the residents. In the same manner animal grazing might be contaminated, and the consequent, indirect exposure route is ingestion of meat and milk (9).

As it can be seen, there are more than one exposure pathways even in a case of above simple example. At the end of the risk assessment process, existing controls should be recorded and further measures may need to be considered to reduce or eliminate the risks identified. Detailed consideration of risk management is beyond the scope of this chapter but, in general terms, risk can be treated by reducing or modifying the source, by managing or breaking the pathway and/or modifying the receptor.

11.4 Risk assessment framework

Risk assessment is a structured process set in the wider frame of risk management (Figure 11.5). Risk assessment generally comprises of three steps: (1) risk identification, (2) risk analysis and (3) risk evaluation, providing the input for risk treatment and the overall risk management process.

RISK ASSESSMENT

risk identification
risk analysis
risk evaluation

RISK ASSESSMENT

Figure 11.5: Risk assessment frame and structure.

According to ISO Risk management – Vocabulary [18], risk is defined as effect of uncertainity on objectives. It should be noted that the objectives can have different aspects such as financial, health and safety, as well as environmental goals. Risk is often expressed in term of combination of the consequences of an event and the associated likelihood of occurrence. Risk identification [18] is a process of finding, recognising and describing the risk. It involves identification of risk sources, hazardous events, their causes and potential consequences. Risk analysis is defined as a process to comprehend the nature of risk and the level of risk. Risk analysis consists of determining consequences and their probabilities for identified scenarios, taking into account existence and effectiveness of control measures. Note that level of risk represents the magnitude of risk expressed in the terms of consequences and their likelihood. Hence, for level of risk eq. (11.1) applies with regard to hazardous event identified in previous step. The purpose of risk evaluation is to determine whether the risk is acceptable or tolerable and provides a base for decisions on how to deal with the undesired, negative consequences in risk treatment as a part of risk management process.

IEC/ISO Risk management – Principles and guidelines standard [19] is intended for organisations of all types and sizes that seek to address uncertainties and increase the likelihood of achieving their objectives by adopting the effective risk management practice. However, the same principles and similar structure of risk assessment is applied within other fields such as environmental risk management.

The objectives of risk assessment include [20] (i) protection of human and ecological health form adverse effects resulting from exposures to hazardous agents or situations (e. g. toxic substances), (ii) balance risks and benefits from usage hazardous substances such as drugs and pesticides, (iii) set target level of risk (e. g. food contaminants and water pollutants), (iv) set priorities from program activities to be used by regulatory agencies, manufacturers and environmental consumers organisations and (v) estimate residual risks and extent of risk reduction after steps are taken to reduce risks.

11.5 Environmental risk assessment

Risk assessment has become a commonly used approach in examining environmental problems mainly caused by human activities, [8, 13] whereas risks under consideration may be of very different natures. ERA [13] is the examination of risks resulting from technology that threaten ecosystems, animals and people, whereas the target/receptors as well as end-points are varied. Accordingly, and taking into account historical development as well as regulatory and policy imperatives and scope, a typology of risk assessment methods breaks ERA into:

- Human health risk assessment
- Ecological risk assessment
- Applied industrial risk assessment

ERA includes following steps: (1) problem formulation, (2) hazard identification, (3) release assessment, (4) exposure assessment, (5) consequence assessment and (6) risk estimation. Risk evaluation (7) is included in assessment of risks of existing and new chemicals that consider emission and the resulting exposure to the environment and humans in all life cycle steps. However, it may be included in other types of risk assessment. It should be noted that the political, economic, legal and social factors influence problem formulation as well as and risk evaluation, and consequently the decisions made within risk management.

As it can be seen ERA steps fits well in above described general risk assessment structure (Figure 11.5). Namely, hazard identification is comprised in risk identification. Release assessment examines likelihood and the magnitude of the release, while exposure assessment examines the pathway and doses that receptor receives, revealing the source–pathway–receptor concept within ERA. Consequence analysis (or dose-response assessment) determines what is the effect on the receptor, while risk estimation gives a quantitative or qualitative measure of the risk. Hence, former steps can be clearly correlated with the risk analysis step within the general risk assessment structure.

Risk assessment and management is used as decision making tool in:

- the design of regulation; in determining societally acceptable risk levels as a base for environmental standards
- providing the basis for site-specific decisions, in land-use planning or sitting of hazardous installations.
- prioritisation of environmental risks; in determination of which chemicals to regulate first
- comparison of the risks between different risk control measures and substitution decisions.

Risk assessment enables the objective decision on whether risks posed by the chemical at certain dose outweigh the benefits of its use, i. e. whether risk is acceptable or not. The term "acceptable risk" might be ambiguous as it heavily depends on risk perception and the question "acceptable to whom" always arises.

Risk perception [13] involves people's beliefs and attitudes, judgments and feelings, as well as the wider social and cultural values that people adopt toward hazard and their benefits. Hence, risk perception will be a major determinant in whether a risk is deemed to be acceptable and whether the risk management measures are seen to resolve the problem.

Risk management is the decision-making process through which choices can be made between a range of options which achieve the "required outcome" specified by environmental standards, determined by a formalised risk-cost-benefit analysis or "industry norms" or "good practice" [13]. Within the risk management process [19] risk can be treated by applying appropriate option(s) for changing the probability or the severity of the consequences related to the unacceptable risks. The residual or acceptable risks may be retained, [18] i. e. the potential benefit of gain, or burden of loss, from a particular risk may be accepted. One of definitions for acceptable risk given by Rausand [10] is: risk considered insignificant and not justifying further effort to reduce it.

European Seveso directive [11] is an example of incorporation of risk assessment in regulation for site-specific problems. The Directive covers establishments where dangerous substances may be present in quantities above a certain threshold. The operators of industrial sites where dangerous substances are produced, used, handled or stored, or may be generated during loss of control of the process, are required to identify the potential major accident hazards and to provide the evidence that the adequate steps have been taken to prevent accidents and to limit the possible consequences to people and environment. Among number of activities used to categorize the Seveso establishments, seven activities contribute to 50 % of establishments: fuel storage (including heating, retail sale, etc.), wholesale and retail storage and distribution (excluding LPG-Liquefied petroleum gas), LPG storage, general chemicals manufacture, production of basic organic chemicals, power generation, supply and distribution, and LPG production, bottling and bulk distribution [21].

As risk assessment combines hazard and exposure data it enables prioritisation based on actual risk rather than the potential hazard [13]. Chemical risks are ranked on the basis of detailed information of risk effects, likelihood of exposure for different exposure scenarios priorities. However, it should be noted that the ability to provide detailed risk assessment is dependent on the availability of relevant toxicological data. Lack of complete data for numerous chemicals represents the limitation for the effective use of risk assessment in as an environmental management tool.

Comparative risk is relative new field in European practice [13]. The aim is to ensure that the most severe risks are dealt with first, and that resources are devoted adequately with regard to the level of risk.

Health risk assessment includes physical risks such as ionising radiation, biological such as pathogens or introduction of genetically modified organisms into the environment and the food chain, and chemical risks such as toxic substance. Different methodologies are applied for human health risk assessment depending on different toxic mechanisms exerted by different classes of chemical and the toxicological end-point being assessed [13]. The end-point being assessed could be death, or a specific pathological condition relating to exposure to a chemical. In some cases, such as immuno-suppressant toxins, specific end-points may be difficult to determine due to the role of other agents and stressors on the body. Hence, the methodology of risk assessment would differ from that applied for some other class of chemicals such as irritants.

Ecological risk assessment [13] involves the assessment of the risks posed by the presence of substances released to the environment by man, on all living organisms in the variety of ecosystems which make up the environment. Methodology for ecological risk assessment is developed form human health assessment. Hence, the main difference between human health and ecological risk assessment can be seen in the scope of the process. Ecological risk assessment is characterised by the complex nature of the potential targets or receptors. Health risk assessment is concerned with individuals and morbidity and mortality, while ecological risk assessment is concerned with populations and communities and the effects of substances on mortality and fecundity. Ecological risk assessment has to deal with a multitude of organisms, all with varying sensitivities to chemicals and various groups have distinct exposure scenarios, such as free swimmers and sediment dwellers [13]. Because of the difficulty in obtaining toxicity data on all organisms in an ecosystem, the recognised practice is to test selected representatives of major taxonomic groups and use these as surrogates for the whole system. This method is questionable as it may not protect the most sensitive species exposed in the environment. Failure to identify the effects of an agent on a potential receptor can result in widespread damage to organisms and ecosystems. A typical example is the use of anti-fouling paints containing tributyltin and the resulting damaging effect on oysters and dog whelks [13].

Risk assessment methodology for industrial risk scenarios is a site specific and includes elements of either, or both, human health or ecological risk assessment [13].

11.6 Risk assessment methodology

The methodologies for human health risk assessment for chemicals are mostly based on so-called NAS model developed by US National Academy of Sciences [22]. It includes: (1) hazard identification, (2) dose-response assessment, (3) exposure assessment and (4) risk estimation/characterisation, Figure 11.6.

The methodology for human health and ecological risk assessment used in the regulation of new and existing substances in the EU [23] consists of the same steps used in above NAS model, although hazard identification and dose-response assessment are combined in the single step called Effects Assessment as presented in Figure 11.7.

```
              ┌─────────────────┐
              │     HAZARD      │
              │ IDENTIFICATION  │
              └─────────────────┘
                 ⇩         ⇩
┌──────────────────┐   ┌──────────────────┐
│  DOSE-RESPONSE   │   │    EXPOSURE      │
│   ASSESSMENT     │   │   ASSESSMENT     │
└──────────────────┘   └──────────────────┘
          ⇩                    ⇩
       ┌──────────────────────────────┐
       │            RISK              │
       │ ESTIMATION/CHARACTERIZATION  │
       └──────────────────────────────┘
```

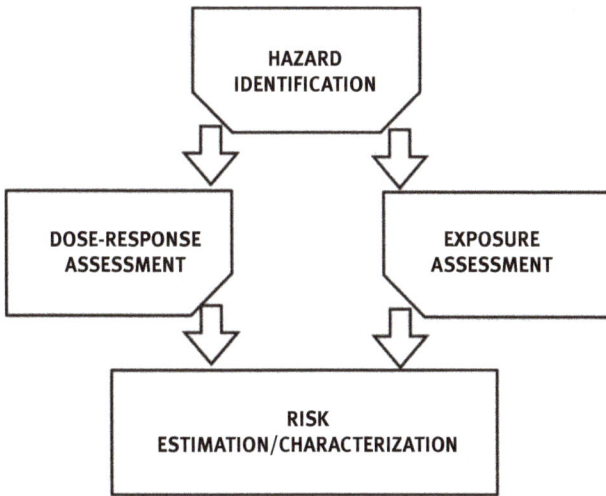

Figure 11.6: Four step risk assessment process, NAS.

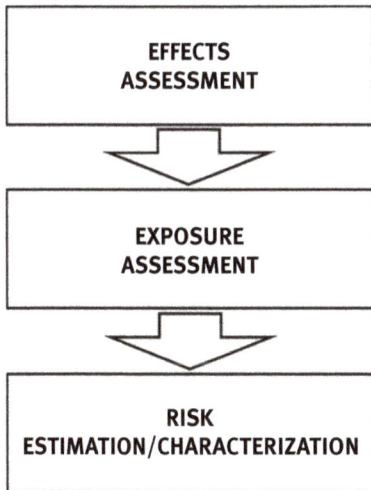

```
┌──────────────────────────┐
│        EFFECTS           │
│      ASSESSMENT          │
└──────────────────────────┘
             ⇩
┌──────────────────────────┐
│        EXPOSURE          │
│      ASSESSMENT          │
└──────────────────────────┘
             ⇩
┌──────────────────────────┐
│          RISK            │
│ ESTIMATION/CHARACTERIZATION │
└──────────────────────────┘
```

Figure 11.7: Three step risk assessment process, EU.

Corresponding research activities are required for each step. Namely, for hazard identification within effect assessment, laboratory and fields observation of adverse health effects and exposures to particular agent are needed in order to determine does the agent cause the adverse effect.

According to technical guidance on risk assessment [23] the risk assessment process, in relation to both human health and the environment, entails a sequence of actions as presented in Figure 11.7:

1. Assessment of effects, comprising
   a. Hazard identification: identification of the adverse effects which a substance has an inherent capacity to cause; and
   b. dose (concentration)-response (effects) assessment: estimation of the relationship between dose, or level of exposure to a substance, and the incidence and severity of an effect, where appropriate.
2. Exposure assessment: estimation of the concentrations/doses to which human populations (i. e. workers, consumers and man exposed indirectly via the environment) or environmental compartments (aquatic environment, terrestrial environment and air) are or may be exposed.
3. Risk characterisation: estimation of the incidence and severity of the adverse effects likely to occur in a human population or environmental compartment due to actual or predicted exposure to a substance, and may include "risk estimation", i. e. the quantification of that likelihood.

To improve the protection of human health and the environment from the risks that can be posed by chemicals, while enhancing the competitiveness of the EU chemicals industry, European Union adopted REACH regulation (Registration, Evaluation, Authorisation and Restriction of Chemicals) [24]. REACH applies to all chemical substances; not only those used in industrial processes but also in our day-to-day lives, for example in cleaning products, paints as well as in articles such as clothes, furniture and electrical appliances. REACH establishes procedures for collecting and assessing information on the properties and hazards of substances. A main task of REACH is to address data gaps regarding the properties and uses of industrial chemicals. Information sets to be prepared include safety data sheets (SDS), chemical safety reports (CSR), and chemical safety assessments (CSA). These are designed to guarantee adequate handling in the production chain, in transport and in use and to prevent the substances from being released to and distributed within the environment. Another important aim is to identify the most harmful chemicals and to set incentives to substitute them with safer alternatives [25].

11.7 Risk assessment for chemicals

Any chemical which could potentially cause harm is considered to be hazardous. Certain chemicals can cause different types of harm, ranging from mild skin irritation to cancer. They can also have significant impacts on the environment, including the air, water and land; and they may adversely affect plants and animals. A chemical can only cause harm to our health or to the environment, if we are exposed to it [26].

The risk assessment for human health addresses the following potential toxic effects and human populations, considering each population's exposure by the inhalation, oral and dermal routes [23]:

Effects:
- acute toxicity;
- irritation;
- corrosivity;
- sensitisation;
- repeated dose toxicity;
- mutagenicity;
- carcinogenicity;
- toxicity for reproduction.

Human population:
- workers;
- consumers;
- humans exposed indirectly via the environment.

The information required for the technical dossier of chemical substance includes [23]:

i. physiochemical properties such as state of the substance at 20 °C and 101,3 kPa, melting-point, boiling-point, relative density, vapour pressure, surface tension, water solubility, partition coefficient n/octanol/water, flash-point, flammability, explosive properties, self-ignition temperature, oxidizing properties, and granulometry;Data on physico-chemical properties are important for both the exposure assessment and the effects assessment. For example, physico-chemical properties are used to estimate emissions and the human exposure scenarios, to assess the design of toxicity tests, and may also provide indications about the absorption of the substance for various routes of exposure.

ii. toxicological studies including: acute toxicity (administered orally, by inhalation and cutaneously, skin irritation, eye irritation and skin sensitisation), repeated dose whereas the route administration should be the most appropriate having regard to the likely route of human exposure, the acute toxicity and the nature of the substance, and other effects including mutagenicity, screening for toxicity related to reproduction and assessment of the toxicokinetic behaviour of a substance to the extent that can be derived from base set data and other relevant information.

iii. ecotoxicological studies including effects on organisms such as acute toxicity for fish, acute toxicity for daphnia, growth-inhibition test on algae and bacterial inhibitor, degradation (biotic and abiotic, and hydrolysis as a function of pH for substances that are not readily biodegradable), adsorption/desorption screening tests.

The principle of the assessment of chemical risks for human health is to compare the concentration of substance to which population is exposed with the concentration at which no adverse (toxic) effects are expected to occur [13, 23].

No Observed Adverse Effect Level (NOAEL) is the outcome of the dose-response assessment. Where it is not possible to establish a NOAEL but a Lowest Observed Adverse Effect Level (LOAEL) can be derived, the latter is compared with the exposure level. The exposure levels can be derived based on available monitoring data and/or model calculations. Both NOAEL and LOAEL values are determined on the basis of results from animal testing, or on the basis of available human data.

The risk characterization involves calculation of the exposure to effect ratio or a qualitative evaluation or the likelihood that an effect will occur at the give exposure. The risk reduction measures may include [13]:

- providing information on the public regarding the safe and responsible handling of substances or products
- the use of emission permits which set limits
- marketing restrictions such as limiting production, import or use
- total ban of a substance or activity

The environmental risk assessment approach [27] address the concern for the potential impact of individual substances on the environment by examining both exposures resulting from discharges and/or releases of chemicals and the effects of such emissions on the structure and function of the ecosystem.

Three approaches are used for this examination [27]:

- quantitative estimation of Predicted Environmental Concentration (PEC) to Predicted No Effect Concentration (PNEC) ratio, i. e. quantitative PEC/PNEC estimation for ERA of a substance comparing compartmental concentrations with the concentration below which unacceptable effects on organisms will most likely not occur. This includes also an assessment of food chain accumulation and secondary poisoning;
- the qualitative procedure for the ERA of a substance for those cases where a quantitative assessment of the exposure and/or effects is not possible;
- the PBT assessment of a substance consisting of an identification of the potential of a substance to persist in the environment, accumulate in biota and be toxic combined with an evaluation of sources and major emissions, whereas PBT stands for Persistent, Bioaccumulative and Toxic.

In principle, human beings as well as ecosystems in the aquatic, terrestrial and air compartment are to be protected. At present, the ERA methodology has been developed for the following compartments [27]:

For inland risk assessment:
- aquatic ecosystem (including sediment);
- terrestrial ecosystem;
- top predators;
- microorganisms in sewage treatment systems;
- atmosphere.

For marine risk assessment:
- aquatic ecosystem (including sediment);
- top predators.

In addition to three primary environmental compartments, effects relevant to the food chain (secondary poisoning) are considered as well as effects on the microbiological activity of sewage treatment systems because proper functioning of sewage treatment plants is important for the protection of the aquatic environment.

Different targets of the risk characterisation and the exposure scenarios, i. e. media of exposure for inland risk assessment and the marine environment as well as of indirect human exposure through the environment are presented in Table 11.2 [27].

Table 11.2: Relationship between different targets of the risk characterisation and medium of exposure.

Target	Media of exposure
Inland compartments	
Aquatic organisms	Surface water
Benthic organisms	Sediment
Terrestrial organisms	Aquatic soil
Fish-eating predators	Fish
Worm-eating predators	Earthworms
Microorganisms	Aeration tank (sewage treatment processes)
Marine compartment	
Aquatic organisms	Seawater
Benthic organisms	Marine sediment
Fish-eating predators	Fish
Top predators	Fish-eaters
Indirect human exposure	
Drinking water production	Surface water, groundwater (annual average)
Inhalation of air	Air (annual average)
Production of crops	Agricultural soil (averaged over 180 days)
Production of meat and milk	Grassland (averaged over 180 days)
Fish for human consumption	Surface water (annual average)

Effects Assessment involves the identification of the hazard based on its physico-chemical properties, ecotoxicity and intended use, and the estimation of a Predicted No Effect Concentration (PNEC), derived from ecotoxicity data and the application of assessment factors.

Exposure assessment involves the calculation of a PEC. This is derived using monitoring data, realistic worst cases scenarios and predictive modelling techniques. It is a complex task and should consider release, degradation, and transport and fate mechanisms.

Risk characterisation involves the calculation of a quotient – the PEC/PNEC ratio. If the ratio is less than 1 the substance is considered to present no risk to the environment in a given scenario.

The Organisation for Economic Co-operation and Development (OECD) developed similar scheme and delivered a set of practical tools on ERA for chemicals [28]. Interactive OECD ERA toolkit describes the general work flow of ERA and provides links to relevant tools developed by OECD and member countries that can be used in each step of the work flow. The toolkit comprises five examples, provide a roadmap of the process, showing the steps involved in each case and the tools which were used on risk assessment. It includes a textile dye, pesticide and metal, end examples on setting an environmental quality standards, and initial screening of substances for persistent, bioaccumulative and toxic properties, as well and on compliance with limits set in a permit regarding ait pollution.

Similarly, a toolkit on human health risk assessment for chemicals is developed by World Health Organisation (WHO) within the International Programme on Chemical Safety (IPCS) project on the Harmonization of approaches to assessment of risks from exposure to chemicals [29].

11.8 Dose–response relationship

Effect assessment comprises of hazard identification and dose–response assessment. In human health risk assessment for chemicals hazard is described as [13] the identification of those substances that are deemed to be hazardous in some concentration or dose. Such assumption is based on physical chemical and toxicological properties of the chemical and on environmental fate mechanism.

A fundamental principle in toxicology is that a relation exists between the dose of an agent (chemical) received and the response produced.

Paracelsus (1493–1541), sometimes called father of toxicology, is frequently quoted for his saying [30, 31]: "What is there that is not poison? All things are poison and nothing (is) without poison; only the dose makes that a thing is no poison." In short: "The dose makes the poison." Paracelsus was the first one who recognised that dose is the most important in determining whether certain chemical is toxic or not. However, he did not have a quantitative notion on how much difference there was between the doses of a substance that separated a poison from a drug. Low doses of certain substances were found to be useful as curative while relatively high doses of the same substance were considered toxic [30]. The body's response to the chemicals depended on the dose received. Consumption of small amounts wine and spirits is considered to be safe. Ethanol blood level of 0.05 % is considered non-toxic, while toxic and lethal dose are determined to be 0.1 % and 0.5 %, respectively [32]. Beneficial dose of aspirin is 2 tablets (0.65 g), while 30 and 105 tablets make toxic and lethal dose, respectively [32].

Dose–response relationship is a relationship between exposure and health effect, which can be established by measuring the response relative to an increasing dose. The dose–response relationship is based on observed data from experimental animal, human clinical, or cell studies. Typical dose–response curve is presented in Figure 11.8.

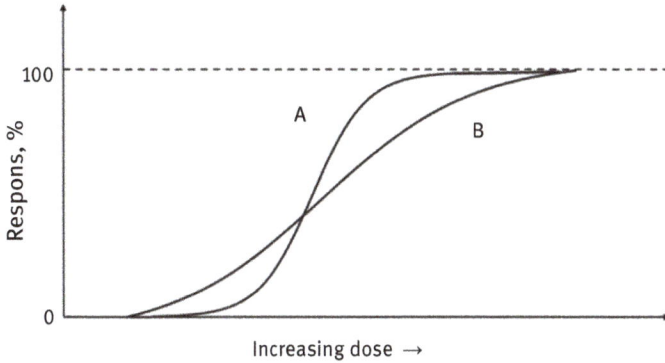

Figure 11.8: Dose–response curve for chemicals A and B.

Such dose–response curve establishes causality that the chemical has in fact induced the observed effects, indicates the lowest dose where an induced effect occurs, and determines the rate at which injury builds up (the slope for the dose response).

The NOAEL (no observable adverse effect level) identified in a particular test will be simply the highest dose level or concentration of the substance used in that test at which no statistically significant adverse effects were observed, i. e. it is an operational value derived from a limited test [23]. For example if the dose levels of 200, 50, 10 and 5 mg/kg/day of a substance have been used in a test and adverse effects were observed at 200 and 50 mg/kg/day/but not at 10 or 5 mg/kg/day, the derived NOAEL will be 10 mg/kg/day. Thus, the NOAEL and LOAEL (lowest observable adverse effect level) values for a given study would depend on the experimental study design, i. e. the selection of dose levels and the spacing between doses.

A threshold for toxic effects occurs at the point where the body's ability to detoxify a xenobiotic or repair toxic injury has been exceeded. For most organs there is a reserve capacity so that loss of some organ function does not cause decreased performance. Humans and other living organisms often can tolerate some degree of exposure to various chemicals before they became toxic or cause ill effect [7]. Hence, many trace metals are toxic in high concentrations but essential in very low quantities. Threshold dosage is a dosage below which there is no increase in adverse effect above natural background rates. It should be noted that for some effects, such as genotoxicity there are no threshold dosage.

Conservative assumption of zero thresholds is commonly used for carcinogens, while nonzero threshold is employed for other types of health effect [7]. The intersection of dose–response curve with the x-axis in Figure 11.7 indicates the threshold dosage. Below that level there is no observable adverse effect.

Knowledge of the shape and slope of the dose–response curve is extremely important in predicting the toxicity of a substance at specific dose levels. Major differences among toxicants may exist not only in the point at which the threshold is reached but also in the percent of population responding per unit change in dose (i. e. the slope). As illustrated in Figure 11.8, toxicant A has a higher threshold but a steeper slope than toxicant B.

Some of the common measures of observed effects include [12]:

LC_{50}: median lethal concentration

LD_{50}: median lethal dose

EC_{50}: median effective concentration

LC_{10}: lethal concentration to 10 % of test organisms

where L stands for lethal, E stands for effective, C is concentration and D is dose.

LD_{50} represents a dose of toxicant at which 50 % of tested organisms die within the specified time. EC_{50} is concentration at which 50 % of tested organisms exhibit the effect of exposure.

Different sublethal effects can be examined such as percentage of organisms that fail to produce eggs or achieve specific height. For example *Vibrio fischeri* luminescence assays are often used as environmental monitoring tool for toxicity of environmental samples. Test is based on measuring changes in the light emitted by luminescent bacteria *Vibrio fischeri* upon exposure to medium containing toxic substance. Resulting EC_{50} is the effective concentration of a sample that causes a 50 % decrease in the light output of the test organisms under controlled conditions. The result is reported in the terms of IC_{50} which stands for inhibitory concentration measuring the quantitative reduction of light production relative to control. LC_{10} is lower than LC_{50} for the same organisms. EC_{50} is lower than LC_{50} as it measure effect other than lethal.

Hence, toxic effects as a function of concentration follow increasing order: $NOEC < LOEC < EC_{50} < LC_{10} < LC_{50}$ [12].

The dose–response assessment or toxicity assessment provides a means of understanding whether exposure to environmental contaminants has the potential to produce adverse effects. The dose–response assessment most often relies on data from animal studies and extrapolates these data to humans. These studies are most often conducted using very high doses relative to the environmental exposures experienced by humans—hence, two extrapolations are required, one from high doses to low doses and the other from animals to humans. Extrapolation of high doses in an epidemiologic study to lower environmental doses is also needed when the dose–response assessment for a chemical is based on high-exposure epidemiology data, such as that in occupational or observational studies [31].

11.9 Risk assessment techniques

As a part of the risk management process, the risk assessment is performed to provide an input to a decision (Figure 11.5). For that purpose relevant and viable information should be obtained. Different techniques might be used depending on the problem in hand and available data. The examples of some commonly used techniques include: brainstorming, interviews, Delphi technique, check lists, primary hazard analysis, hazard analysis and critical control points, scenario analysis, fault tree analysis, event tree analysis, bow-tie analysis, Bayesian statistics and networks, risk indices, consequence/probability matrix (risk matrix) and multi-criteria decision analysis. According to ISO 31010 standard [19] risk assessment techniques can be classified depending on phase of the assessment process; risk identification, analysis or evaluation. Some of the listed techniques are applicable in just one of the assessment phases, while others can be used for entire process. For example the brain storming and check-lists are applicable only for the risk identification, while risk matrices can be used in all phases of risk assessment.

The methods used for describing and estimating the risk levels can be characterized as qualitative, quantitative or semi-quantitative. Qualitative methods describe probabilities, consequences and risk in words with different significance level, while quantitative methods express the risk in numerical values. The methods that are basically qualitative but the numerical values, i.e. rating scales, are assigned to probability and consequence classes can be categorised as semi-quantitative [15]. Scales, which may be linear or logarithmic, are combined to produce a level of risk using different formula. Risk matrices can be used for the qualitative but also semi-quantitative risk description. Logic trees can be applied both qualitatively and quantitatively. The main characteristics of these techniques, as a simple representatives of above types, are further described in more details.

11.9.1 Risk matrices

Risk can be expresses as a combination of likelihood (frequency or probability) and the consequences of hazardous event (eq. (11.1)). It can be visualised in a two-dimensional space constructed with axes of likelihood and severity of consequence [31] (Figure 11.9). The risk is lowest at left bottom and increases toward right top.

A risk matrix is overlaid on this continuum with cells defined in the terms of consequence columns and likelihood rows representing the tabular illustration of the risk [10, 31].

Risk matrices are used to screen the significance of the risk, to compare risks relative to one another, to prioritize risks based on their ranks, and to highlight the areas for further quantitative analysis [31]. The risk matrices can be of different sizes, ranging from the simplest matrix 2 × 2 to highly complex matrices comprising a range of different severities on the same risk continuum [31]. However, in most risk

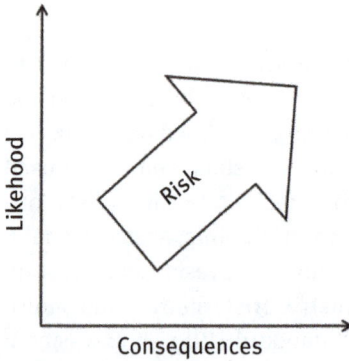

Figure 11.9: Visualisation of the risk continuum.

matrices the frequencies/probabilities and the severity of the consequences are divided into three to six categories [10]. Each cell in the matrix corresponds to the specific combination of frequencies/probabilities and consequences which can be defined descriptively, i. e. qualitatively or semi-quantitatively, whereas descriptive definitions are given numerical interpretations. It should be noted that different terms are used to describe probabilities, severity of consequences and the resultant risk. Murray et al. [32] reported consequence categories used in different studies such as "catastrophic", "critical", "marginal", and "negligible" or "critical", "serious", "moderate", "minor", and "negligible", or "catastrophic", "major", "moderate", "minor", and "insignificant". Terms "frequent", "probable", "occasional", and "remote", or "very likely", "probable", and "improbable", or "almost certain", "likely", "possible", "unlikely", and "rare" are used to describe the extent of probability. The matrix can be further extended by adding terms "very", "highly" and "extremely". However, both probabilities and severity of the consequences can be simply described in terms "low", "medium" and "high" as presented in Table 11.3.

Table 11.3: Example of 3 × 3 risk matrix.

Consequences Probability	Low	Medium	High
High	Moderate risk	High risk	High risk
Medium	Low risk	Moderate risk	High risk
Low	Low risk	Low risk	Moderate risk

In this matrix, for example, each category may be assigned with an arbitrary score from 1 to 3, and the risk may be simply calculated by multiplying two factors to obtain a comparative number representing the level of resultant risk [33].

The risk matrix is widely adopted technique in various fields of risk assessment because matrices are easy to use and they provide a rapid ranking of risks into different significance levels [18]. However, there are some limitations regarding the need to be designed for specific circumstances (scenarios) so it may be difficult to have a common system across the range of circumstances [18]. Also, scaling of verbally described probabilities and consequences is always ambiguous to some extent. The use of matrix is subjective so differences may arise between the assessors. Further limitation of a risk matrix is related to the difficulties in comparing the resulting level of risks for different consequence categories.

European Chemical Agency (ECHA) describes application of risk matrices in qualitative risk assessment for human health [34], required for substances for which a threshold cannot be established such as irritants/corrosives, sensitizers, carcinogens, mutagens and reproductive toxicants. Risk matrix is constructed by describing the resultant risk for each combination of the likelihood of exposure and the substance hazard band. The substance hazard band is assigned as low, moderate or high hazard, based on the substance classification and hazard statements [35]. The likelihood of exposure is assigned to the same three categories. Hence, when combining low likelihood of exposure with low substance hazard band the resultant risk is low, while high likelihood of exposure and high substance hazard band result with the high risk, according to the example given in Table 11.3. It should be noted that in the case when either substance hazard band or likelihood of exposure is high, the resultant risk cannot be low.

11.9.2 Logic trees

Logic trees are diagrams that link all the events and processes within the considered system that could lead to the hazard or could develop form the hazard [12]. Hence, they can be considered as conceptual models (see subchapter 3.) Basically, there are two types of the logic trees depending on the type of applied logic approach. Fault trees are deductive, as they "look backward", i. e. works from the top down, linking chains of events to the outcome. Event trees are inductive, as they "look forward", i. e. works from the bottom up, starting from triggering event and following all possible outcomes to the final consequences [12].

Fault tree is a deductive logic diagram that displays the interrelationships between a potential hazardous event called top event and the causes of that event [10, 19]. Hence, fault trees identify what has to happen [12] i. e. what is the scenario of hazardous event (see subchapter 2.). Top events might be toxic release, explosion, failure of technological system, injury or death. The fault tree is constructed using AND and OR logic gates determining preconditions for the top event and all intermediate events until the basic causes are identified [36]. The events at the lowest level are so-called basic events such as component failure, environmental conditions, or human errors. An example of the fault tree diagram is presented in Figure 11.10.

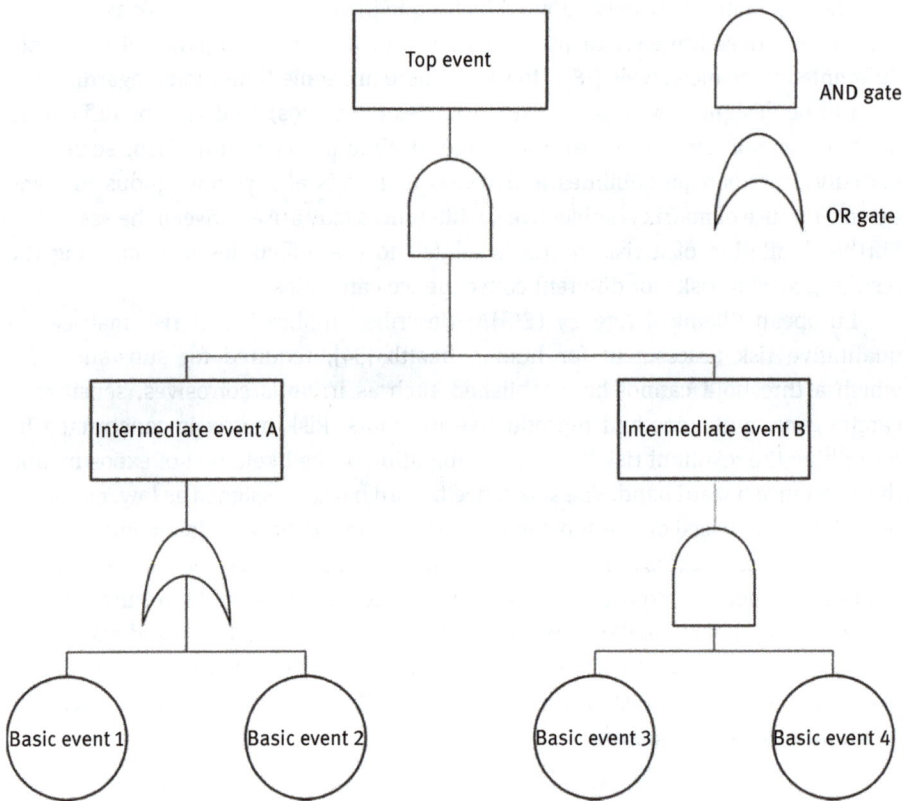

Figure 11.10: An example of a fault tree diagram.

The output event of an AND gate occurs only if all input events occur simultaneously and the output event of an OR gate occurs if at least one of the input events occur.

Fault tree may be used both qualitatively and quantitatively. In former case fault tree are used to identify potential causes and pathways to a failure, while later case the probability of the top event is calculate based on the known (or assumed) probabilities of basic events [19].

If identified events are mutually exclusive, the probability that one or the other will occur is presented by eq. (11.4), while the probability that both events would occur simultaneously is presented by eq. (11.5) [12].

$$p(B1 \cup B2) = p(B1) + p(B2) \tag{11.4}$$

$$p(B1 \cap B2) = 0 \tag{11.5}$$

whereas p stands for probability, while B1 and B2 stand for basic events 1 and 2, as presented in Figure 11.10.

Assuming that events B3 and B4 are independent, the probabilities in related OR and AND gates can be calculated using eqs (11.6) and (11.7), respectively.

$$p(\text{B3} \cup \text{B4}) = p(\text{B3}) + p(\text{B4}) - p(\text{B3} \cap \text{B4}) \qquad (11.6)$$

$$p(\text{B3} \cap \text{B4}) = p(\text{B3}) \times p(\text{B4}) \qquad (11.7)$$

Calculated probabilities in the gates represent the probabilities of intermediate events and can be used as inputs to the next level, up to the top event.

11.9.3 Event trees

Event trees are graphical presentation of the possible outcomes following an initiating event also called triggering event. It is used to identify and evaluate sequence of events in accident scenario under consideration (see subchapter 2). Event trees are generally based on binary-logic in which an event either has happened or not [36]. Hence, the diagram is constructed as a series of dichotomies as presented in Figure 11.11. Intermediary events (also called pivotal events) are events between the initial event and the end event or final state, i. e. outcome. Event tree splits intermediate events in two branches representing two mutually exclusive options; the phenomena either occurs or not, the system component either fails or not. It should be noted that in

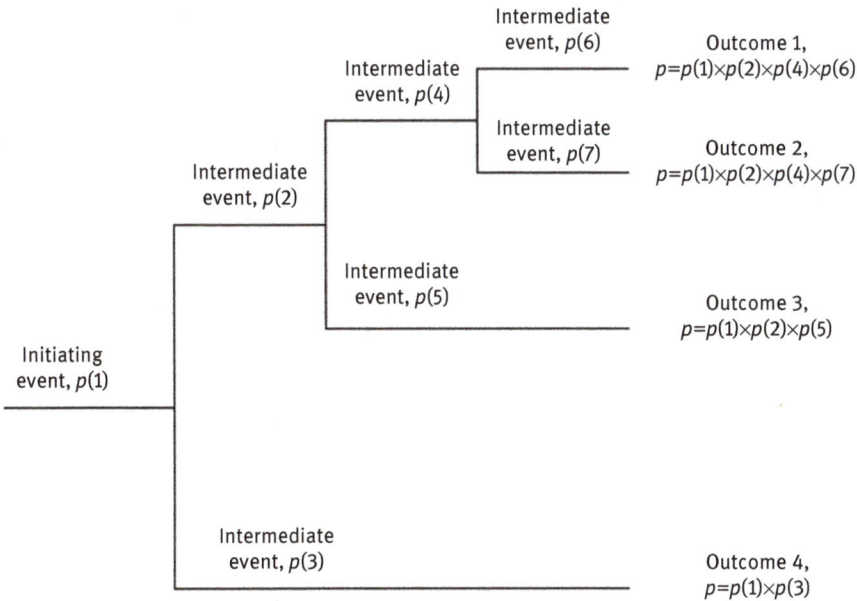

Figure 11.11: The example of an event tree.

some cases intermediate event may be split in more than two branches, but they should be mutually exclusive and the sum of probabilities should be 1.

According to eq. (11.7), the probability of each of possible outcome is obtained by multiplying probabilities of all preceding events.

The above described methodology of event tree can be illustrated by an simple example. Consider the start of fire as an initiating event. Activation or failure of the sprinkler system is the first, while call to the fire department is the second intermediate event. What are the possible outcomes? When fire starts and then the subsequent intermediate events are successful (yes) there won't be any material losses, injuries or environmental damage. In the case when the sprinkler system is activated (yes) but the call to fire department fails (no), the partial damage would occur. The similar would happen when the sprinkler system fails (no) but the call to the fire department is successful (yes). The worst scenario is when both sprinkler system and call to the fire department fails (no) resulting with the total damage. The event tree describes conditional probabilities, e. g. the probability of an activated sprinkler system given there is a fire. Knowing the probabilities of initial and intermediate events, the probabilities of the each outcome can be calculated.

11.10 Conclusion marks

Risk assessment is used as a tool in wide range of professional and scientific areas. Hence, different definitions of terms used and methodologies applied can be found in the relevant literature. However, the core principles and the key stages including risk identification, risk analysis and risk evaluation of the process are fundamentally the same in each case. Different qualitative, quantitative and semi-quantitative techniques are available to support the risk assessment and management process in all stages. Generally, the risk can be eliminated, modified and mitigated, or accepted. ERA is very much developing field. The main purpose of ERA is to provide that well-informed decisions can be made and to protect human health and the environment.

References

[1] Merriam-Webster Dictionary. Risk. https://www.merriam-webster.com/dictionary/risk. Accessed: 1 Apr 2017.
[2] Collins Dictionary. Risk. https://www.collinsdictionary.com/dictionary/english/risk. Accessed: 1 Apr 2017.
[3] Cambridge Dictionary. Risk. http://dictionary.cambridge.org/us/dictionary/english/risk. Accessed: 1 Apr 2017.
[4] The Free Dictionary by Farlex. Risk. http://www.thefreedictionary.com/risk. Accessed: 1 Apr 2017.
[5] English Oxford Living Dictionaries. Risk. https://en.oxforddictionaries.com/definition/risk. Accessed: 1 Apr 2017.
[6] Sušnik J, Vamvakeridou-Lyroudia L, Kapelan Z, Savić DA. Major risk categories and associated critical risk event trees to quantify, PREPARED 2011.003, 2011. http://www.prepared-fp7.eu/viewer/file.aspx?fileinfoID=161. Accessed: 8 Mar 2017.

[7]   Rubin S. Introduction to engineering and the environment. New York: McGraw-Hill, 2001.
[8]   Muralikrishna IV, Manickam V. Environmental management-science and engineering for industry. Oxford: Butterworth-Heinemann, BSP books Pvt, Published by Elsevier Inc., 2017.
[9]   Technical Workbook on Environmental Management Tools for Decision Analysis. United nations environment programme, division of technology, industry and economics, 1999. http://www.unep.or.jp/ietc/publications/techpublications/techpub-14/index.asp. Accessed: 8 Mar 2017.
[10] Rausand M. Risk assessment; Theory, methods, and applications. Hoboken, NJ: John Wiley & Sons, Inc., 2011.
[11] Directive 2012/18/EU of the European Parliament and of the Council on the control of major-accident hazards involving dangerous substances. Off J Eur Union. L 197.
[12] Burgman M. Risks and decisions for conservation and environmental management. Cambridge UK: Cambridge University Press, 2007.
[13] Fairman R, Mead CD, Williams WP. Environmental risk assessment: approaches, experiences and information sources, European Environmental Agency, 1998. https://www.eea.europa.eu/publications/GH-07-97-595-EN-C2. Accessed: 1 Apr 2017.
[14] Kaplan S, Garrick BJ. On the quantitative definition of risk. Risk Anal. 1981;1:11–27.
[15] Lindhe A. Risk assessment and decision support for managing drinking water systems, PhD Thesis. Gothenburg, Sweeden: Chalmers University of Technology, 2010.
[16] EHSC note on environmental risk assessment environment health and safety committee, Royal Society of Chemistry, 2013.
[17] Hester RE, Harrison RM. Risk assessment and risk management. Oxford, UK: Royal Society of Chemistry, Information services, 1998.
[18] ISO Guide 73:2009. Risk management –vocabulary; ISO Technical Management Board Working Group on risk management.
[19] IEC/ISO 31010:2009. Risk management – principles and guidelines, IEC technical committee 56 and ISO Technical Management Board Working Group on risk management.
[20] Klaassen CD. Casarett & Doull's toxicology; The basic science of poisons. 8th ed. New York, NY: McGraw-Hill Education, 2013.
[21] Report on the Application in the Member States of Directive 96/82/EC on the control of major-accident hazards involving dangerous substances for the period 2009-2011, C (2013) 4035 final.
[22] National Research Council. Risk assessment in the federal government, managing the process. Washington, DC: National Academy Press, 1983.
[23] Technical guidance document on risk assessment, in support of Commission Directive 93/67/EEC on Risk Assessment for new notified substances, Commission Regulation (EC) No 1488/94 on Risk Assessment for existing substances, and Directive 98/8/EC of the European Parliament and of the Council concerning the placing of biocidal products on the market, Part 1. Institute for Health and Consumer Protection, European Chemicals Bureau, European Commission Joint research centre, EUR 20418 EN/1, European Communities, 2003
[24] Understanding REACH, ECHA European Chemicals Agency. https://echa.europa.eu/regulations/reach/understanding-reach. Accessed: 7 Dec 2017.
[25] Foth H, Hayes A. Concept of REACH and impact on evaluation of chemicals. Hum Exp Toxicol. 2008;27(1):5–21.
[26] ECHA European Chemical Agency. Chemicals in our life. https://echa.europa.eu/chemicals-in-our-life/which-chemicals-are-of-concern/svhc. Accessed: 25 May 2017.
[27] Technical guidance document on risk assessment, in support of Commission Directive 93/67/EEC on Risk Assessment for new notified substances, Commission Regulation (EC) No 1488/94 on Risk Assessment for existing substances, and Directive 98/8/EC of the European Parliament and of the Council concerning the placing of biocidal products on the market, Part 2. Institute

for Health and Consumer Protection, European Chemicals Bureau, European Commission Joint research centre, EUR 20418 EN/1, European Communities, 2003.

[28] OECD environmental risk assessment toolkit: tools for environmental risk assessment and management. http://www.oecd.org/chemicalsafety/risk-assessment/environmental-risk-assessment-toolkit.htm Accessed: 14 Feb 2017.

[29] WHO human health risk assessment toolkit for chemical hazards. Geneva, Switzerland: WHO Press, World Health Organization, 2010. http://www.who.int/ipcs/methods/harmonization/areas/ra_toolkit/en/. Accessed: 14 Feb 2017.

[30] Simon T. Environmental risk assessment; A toxicological approach. Boca Raton, FL: CRC Press, Taylor & Francis group, 2014.

[31] Wilkinson G, David R. Back to basics: risk matrices and ALARP. In: Dale C, Anderson T, editors. Safety-critical systems: problems, process and practice. London, UK: Springer, 2009.

[32] Murray SL, Grantham K, Damle SB. Development of a generic risk matrix to manage project risks. J Ind Syst Eng. 2011;5(1):35–51.

[33] Robu BM, Căliman FA, Beţianu C, Gavrilescu M. Methods and procedures for environmental risk assessment. Environ Eng Manag J. 2007;6(6):573–92.

[34] How to undertake a qualitative human health assessment and document it in a chemical safety report, European Chemicals Agency, ECHA-12-B-49-EN, 2012. https://echa.europa.eu/documents/10162/13655/pg_15_qualitative-human_health_assessment_documenting_en.pdf. Accessed: 14 Feb 2017.

[35] Guidance on information requirements and chemical safety assessment, Part E: risk characterisation, European Chemicals Agency ECHA-2016-G-04-EN. https://echa.europa.eu/web/guest/guidance-documents/guidance-on-reach. Accessed: 14 Feb 2017.

[36] Marhavilas P, Koulouriotis D, Mitrakas C. Fault and event-tree techniques in occupational health and safety systems – part I: integrated risk evaluation scheme. Environ Eng Manag J. 2014;8:2097–108.

Study questions and tasks

-   What is the difference between the risk and hazard?
-   Make a list of all "risky" situations that you encountered this week. Identify the sources of hazard and actual as well as potential undesired outcomes. Consider different scenarios and risk treatment option.
-   Rank above listed risks based on your personal perception. Discuss with the class and consider arguments for different rankings.
-   Explain "set of triplets" idea underlying the risk assessment.
-   What are conceptual models and what is their role in risk assessment?
-   Sketch and describe the conceptual model for Student Chemistry Lab. Besides chemicals, what hazards did you identify? Describe different exposure pathways. Classify safety measures regarding their influence on source, pathway or receptor.
-   What are the main characteristics of the Human health risk assessment, Ecological risk assessment and Applied industrial risk assessment within ERA? Additional reading is recommended; available at https://www.eea.europa.eu/publications/GH-07-97-595-EN-C2)

- Define general risk assessment framework and describe the steps involved and make the correlation with risk assessment steps within ERA, NAS model and methodology for human and ecological risk assessment used in regulations for new and existing substances.
- Which potential toxic effects are considered within the human health risk assessment for chemicals? Explain the importance of knowledge on physio chemical properties of the hazardous substances.
- Explain terms NOAEL and LOAEL and their correlation with the underlying principle of human health assessment for chemicals.
- Explain terms PEC and PNEC and their usage in risk characterization within the ERA for chemicals.
- Read carefully the example on ERA for a textile dye using OECD interactive toolkit (http://envriskassessmenttoolkit.oecd.org/default.aspx.). Assuming the linear correlations within the system set by described scenario; assess the maximal increase in textile processing that would not pose an environmental risk regarding the toxicity on *Daphina magna*. Note that 30 % of the textile is treated with the reactive dye in question.
- What information can be obtained from the dose-response curve?
- What is the meaning of the threshold dosage? Is there a threshold dosage for each type of a health effect (why)? Sketch the dose-response curve for the zero and nonzero threshold.
- What are risk matrices and how are they used in risk assessment?
- Describe similarities and differences between fault and event trees.
- Based on fault tree diagram presented in Figure 11.10), calculate the probability of top event assuming that all events (basic and intermediate) are independent and that the probabilities of all basic events are the same $p = 0.1$.
- For above example, evaluate the influence of applied risk mitigation measures, which lowered probabilities of all basic events from $p = 0.1$ to $p = 0.09$.
- Using the same diagram (Figure 11.10), calculate the probability of the top event assuming that B1 and B2 are mutually exclusive while all other events are independent, and the probabilities are the same $p = 0.1$.
- Translate Figure 11.1, representing the sequence of events, into the event tree. Initiating event: a cat pushes the 1st tile; intermediate events 1, 2, 3 and 4: tile falls and pushes the next one (yes/no), intermediate event 5: tile falls and pushes the glass with milk(yes/no). The probability of initial event and all intermediate events are $p = 0.5$. What is the probability for the spilled milk?

Index

https://doi.org/10.1515/9783110468038-012

www.ingramcontent.com/pod-product-compliance
Lightning Source LLC
Chambersburg PA
CBHW080656220326
41598CB00033B/5223